AF567054

# PHYSICS OF ACCRETION DISKS

## Advances in Astronomy and Astrophysics

A series of books and monographs which aims to supply professionals and serious graduate students with the intellectual tools necessary for the appreciation of the present status of topics at the forefront of current research, and provides a framework upon which future developments may be based.

Series Editors:
*V.G. Gurzadyan*, Yerevan Physics Institute, Armenia
*S. Inagaki*, Kyoto University, Japan
*G. Meylan*, European Southern Observatory, Garching bei München, Germany

**Volume 1**
Highly Evolved Close Binary Stars: Catalog
Highly Evolved Close Binary Stars: Finding Charts
*A.M. Cherepashchuk, N.A. Katysheva, T.S. Khruzina, S.Yu. Shugarov*

**Volume 2**
Physics of Accretion Disks: Advection, Radiation and Magnetic Fields
*Edited by S. Kato, S. Inagaki, S. Mineshige, J. Fukue*

This book is part of a series. The publisher will accept continuation orders which may be cancelled at any time and which provide for automatic billing and shipping of each title in the series upon publication. Please write for details.

# PHYSICS OF ACCRETION DISKS
## ADVECTION, RADIATION AND MAGNETIC FIELDS

**Edited by**

**S. Kato, S. Inagaki, S. Mineshige**

*Department of Astronomy, Kyoto University, Japan*

**J. Fukue**

*Astronomical Institute, Osaka-Kyoiku University, Japan*

GORDON AND BREACH SCIENCE PUBLISHERS

Australia • Canada • China • France • Germany • India • Japan
Luxembourg • Malaysia • The Netherlands • Russia • Singapore
Switzerland • Thailand • United Kingdom

Copyright © 1996 OPA (Overseas Publishers Association) Amsterdam B.V.
Published in The Netherlands under license by Gordon and Breach Science Publishers.

All rights reserved.

No part of this book may be reproduced or utilized in any form or by any means, electronic or mechanical, including photocopying and recording, or by any information storage or retrieval system, without permission in writing from the publisher. Printed in Singapore.

Amsteldijk 166
1st Floor
1079 LH Amsterdam
The Netherlands

---

**British Library Cataloguing in Publication Data**

Physics of accretion disks. — (Advances in astronomy and astrophysics; v. 2)
1. Accretion (Astrophysics) 2. Disks (Astrophysics)
I. Kato, S.
523′.01

ISBN 90-5699-016-0

# CONTENTS

Photographs of the conference in progress.

# PREFACE

The international workshop on Basic Physics of Accretion Disks was held at Kyoto during 22–27 October, 1995. The main purpose of this workshop was to discuss a "next-generation standard disk model" which can satisfactorily explain general behavior of accreting objects. The so-called standard $\alpha$-disk model has been proposed by Shakura and Sunyaev, and by Novikov and Thorne in 1973. The standard model assumes that the disk is axisymmetric, in steady state, optically thick, thereby emitting blackbody radiation, and is geometrically thin. By adopting the $\alpha$ prescription for disk viscosity, one can construct the disk structure.

Although the standard disk model has contributed much towards the physical understanding of the astrophysical disks, recent observations have revealed a variety of accretion phenomena that cannot always be accounted for by the standard picture of accretion disks. Good examples are the presence of high-energy radiation, rapid variabilities, outburst behaviors, and so on. Distinct kinds of accretion disk models have been proposed, therefore, in place of or to complement the standard disk model. One possibility is a cooling-dominated, optically thin disk. This is, however, thermally unstable and is thus unable to explain persistent hard radiation. Another possibility is an advection-dominated, optically thin disk, which was the main theme of the workshop.

Why do we focus on the advection-dominated disks? There are two main reasons for this. First, advection-dominated disks are thought to exist for a wide range of mass-input rates (and disk luminosities), just as the standard-type disks do. Second, the properties of the advection-dominated disk are distinct from those of the standard-type disk. Unlike the standard disk, the advection-dominated disk is characterized by a high temperature, rapid accretion, sub-Keplerian rotation, and inefficient radiation. The advection-dominated disks are rather common, but their nature is rather unique; this motivated us to plan a workshop. This is, to our knowledge, the first international workshop, in which various aspects of the advection-dominated disks have been extensively discussed.

The workshop began with two long reviews by Abramowicz and Narayan, followed by other invited talks discussing many of the outstanding issues associated with the advection-dominated disk model. The most interesting issue would be the relationship between two types of disks (standard disks and advection-dominated disks); are they bi-modal or coexistent? Optically thin and thick portions of the disk may coexist with either vertical separation (i.e., a disk-corona structure) or radial separation (i.e., an inner hot disk plus an outer cool disk). The hard-soft spectral transitions observed in the black hole candidate, on the other hand, suggest that the black-hole disk may alternate between the two types with time. Other interesting issues include the global disk structure, the thermal stability, two-temperature regimes, electron-positron pair production, disk spectra, the relation to the outburst behavior of X-ray transients and hard-soft spectral transitions, the effects of radiation drag force, jet formation, and possible nucleosynthesis in the advection-dominated disks. As readers will find in the proceedings, other fundamental issues concerning the basic physics of accretion disks were also extensively discussed, which include the origin of disk viscosity (or turbulence), the observational appearance of the tidal instability in cataclysmic variables, and the role of self-gravity of accretion disks in relation to stellar systems and to star/planet formation.

This workshop was financially supported by the Ministry of Education, Science, Sports and Culture of Japan. We also gratefully acknowledge the Kyoto School of Computer Science, which kindly provided us with the places for the scientific sessions and poster discussion, offered the shuttle bus service, and sponsored the Koto concert. We are indebted to the Department of Astronomy, Kyoto University for assistance. Finally, we wish to express heartfelt thanks to all the participants for their contributions to the workshop.

Shoji Kato
Shogo Inagaki
Shin Mineshige
Jun Fukue

# LIST OF PARTICIPANTS

| | |
|---|---|
| Abramowicz, Marek A. | Göteborg University, Sweden |
| Arai, Kenzo | Kumamoto University, Japan |
| Arimoto, Jun'ichi | Osaka-Kyoiku University, Japan |
| Balbus, Steven A. | University of Virginia, Charlottesville, USA |
| Beckert, Thomas | University of Heidelberg, Germany |
| Brandenburg, Axel | NORDITA, Copenhagen, Denmark |
| Chen, Xingming | Göteborg University, Sweden |
| Duschl, Wolfgang J. | University of Heidelberg, Germany |
| Fukue, Jun | Osaka-Kyoiku University, Japan |
| Furuya, Natsumi | Kyoto University, Japan |
| Godon, Patrick | NASA/Jet Propulsion Laboratory, Pasadena, CA, USA |
| Hagio, Fumihiko | Kumamoto Institute of Technology, Japan |
| Hanawa, Tomoyuki | Nagoya University, Japan |
| Hashimoto, Masa-aki | Kyushu University, Japan |
| Hawley, John F. | University of Virginia, Charlottesville, USA |
| Hayashi, Masahiko | National Astronomical Observatory of Japan, Tokyo, Japan |
| Hirotani, Kouichi | National Astronomical Observatory of Japan, Tokyo, Japan |
| Honma, Fumio | Kyoto University, Japan |
| Hozumi, Shunsuke | Shiga University, Otsu, Japan |
| Inagaki, Shogo | Kyoto University, Japan |
| Inutsuka, Shu-ichiro | National Astronomical Observatory of Japan, Tokyo, Japan |
| Kaburaki, Osamu | Tohoku University, Sendai, Japan |
| Kato, Shoji | Kyoto University, Japan |
| Kato, Taichi | Kyoto University, Japan |
| Kikuchi, Nobuhiro | University of Tokyo, Japan |
| Kim, Soon-Wook | University of Texas at Austin, USA |
| Kojima, Yäsufumi | Hiroshima University, Japan |
| Korchagin, Vladimir | National Astronomical Observatory of Japan, Tokyo, Japan |
| Kudoh, Takahiro | National Astronomical Observatory of Japan, Tokyo, Japan |
| Kusunose, Masaaki | University of Texas at Austin, USA |
| Lamb, Frederick K. | University of Illinois at Urbana-Champaign, USA |
| Lasota, Jean-Pierre | Observatoire de Paris, France |
| Manmoto, Tadahiro | Kyoto University, Japan |
| Maddison, Sarah | Monash University, Clayton, Australia |
| Matsumoto, Ryoji | Chiba University, Japan |
| Matsuzaki, Takaaki | Chiba University, Japan |
| Meyer, Friedrich | Max-Planck-Institute for Astrophysics, Garching, Germany |

| | |
|---|---|
| Mineshige, Shin | Kyoto University, Japan |
| Miyama, Shoken M. | National Astronomical Observatory of Japan, Tokyo, Japan |
| Miyamoto, Sigenori | Osaka University, Japan |
| Nakao, Yasushi | Kyoto University, Japan |
| Nakamura, Kaori | University of Tokyo, Japan |
| Nakamura, Kenji E. | Kyoto University, Japan |
| Nakayama, Kunji | Kochi University, Japan |
| Narayan, Ramesh | Center for Astrophysics, Cambridge, MA, USA |
| Negoro, Hitoshi | Institute of Space and Astronautical Science, Sagamihara, Japan |
| Nobuta, Koji | Nagoya University, Japan |
| Nogami, Daisaku | Kyoto University, Japan |
| Ohna, Eetsuko | Osaka-Kyoiku University, Japan |
| Osaki, Yoji | University of Tokyo, Japan |
| Saigo, Kazuya | Nagoya University, Japan |
| Sano, Takayoshi | University of Tokyo, Japan |
| Schandl, Susanne | Max-Planck-Institute for Astrophysics, Garching, Germany |
| Sellwood, Jerry A. | Rutgers University, N.J., USA |
| Shimizu, Souichi | University of Tokyo, Japan |
| Szuszkiewicz, Ewa | SISSA, Trieste, Italy |
| Taam, Ronald E. | Northwestern University, Evanston, IL, USA |
| Tajima, Yukiko | Osaka-Kyoiku University, Japan |
| Takeuchi, Mitsuru | Kyoto University, Japan |
| Tsugawa, Motohiko | University of Tokyo, Japan |
| Tsuribe, Tohru | University of Tsukuba, Japan |
| Umemura, Masayuki | University of Tsukuba, Japan |
| Watanabe, Yoichi | Osaka-Kyoiku University, Japan |
| Wheeler, J. Craig | University of Texas at Austin, USA |
| Yamada, Tatsuya T. | Osaka University, Japan |
| Yamasaki, Tatsuya | Kyoto University, Japan |
| Yi, Insu | Princeton Institute for Advanced Study, N.J., USA |
| Yokosawa, Masayoshi | Ibaraki University, Japan |

# Basic Physics of Advection in Black Hole Accretion Disks

Marek ABRAMOWICZ[1,2]
*1. Department of Astronomy & Astrophysics, Göteborg University and Chalmers University of Technology, S-412 96 Göteborg, Sweden*
*2. Nordita, Blegdamsvej 17, DK-2100, Copenhagen Ø, Denmark*

**Abstract**

I describe from my personal perspective the history of the progress in understanding the role of advective cooling in hot accretion disks around black holes. I argue that advection strongly influences equilibria and stability of this type of disk .

## 1. Prologue: Accretion Flows with Low Radiative Efficiency

Rees, Begelman, Blandford & Phinney (1982) suggested that at low (very sub Eddington) accretion rates starved black holes have low radiative efficiency. They argued that this is relevant for radio galaxies, the $\sim 10^6 M_\odot$ black hole in the Galactic Center and stellar mass black holes in our Galaxy (see also Rees, 1982).

The low radiative efficiency is also characteristic of spherical accretrion at high (near Eddington) accretion rates, for the reason first explained by Katz (1977). Radiation diffuses outward at a speed $v_{\rm diff} = c/\tau$, where $\tau$ is the optical depth with $v$ being the velocity of matter, $v_{\rm diff}/v = \dot{M}_{\rm E}/2\dot{M}$. Thus, as $\dot{M}$ approaches $\dot{M}_{\rm E}$, radiation is trapped within the accreting matter and cannot escape: $\dot{M}$ may become very large without $L$ exceeding $L_{\rm E}$. Here $\dot{M}$ denotes accretion rate, $L$ luminosity, and the corresponding Eddington quantities are $\dot{M}_{\rm E} = L_{\rm E}/c^2$. This idea was later worked out in more details by Rees (1978) and Begelman (1978, 1979).

My review describes accretion disks with strong advective cooling that provide a still different example of flows with low radiative efficiency. Understanding of their nature has been gradually progressing thanks to studies on stability of thin disks (1978), discovery of stabilization of thick disks by the Roche lobe overflow (1978, 1981) and research on the transonic accretion

*S. Kato et al. (eds.), Physics of Accretion Disks, 1–14.*
© 1996 OPA (Overseas Publishers Association) Amsterdam B.V.

with non-zero angular momentum (1980). The unification of the subject was provided in terms of slim accretion disks — first optically thick (1986, 1988), and later optically thin (1995).

## 2. The Standard Model of Accretion Disks

The standard model of accretion disks has been developed in the early seventies in the classic papers of Lynden-Bell (1969), Shakura (1972), Pringle & Rees (1972), Shakura & Sunyaev (1973), and Novikov & Thorne (1973). It provided remarkably successful contributions to understanding of quasars, X-ray binaries and other high-energy astronomical objects and generated a progress in black hole astrophysics, theory of radiative processes, understanding of turbulence, and development of relativistic numerical hydrodynamics. The standard model was presented in several excellent reviews, including the classic one by Pringle (1981). It was also the main subject of a textbook (Frank, King & Raine, 1992), and numerous conferences (e.g., Bertout, Collin-Soufrin & Lasota 1991; Meyer, Duschl, Frank & Meyer-Hofmeister 1989). The standard model's most obvious success was in explaining the recurrent outbursts and other properties of dwarf novae (see reviews by Cannizzo 1993, and Lasota 1995, 1996).

The assumption on which the standard model rests is that the vertical thickness of the disk $H$ is much smaller than the corresponding radius $r$. This allows a separate treatment of the radial and vertical structures. Till recently, an overwhelming majority had the view that the radial structure is already well understood and that for astrophysically relevant accretion disks it should be similar to that in the standard model: with negligible radial gradients of pressure, density, temperature, and entropy, with almost Keplerian rotation, and no radial transport of heat. Thick $H/r > 1$ and slim $H/r \sim 1$ accretion disks with significant radial gradients of pressure, density and entropy, non-Keplerian rotation and heat balance strongly influenced by radial advection, had less obvious astrophysical applications. For this reason they have been considered as interesting perhaps, but mostly irrelevant, theoretical curiosities.

The opinion that there is little to be seriously changed in the radial structure of the standard disk model is still persistent, but it starts to erode, thanks mostly to recent works of Narayan, Lasota and their collaborators. They have convincingly demonstrated that hot slim disks explain much better than the standard model detailed observations of sources that appear hot and underluminous, like Sgr A* (Narayan, Yi & Mahadevan 1995) soft X-ray transients (Narayan & Yi 1995b; Lasota, Narayan & Yi 1995; Lasota 1996), and the dwarf active galactic nucleus NGC 4258 (Lasota, Abramowicz, Chen, Krolik, Narayan & Yi 1996). These solutions may also possibly be relevant to somewhat luminous sources, both black hole X-ray binaries and active galactic nuclei (Narayan 1996). In a parallel development, Rees (1993) and

Fabian & Rees (1995) discussed properties of underluminous active galactic nuclei in more general terms of accretion flows with low radiative efficiency.

## 3. Stability of the Standard Model

Piran (1978) introduced four convenient parameters,

$$\mathcal{K} = \left(\frac{\partial \ln Q^-}{\partial \ln H}\right), \ \mathcal{L} = \left(\frac{\partial \ln Q^-}{\partial \ln \Sigma}\right), \ \mathcal{M} = \left(\frac{\partial \ln Q^+}{\partial \ln H}\right), \ \mathcal{N} = \left(\frac{\partial \ln Q^+}{\partial \ln \Sigma}\right). \quad (1)$$

Here $\Sigma$ is the surface density, $Q^+$ is the rate of heat generation by viscous processes, and $Q^-$ is the rate of heat losses. The necessary conditions for thermal and viscous stability are,

$$\mathcal{K} - \mathcal{M} > 0, \quad \frac{\mathcal{N}\mathcal{K} - \mathcal{M}\mathcal{L}}{\mathcal{K} - \mathcal{M}} > 0. \quad (2)$$

The criterion for thermal stability was first derived by Pringle, Rees & Pacholczyk (1973), (see also Shakura & Sunyaev 1976), while the condition for viscous stability was derived by Lightman & Eardley (1974) and Lightman (1974). In the optically-thick standard disk model, assuming that kinematic viscosity is proportional to the *total* pressure, $\mathcal{K} = 4(1+\beta)/B$, $\mathcal{L} = -\beta/B$, $\mathcal{M} = 2$, $\mathcal{N} = 1$. Here $\beta = P_{\rm gas}/P$ is the ratio of gas to total pressure and $B = 4 - 3\beta$. Thus, the condition for thermal and viscous stability is $\beta > 2/5$. Therefore, in the standard model, radiation-pressure supported parts (i.e., with $\beta \sim 0$) are unstable.

This is a problem, because the innermost parts of the standard optically thick black hole accretion disks *are* radiation-pressure supported. Thorne & Price (1975) have proposed that the instability may swell the inner part into much hotter, gas-pressure dominated, optically-thin state. This idea was further developed in the classic paper by Shapiro, Lightman & Eardley (1976) who constructed a hot, optically-thin, two-temperature model of accretion disk. The instability problem, was not solved, however. The two-temperature hot disks were still violently thermally unstable.

## 4. Thick Disks and the Roche Lobe Overflow

Results of the Warsaw group research on thick accretion disks, started by Abramowicz, Jaroszyński & Sikora (1978) and summarized by Abramowicz, Calvani & Nobili (1980), have been different in a rather crucial point from these obtained earlier by Fishbone & Moncrief (1976). The Warsaw group found that thick disks have radiative efficiency smaller than the standard disks with the same accretion rate because of mass and heat losses at the inner edge of the disk, induced by the Roche lobe overflow similar to that in close binaries (Paczyński 1978).

Let $H_0$ and $\Sigma_0$ be the vertical thickness and surface density at the "cusp" where the Roche lobe crosses itself. The mass loss and the advective heat flux due to the relativistic Roche lobe overflow are, independently of any particular details (Abramowicz 1981),

$$\dot{M} \sim H_0\Sigma_0, \quad Q^{-}_{\rm adv} \sim H_0^3\Sigma_0. \tag{3}$$

This means that at the inner edge the Piran's coefficients are $\mathcal{K} = 3$, $\mathcal{L} = 1$, $\mathcal{M} = 2$, $\mathcal{N} = 1$, and thus that both thermal and viscous stability conditions (2) are fulfilled: *The inner edge of* any *disk, standard or not, is thermally and viscously stable* (Abramowicz, 1981). Among thick disks models constructed at that time, particularly important for understanding of the advective cooling was the self-similar model of Begelman & Meier (1982). Later, advanced numerical works by Hawley, Smarr, & Wilson (1984), Eggum, Coroniti & Katz (1987) provided a sophisticated and detailed treatment of physical conditions in the flow. They have been constructed independently of the analytic theory developed earlier by the Warsaw group. However, all these models are advectively cooled by the Roche lobe overflow *exactly* according to the description given by the Warsaw group (see a demonstration of this point in Igumenshchev, Chen & Abramowicz 1996).

## 5. Transonic Part of Black Hole Accretion Flows

Properties of quasi-spherical and disk-like transonic accretion are qualitatively different: flow properties jump along sequences with continuously changing angular momentum depending on two parameters: radius of the sonic point $r_{\rm S}$ and value of angular momentum there, $\ell_{\rm S}$ (Abramowicz & Zurek 1981). For the disk-like accretion $r_{\rm S} < r_{\rm MS}$ and $\ell_{\rm S} > \ell_{\rm K}(r_{\rm MS})$. Here $r_{\rm MS}$ is the radius of the last stable Keplerian orbit ($r_{\rm MS} = 3r_{\rm G}$ for a non-rotating black hole) and $\ell_{\rm K}$ denotes the Keplerian angular momentum. With fixed outer boundary conditions these parameters depend mostly on viscosity in the disk. High viscosity usually corresponds to low values of $\ell_{\rm S}$ and high values of $r_{\rm S}$, consistent with the quasi-spherical case. For this reason, properties of transonic flows are different for high and low values of the viscosity parameter $\alpha$ (Muchotrzeb 1983).

Moncrief (1980) proved that a transonic, spherically symmetric, adiabatic black hole accretion is globally stable against small perturbations, because the perturbations are advected inside the sound horizon and therefore cannot grow. In the case of a transonic, viscous accretion with non-zero angular momentum advection stabilizes local, axially symmetric thermal modes (Abramowicz 1981; Honma, Matsumoto, Kato & Abramowicz 1991; Kato, Abramowicz & Chen 1996), and global, dynamical, non axially symmetric Papaloizou-Pringle (1984) modes (Abramowicz, Blaes & Lu 1986; Blaes 1987). However, for a sufficiently high value of the viscosity parameter $\alpha$, a global unstable acoustic standing mode is present at the sonic surface (Muchotrzeb 1983; Matsumoto, Kato, Fukue & Okazaki 1984; Abramowicz & Kato 1989). In relativistic flows acoustic modes may be trapped in the region where the epiciclic frequency has its maximum (Okazaki, Kato & Fukue 1987; Kato & Honma 1991). The subject was reviewed by Wallinder, Kato & Abramowicz (1993).

Abramowicz & Zurek (1981) suggested that the non-uniquness of the local transonic solutions may result in shocks that discontinuously join different multiple solutions. This idea was explored by Fukue (1987), Abramowicz & Chakrabarti (1990), and later by other authors.

Either density $\rho$ or velocity $v$ are infinite at the inner edge of the standard accretion disk. The artificial nature of this singularity has been recognized quite early by Stoeger (1975), but the true remedy for it was found only much later by Paczyński & Bisnovatyi-Kogan (1981) who demonstrated that one should assume the zero torque boundary condition at the black hole's horizon, and include the advective transport of heat at the inner edge. In this pioneering paper, Paczyński and Bisnovatyi-Kogan have derived and solved accretion disk equations accurate in second order of $H/r$ that described radial gradients of pressure and entropy, the $v(dv/dr)$ term in the radial momentum equation, possibility of a non-Keplerian rotation, and the advective cooling. More sophisticated solutions of these equations, smoothly passing through the sonic point, have been found by Muchotrzeb & Paczyński (1982), Muchotrzeb (1983), and Muchotrzeb-Czerny (1986). These works have provided the necessary numerical know-how for the next crucial step: slim accretion disks.

## 6. Unification: Slim Accretion Disks

The unification of the ideas and developments described above has happened at a precise time and place: in the early evening on 5th of October 1984 at Dottore Montanari's apartment, 12, via Tor San Piero, Trieste, Italy. Jean-Pierre Lasota and I have been trying to understand implications of an interesting unpublished observational result by Danielle Alloin and others indicating that a particular active galactic nucleus NGC 1566 varies quasi-periodically. During this discussion (at various moments disturbed by Italian *dolce vita*), we invented slim accretion disks. I still do remember the feeling of illumination and joy when, finally, we understood how advection influences thermal equilibria and stability of optically thick disks. Everything snapped in place: Piran's stability criteria, Roche lobe overflow, transonic solutions found by Paczyński, Bisnovatyi-Kogan & Muchotrzeb, inner boundary condition, $S$-shaped equilibrium curves... The paper that we wrote (Abramowicz, Alloin, Lasota & Pelat) has never been published, however. For two expert referees, our challenge of the standard model appeared at that time not sufficiently convincing.

Approximate semi-analytic models of slim disk have been reported by us later in a conference proceedings (Abramowicz, Lasota & Xu 1986, hereafter ALX). Detailed numerical models have been published by Abramowicz, Czerny, Lasota & Szuszkiewicz (1988, hereafter ACLS). Difficult numerical calculations performed in ACLS by our co-workers Bożena Czerny and Ewa Szuszkiewicz have been based on a modification of the method developed earlier by Paczyński. They fully confirmed our thoughts of that evening at Montanari's.

In contrast to the standard disks, slim disks have significant radial gradients of pressure and entropy, non-Keplerian rotation, and a significant radial heat transport (advection). ALX suggested that due to non-linearities connected with transitions from gas to radiation pressure and from radiative to advective cooling, sequences of optically-thick accretion disk equilibria should bend twice in the $\ln \dot{M}$ versus $\ln \Sigma$ plane and form an $S$-curve. This was fully confirmed by exact numerical models calculated later by ACLS. The lower branch of the $S$-curve corresponds to stable, gas pressure dominated standard disk models. The middle branch corresponds to unstable, radiation-pressure dominated standard models. Both lower and middle branches exist for sub-Eddington accretion rates. The upper branch exists at moderately super-Eddington accretion rates and corresponds to stable slim disks. Slim disks are optically-thick, radiation-pressure dominated, and have $H$ slightly smaller than $r$. They are thermally stable, because the timescale of thermal vertical expansion is shorter than the timescale of advection. Thus, perturbations are advected inward before they could grow. Calculations of slim disks models have been immediately after performed by the Kyoto group and reported in a short paper by Abramowicz, Kato & Matsumoto (1989) (see also Kato, Honma & Matsumoto 1988) and later by Chen & Taam (1993). They have confirmed the results of ALX and ACLS, in particular the existence of the $S$-shaped equilibrium sequences. Electromagnetic spectra of slim disks are discussed by Szuszkiewicz (1996) in this volume (see also Szuszkiewicz, Malkan & Abramowicz 1996).

Having in mind the analogy with the dwarf novae $S$-curved equilibrium sequences, ALX suggested that slim disks should experience a limit cycle behaviour. Numerical simulation of Matsumoto, Honma & Kato (1989) and Honma, Matsumoto & Kato (1991) confirmed this suggestion. However, it would be premature to attempt detailed explanations of the observed variability of some real objects as a slim disk limit cycle. Two problems are still unsolved: (i) results of simulations depend strongly on unknown and thus arbitrarily assumed properties of viscosity, (ii) in the non-stationary state the flow experiences optically thick-thin transitions which (in absence of a full solution of the radiative transfer in disks) are still treated by a phenomenological approach (Lasota & Pelat 1991; see also Taam & Lin 1984).

The slim disks provided, for the first time, a physically complete, and mathematically self-consistent model of seriously non-standard accretion disks. Being optically thick, they could not have answered the question of what happens to the unstable, optically-thin, Shapiro, Lightman & Eardley disks. However, they have offered a helpful clue that I describe in the next Section.

## 7. Advectively Dominated, Hot Disks

It was initially thought that the effect of the Roche lobe overflow stabilization could be relevant only at the immediate vicinity of the cusp (Wheeler 1981; Shapiro & Teukolsky 1983). Later, in the context of slim disks, Lasota and I have realized that the reason for stabilization of thermal modes there was not directly connected to the Roche lobe overflow, but rather to the fact that the thermal timescale of vertical expansion $\tau_{\rm th} \sim 1/(\alpha\Omega)$ was of the same order as the timescale of inward advection $\tau_{\rm adv} \sim r/v_r$. When this happens, the instability is advected inwards before it has a chance to grow, in an analogous way to what stabilizes the spherical accretion (Moncrief 1980).

In standard thin accretion disks the timescale for radial advection is given by the viscous timescale $\tau_{\rm adv} \sim \tau_{\rm vis} \sim \tau_{\rm th}/(H/r)^2$. Thus, stabilization of the thermal modes by advective *cooling* should occur when $H \sim r$. This is consistent with the fact that in general

$$Q^-_{\rm adv} \sim \left(\frac{H}{r}\right)^2 Q^+, \tag{4}$$

and therefore when $H \sim r$, the cooling is dominated by advection: $Q^-_{\rm adv} \sim Q^+$, and radiative cooling is thus insignificant in the heat balance. Note that advection could be locally a *heating* process. In this case $Q^-_{\rm adv} + Q^+ = Q^-_{\rm rad}$ and therefore $Q^-_{\rm adv} < Q^-_{\rm rad}$, which means that radiative cooling must be important, and radiative efficiency must be high. This happens in the Shapiro, Lightman & Eardley solution.

In deriving the condition $H \sim r$ we have not assumed any particular regime of optical depth. Thus, the condition should also be valid for *optically thin* disks. This offers a clue that I now describe. The dashed line in figure 1 shows a (i)-(ii)-(iii) sequence of the "classic" (i.e., with no advective cooling) accretion disk equilibria. The three segments correspond to: (i) standard, cool, optically-thick, gas-pressure dominated equilibria, (ii) standard, cool, optically-thick, radiation-pressure dominated equilibria, and (iii) non-standard, hot, two-temperature, optically-thin, gas-pressure dominated equilibria of Shapiro, Lightman & Eardley. The shaded region corresponds to $H/r > 1$, and the unshaded region to $H/r < 1$. The slopes of the three segments and the $H/r = 1$ line do not depend much on the viscosity parameter $\alpha$, but their relative locations in the plane strongly depend on $\alpha$. When $\alpha$ is sufficiently low, the $H/r = 1$ line and the (i)-(ii)-(iii) sequence do cross each other, as shown in figure 1. For high values of $\alpha$ they do not cross. I discuss now the case of low $\alpha$.

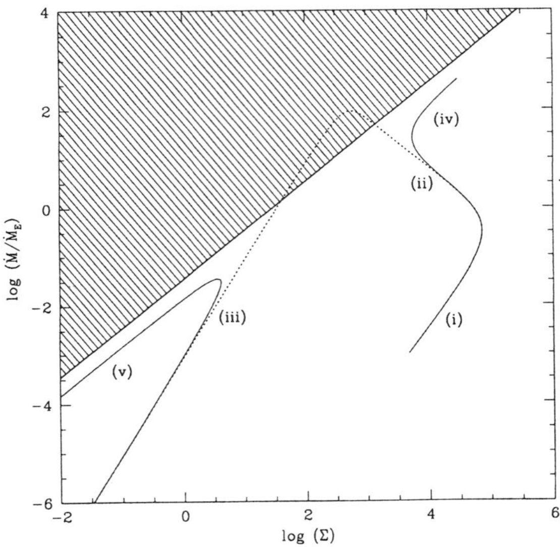


Fig. 1. Some of the low viscosity standard disk models cannot be self-consistent. All models correspond to $M = 10M_{\odot}$, $\alpha = 0.01$ and $r = 5r_G$.

In the shaded region, the (i)-(ii)-(iii) sequence cannot represent a self-consistent solution: advection in the shaded region is a *dominant* cooling process, but it was neglected in calculating the (i)-(ii)-(iii) models. Obviously, for the self-consistency, solutions must follow the path shown by the heavy lines. We see that there are two separate self-consistent sequences of thermal equilibria in the low viscosity case. The one on the right corresponds to *optically thick* accretion disks. It is characteristically $S$-shaped. The lower and middle branches of the $S$ corresponds to the standard accretion disk branches (i) and (ii). The upper branch (iv) is non-standard, and corresponds to slim disks: thermally and viscously stable, radiation-pressure dominated, and with a significant advective cooling. There are two branches of *optically-thin* self-consistent solution: the lower branch (iii) is the classic Shapiro, Lightman & Eardley one, and the upper branch (v) corresponds to the newly discovered hot, optically-thin, advectively cooled, thermally and viscously stable disks (Abramowicz, Chen, Kato, Lasota & Regev 1995; Narayan & Yi 1994, 1995a, b; Chen, Abramowicz, Lasota, Narayan & Yi 1995, hereafter CALNY). Properties of the newly discovered solution are discussed in great details by Narayan (1995) and also by Narayan (1996) in this volume.

CALNY constructed a topological map of all possible thermal equilibria of accretion disks in the whole range of $\dot{M}$, $\alpha$ and $r$. The map provides a unification of all the accretion disk models discussed before in my review. Models in CALNY's topological map include advective cooling, black body radiation for optically thick flows, Comptonization of soft bremsstrahlung and synchrotron photons for optically thin ones, and a phenomenological treatment of the gray regime. Later, Björnsson, Abramowicz, Chen & Lasota (1996, hereafter BACL) included also the presence of the $e^+e^-$ pairs and the relevant physics of hot plasma, using the "mapping method" of Björnsson & Svensson (1991). One should note that physics of hot plasma with advective cooling in hot boundary layers has been already discussed before by Narayan & Popham (1993).

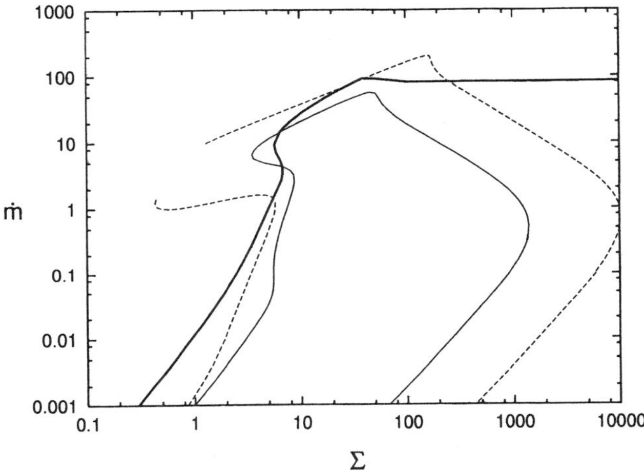


Fig. 2. All possible accretion disk solution at $r = 5r_G$ for $M = 10M_\odot$ and $\alpha = 0.01$ (according to BACL).

It was found by CALNY and confirmed by BACL that, at a given radius $r$, there exist exactly four physically distinct types of accretion disks, as shown in figure 2. Two of these types correspond to a low viscosity, $\alpha < \alpha_{cr}(r)$, and the two other types correspond to a high viscosity, $\alpha > \alpha_{cr}(r)$. The critical viscosity $\alpha_{cr}$ depends strongly on the radius and weakly on the mass of the central accreting object. According to CALNY, in the inner part of the disk $\alpha_{cr} \sim 0.1$. This value was changed by the more accurate treatment of microphysics and inner boundary condition by BACL who found $\alpha_{cr} > 1$. The two low viscosity type of flows are further differentiated by whether the flow is optically thin or optically thick, and the two high viscosity flow solutions by whether advection is negligible or dominant. BACL found that the electron-positron pairs are much less relevant for the hot disk equilibria than previously thought. The same results have been independently obtained by Kusunose & Mineshige (1996). See also Kusunose (1996) in this volume.

In all the above described papers (see also Chen 1995a; Esin, Narayan, Ostriker & Yi 1996), thermal equilibria of hot optically-thin disks have been calculated either locally, or by assuming that solution are self-similar. Mathematically consistent global solution that obey realistic boundary conditions and smoothly pass through the sonic point have been constructed only most recently by Chen, Abramowicz & Lasota (1996), Narayan, Honma, & Kato (1996), and Abramowicz, Chen, Granath & Lasota (1996). In the last paper, fully relativistic equations in Kerr geometry have been solved.

## 8. Real Starvers and Secret Guzzlers

In thick disks, the dominant energy source is the binding energy of matter in the disk, and the dominant energy sink is radiation and advection. Binding energy is converted to heat by viscous dissipation and radiated away, or advected into the black hole. Low radiative efficiency of thick accretion disks is due to significant advective losses of heat. In the ion tori (Rees, Begelman, Blandford & Phinney 1982) on the other hand, the dominant energy source is the rotational energy of the black hole, and the dominant energy sink is the Poynting flux carried away by jets emerging along the funnels. Radiation is insignificant. The energy is tapped from the hole and converted to jets by an electromagnetic analogue of the Penrose process. The low radiative efficiency is not due to advective losses but to the fact that the ion tori are optically thin. Indeed, Rees et al. (1982) have explicitly stated that accretion (and therefore advection) is not relevant at all for their model: all that is needed is the inertia of the torus. Abramowicz & Lasota (1995) have summarized these differences for models with *small* luminosities by calling the black holes with ion tori of the type discussed by Rees et al. (1982) "the real starvers" (accretion rate *must be* low), while the black hole with advectively dominated thick accretion disks "the secret guzzlers" (accretion rate could be relatively high, but the observed evidence for it is hidden: the luminosity is low).

Is there a way to observationally distinguish between a real starver and a secret guzzler? My answer is that, for small luminosities, stability properties are most likely to provide such a way. I have discussed stabilization of several types of unstable modes by advection. Secret guzzlers are advectively dominated, while the real starvers with small luminosities are not, and it has not yet been demonstrated that the starvers are stable. If they are, it would still be interesting to compare time behaviour of these two types of models: one guess that guzzlers should be more quiet.

On the other hand, their electromagnetic spectra must be rather similar owing to the fact that both guzzlers and starvers are optically thin and very hot — the spectra are probably determined in both cases mostly by Comptonization. Still, some differences could be induced by the (different) low temperature parts of the models.

Note, however, that at higher luminosities one may imagine a hybrid model that mixes properties of the ion tori and advectively cooled thick disks: these disks are guzzlers on a double diet! One should also keep in mind that optically *thick* slim accretion disk do explain the UV-optical part of active galactic nuclei spectra *much better* than the standard model does (Szuszkiewicz 1996, this volume).

## 9. Concluding Remarks

The present understanding of the role of advective cooling in hot accretion disks has benefited, to different degree, from many theoretical ideas: thick disks and the Roche lobe overflow, trapping radius, ion tori, stability conditions for accretion flows, discussion of the inner boundary condition and transonic flows, slim disks, boundary layers, physics of hot plasma and two-temperature disks, hot disks with advective cooling, topological unification of the local solutions... These theoretical developments have often been directly motivated by observations of objects with particularly interesting properties, suggestive of low radiative efficiency: underluminous active galactic nuclei, Galactic centre, soft X-ray transients.

What one should expects in the future? Certainly, present works on improvements in microphysics will continue, and significant results should be expected from simulations of turbulent viscosity (Balbus & Hawley 1996; Brandenburg et al. 1996; Kato 1996, in this volume), and from works on radiative transfer and convective instability. Global methods should be improved, and global numerical models (1-D, 2-D and 3-D) of accretion flows constructed (Chen 1996 in this volume). Serious models will be calculated directly in the Kerr geometry. Of a particular interest are (i) understanding which of the two possible stable solutions that exist for the low accretion rates is prefered in real flows and (ii) understanding of the optically thin-thick transitions (Narayan 1996; Yi 1996, in this volume). The transitions could be either radial (e.g., outer part of a disk optically thick, the inner part optically thin), vertical (optically thin corona above the optically thick disk as in Chen 1995b and Abramowicz, Chen & Taam 1995), or temporal (optical depths of the disk changing in time as in Lasota & Pelat 1991).

In the rapidly developing subject one should also expect unexpected: new, brilliant theoretical ideas, and new mind-opening observational discoveries. Indeed, I expect a lot of beautiful surprises ahead. Spiritus flat ubi vult.

## Acknowledgements

I wish to thank S. Kato, J.-P. Lasota, R. Narayan, and M. J. Rees for their helpful comments on the draft of this review.

## References

Abramowicz M. A. 1981, Nature 294, 235

Abramowicz M. A., Blaes O. M., Lu J. 1986, in Structure and Evolution of Active Galactic Nuclei, eds Giuricin et al. (Reidel, Dordrecht)

Abramowicz M. A., Calvani M., Nobili L. 1980, ApJ 242, 772
Abramowicz M. A., Chakrabarti S. 1990, ApJ 350, 281
Abramowicz M. A., Chen X., Granath M., Lasota J.-P. 1996, ApJ submitted
Abramowicz M. A., Chen X., Kato S., Lasota J.-P., Regev O. 1995, ApJL 438, L37
Abramowicz M. A., Chen X., Taam R. 1995, ApJ in press
Abramowicz M. A., Czerny B., Lasota J.-P., Szuszkiewicz E. 1988, ApJ 332, 646
Abramowicz M. A., Jaroszyński M., Sikora M. 1978, A&A 63, 221
Abramowicz M. A., Kato S. 1989, ApJ 336, 304
Abramowicz M. A., Kato S., Matsumoto R. 1989, PASJ 41, 1215
Abramowicz M. A., Lasota J.-P. 1995, Comments Astrophys 18, 141
Abramowicz M.A., Lasota J.-P., Xu C. 1986, in Quasars, eds G. Swarup, V.K. Kapachi (D. Reidel Pub. Co., Dordrecht)
Abramowicz M. A., Zurek W. H. 1981, ApJ 264, 314
Balbus S., Hawley J. F. 1996, in this volume
Begelman M. C. 1978, MNRAS 184, 53
Begelman M. C. 1979, MNRAS 187, 237
Begelman M. C., Meier D. L. 1982, ApJ 253, 873
Bertout C., Collin-Soufrin S., Lasota J.-P., (eds), 1991, Structure and Emission Properties of Accretion Disks (Editions Frontieres, Paris)
Björnsson G., Abramowicz M. A., Chen X., Lasota J.-P. 1995, ApJ submitted
Björnsson G., Svensson R. 1991, ApJL 371, L69
Blaes O. M. 1987, MNRAS 227, 975
Brandenburg A., Nordlund A., Stein R. F., Torkelsson U. 1996, in this volume
Cannizzo J. K. 1993, in Accretion Disks in Compact Stellar Systems ed J. C. Wheeler (World Scientific, Singapore), p6
Chen X. 1995a, MNRAS 275, 641
Chen X. 1995b, ApJ 448, 803
Chen X. 1996, in this volume
Chen X., Abramowicz M. A., Lasota J.-P. 1995, ApJ submitted
Chen X., Abramowicz M. A., Lasota J.-P., Narayan R., Yi I. 1995, ApJL 443, L61
Chen X., Taam R. E. 1993, ApJ 412, 254
Eggum G. E., Coroniti F. V., Katz J. I. 1987, ApJ 323, 634
Esin A. A., Narayan R., Ostriker E., Yi I. 1996, ApJ submitted
Fabian A. C., Rees M. J. 1995, MNRAS 277, L5
Fishbone L. G., Moncrief V. 1976, ApJ 207, 962
Frank J., King A. R., Raine D. 1992, Accretion Power in Astrophysics (Cambridge University Press, Cambridge)
Fukue J. 1987, PASJ 39, 309
Hawley, J. F., Smarr L. L., Wilson J. R. 1984, ApJ 277, 296
Honma F., Matsumoto R., Kato S. 1991, PASJ 43, 147

Igumenshchev I. V., Chen X., Abramowicz M. A. 1996, MNRAS 278, 236
Jaroszyński M., Abramowicz M. A., Paczyński B. 1980, ActaAstr 30, 1
Kato S., Abramowicz M. A., Chen X. 1996, PASJ in press
Kato S., Honma F. 1991, PASJ 43, 95
Kato S., Honma F., Matsumoto R. 1988, PASJ 40, 709
Kato S. 1996, in this volume
Katz J. I. 1977, ApJ 215, 265
Kozłowski M., Jaroszyński M., Abramowicz M. A. 1978, A&A 63, 209
Kusunose M. 1987, ApJ 321, 186
Kusunose M. 1996, in this volume
Kusunose M., Mineshige S. 1996, ApJ submitted
Lasota J.-P. 1995, in Compact Stars in Binaries. IAU Symp. 165, eds E. P. J. van den Heuvel, J. van Paradijs (Kluwer, Dordrecht)
Lasota J.-P. 1996, in Cataclysmic Variables and Related Objects. IAU Coll. 158, eds J. H. Wood et al. (Kluwer, Dordrecht)
Lasota J.-P. 1996, in this volume
Lasota J.-P., Abramowicz M. A., Chen X., Krolik J., Narayan R., Yi I. 1995, ApJ in press
Lasota J.-P., Narayan R., Yi I. 1995, A&A submitted
Lasota J.-P., Pelat D. 1991, A&A 249, 574
Lightman A. P. 1974, ApJ 194, 429
Lightman A. P., Eardley D. M. 1974, ApJL 187, L1
Lynden-Bell D. 1969, Nature 223, 690
Matsumoto R., Kato S., Fukue J., Okazaki A. T. 1984, PASJ 36, 71
Matsumoto R., Kato S., Honma F. 1989, in Theory of Accretion Disks, eds F. Meyer et al. (Kluwer, Dordrecht), p167
Meyer F., Duschl W. J., Frank J., Meyer-Hofmeister E. (eds) 1989, Theory of Accretion Disks (Kluwer, Dordrecht)
Moncrief V. 1980, ApJ 235, 1038
Muchotrzeb B. 1983, ActaAstr 33, 79
Muchotrzeb-Czerny B. 1986, ActaAstr 36, 1
Muchotrzeb B., Paczyński B. 1982, ActaAstr 32, 1
Narayan R. 1995, in Unsolved Problems in Astrophysics, eds J. Bahcall, J. P. Ostriker (Princeton University Press, Princeton) in press
Narayan R. 1996, in this volume
Narayan R. 1996, ApJ submitted
Narayan R., Kato S., Honma F. 1996, ApJ submitted
Narayan R., McClintock J. E., Yi I. 1996, ApJ in press
Narayan R., Popham R. 1993, Nature 362, 820
Narayan R., Yi I. 1994, ApJL 428 L13
Narayan R., Yi I. 1995a, ApJ 444, 231
Narayan R., Yi I. 1995b, ApJ 452, 710
Narayan R., Yi I., Mahadevan R. 1995, Nature 374, 623

Novikov I. D., Thorne K. S. 1973, in Black Holes, eds C. DeWitt, B. DeWitt (Gordon and Breach, New York), 343
Okazaki A.T., Kato S., Fukue J. 1987, PASJ 39, 457
Paczyński B. 1978, quoted in Kozłowski et al. (1978)
Paczyński B., Bisnovatyi-Kogan G. 1981, Acta Astron 31, 283
Papaloizou J. C. B,, Pringle J. E. 1984, MNRAS 208, 721
Piran T. 1978, ApJ 221, 652
Pringle J. E. 1981, ARA&A 19, 137
Pringle J. E., Rees M. J. 1972, A&A 21, 1
Pringle J. E., Rees M. J., Pacholczyk 1973, A&A 29, 179
Rees M. J. 1978, PhysScript 17, 193
Rees M. J. 1982, in The Galactic Center, eds G. Riegler, R. Blandford (AIP, New York)
Rees M. J. 1993, in The Renaissance of General Relativity and Cosmology, eds G. Ellis et al. (Cambridge University Press, Cambridge)
Rees M. J., Begelman M. C., Blandford R. D., Phinney E. S. 1982, Nature 295, 17
Shakura N. I. 1972, AstrZh 49, 921
Shakura N. I., Sunyaev R. A. 1973, A&A 24, 337
Shakura N. I., Sunyaev R. A. 1976, MNRAS 175, 613
Shapiro S. L., Lightman A. P., Eardley D. M. 1976, ApJ 204, 187
Shapiro S. L., Teukolsky S. A. 1983, Black Holes, White Dwarfs and Neutron Stars (Wiley, New York)
Stoeger W. R. 1976, A&A 53, 267
Szuszkiewicz E. 1996, in this volume
Szuszkiewicz E., Malkan M., Abramowicz M. A. 1996, ApJ in press
Taam R. E., Lin D. N. C. 1984, ApJ 287, 761
Thorne K. S., Price R. H. 1975, ApJL 195, L101
Wheeler J. C. 1981, Nature 294, 230
Wallinder F., Kato S., Abramowicz M. A. 1992, A&ARev 4, 79
Yi I. 1996, in this volume

# Advection-Dominated Accretion Flows: Optically Thin Solutions

Ramesh NARAYAN[1,2]
*1. Harvard-Smithsonian Center for Astrophysics, Cambridge, MA 02138, USA*
*2. Institute for Theoretical Physics, University of California Santa Barbara, CA 93106, USA*


### Abstract

General properties of advection-dominated accretion flows are discussed. Special emphasis is given to the optically thin branch of solutions, which has very high ion and electron temperatures and is thermally stable. This solution branch has been applied to a number of low-luminosity accreting black holes. The models have resolved some puzzles and have provided a straightforward explanation of the observed spectra. The success of the models confirms that the central objects in these low-luminosity sources are black holes. There is some indication that advection-dominated models may be relevant also for higher luminosity systems. The properties of the Low state of accreting black holes, the transition from the Low state to the High state, and the similarity of hard X-ray/$\gamma$-ray spectra of black hole X-ray binaries and active galactic nuclei, are explained.


## 1. Introduction

An advection-dominated accretion flow (ADAF) is defined as one in which a large fraction of the viscously generated heat is advected with the accreting gas, and only a small fraction of the energy is radiated. Although the basic idea of advection goes back a number of years (see the accompanying review by Abramowicz for a history of the subject), the relevance of ADAFs to real astrophysical systems was recognized only recently and much of the work in the subject dates back less than two years.

Advection-dominated accretion can occur in two different limits:

*S. Kato et al. (eds.), Physics of Accretion Disks, 15–28.*
© 1996 OPA (Overseas Publishers Association) Amsterdam B.V.

1. At very high mass accretion rates, radiation is trapped in the accreting gas because of the large optical depth and is advected with the flow. This limit of advection-domination, which typically occurs for mass accretion rates $\dot{M} > \dot{M}_{\rm Edd}$ (the Eddington rate), was considered initially by Begelman (1978) and Begelman & Meier (1982). Abramowicz et al. (1988) carried out a detailed study of the solutions and discussed their properties, including stability.

2. At sufficiently low $\dot{M}$, the accreting gas can become optically thin. The cooling time of the gas is then longer than the accretion time, and once again we have an ADAF. This regime was first mentioned by Rees et al. (1982) in the context of their "ion torus" model, and has been studied in detail in recent papers by Narayan & Popham (1993), Narayan & Yi (1994, 1995a, b), Abramowicz et al. (1995), Chen (1995), and Chen et al. (1995).

In section 2 of this review, we discuss the current status of our understanding of ADAFs in general. Section 3 then proceeds to a discussion of the particular features of the optically thin branch of solutions. Section 4 reviews applications of optically-thin ADAFs to various astrophysical systems, and section 5 concludes with some final remarks.

## 2. Structure of ADAFs

Ideally, we would like to understand the structure and properties of an ADAF in four dimensions, namely three coordinates $(R\theta\phi)$ and time $(t)$. Numerical simulations are the only way to achieve this goal, and given the enormous dynamic range involved and the messy radiation processes that one should include for a realistic calculation, no work has been done on so far the full 4D problem.

If we assume that the accretion flow is axisymmetric and in steady state, then we have a 2D problem $(R\theta)$. Even this is a hard problem, because it leads to partial differential equations with boundary conditions. No solutions are available yet.

The work done so far on steady state ADAFs is limited to 0D and 1D models, with some attempt to approximate the 2D solution via a $(1+1)$D approach. The following subsections summarize the results. In addition, there have been preliminary studies of time-dependent flows in 2D ($Rt$, Manmoto et al., this volume) and 3D ($R\theta t$, Igumenshchev, Abramowicz & Chen 1995).

ADAF studies are currently at a stage roughly similar to that achieved in the study of thin disks over the last 20 years. The bulk of the work on thin disks is based on the Shakura-Sunyaev (1973) model or its variants (cf. Frank, King & Raine 1992), which is a 0D solution in the language adopted here. Some attempts have been made to do 1D calculations to obtain the local vertical structure of thin disks (e.g., Shaviv & Wehrse 1991) or the global radial structure (e.g., Narayan & Popham 1993). In addition, there have been time-dependent 2D ($Rt$) calculations of disk instabilities and 3D ($R\theta t$) computations of boundary layers in thin disks.

### *2.1. 0D — The Basic Self-Similar Solution of ADAFs*

Narayan & Yi (1994) simplified the 2D axisymmetric steady state ADAF problem by assuming that the dynamical variables of the flow have power law dependences on $R$. This allowed them to evaluate directly all radial derivatives in the equations. Further, they eliminated $\theta$ by considering a height-integrated set of equations (the Slim Disk Equations, cf. Abramowicz et al. 1988). This results in a set of algebraic equations. The equations have an analytic self-similar solution which depends only on three parameters: the viscosity parameter $\alpha$, the ratio of specific heats of the accreting gas $\gamma$, and an advection parameter $f$ which is defined to be the ratio of the advected energy to the viscously generated energy. (Equivalently, the radiative efficiency is $1-f$.) A variant of the Narayan & Yi solution had been derived earlier by Spruit et al. (1987) in a different context.

The self-similar solution reveals many of the basic properties of ADAFs:
(i) The isothermal sound speed is very large, $c_s \gtrsim v_{\rm ff}/2$, where $v_{\rm ff} = (GM/R)^{1/2}$ is the free-fall velocity, with $M$ being the mass of the central object and $R$ the radius.
(ii) The radial velocity is $v \sim \alpha v_{\rm ff}$, where $v_{\rm ff} = (GM/R)^{1/2}$ is the free-fall velocity, with $M$ being the mass of the central object and $R$ the radius. The radial velocity is much larger than in thin disks. This follows from point (i).
(ii) The angular velocity is distinctly sub-Keplerian, $\Omega \lesssim \Omega_{\rm K}/2$, the exact value depending on $\gamma$. Here $\Omega_{\rm K} = (GM/R^3)^{1/2}$ is the Keplerian $\Omega$.
(iv) The Bernoulli parameter of the accreting gas, viz. the sum of the kinetic energy, the potential energy and the enthalpy, is positive. Since the Bernoulli parameter is conserved in an adiabatic flow, if the gas in an ADAF moves adiabatically to a large radius it will have a bulk velocity comparable to the escape velocity at its point of origin. This suggests a possible connection between ADAFs and jets (Narayan & Yi 1995a) which needs to be explored.
(v) Entropy increases inward and, therefore, the accreting gas is convectively unstable. This has been confirmed by Igumenshchev et al. (1995) using numerical experiments (see Chen, this volume). It has been suggested that convection may possibly enhance the effective viscosity in ADAFs (Narayan & Yi 1994, 1995a).

Point (i) implies that the temperature of the gas is nearly virial so that the vertical height of the gas is very large: $H \sim c_s/\Omega_{\rm K} \sim R$. This leads one to ask the following important question: Is it valid to use height-integrated equations when the flow has such a thick morphology?

### *2.2. 1D Solution in $\theta$ — Vertical Structure of ADAFs*

In a follow-up study, Narayan & Yi (1995a) avoided height-integration and considered an exact set of equations for a viscous axisymmetric steady state accretion flow. Assuming self-similarity in $R$, as had Begelman & Meier

(1982) earlier, they calculated exact numerical solutions of the resulting differential equations in $\theta$.

The solutions reveal that an ADAF is not at all disk-like in morphology, but nearly spherical. In fact, the flow configuration is similar to a differentially-rotating settling star, and there are no funnels. The radial velocity is maximum at the equator and goes to zero at the rotation pole, while the angular velocity and sound speed are nearly constant on spherical shells.

Despite the quasi-spherical morphology, the height-integrated analytic 0D solutions (section 2.1) are in excellent agreement with the exact 1D solutions. This implies that the height-integration approximation is quite accurate even for nearly spherical flows. How is this possible? The answer is that height-integration should not be interpreted as a cylindrical average over $z$ at a given cylindrical radius $R$, but rather as a spherical average over $\theta$ at a given spherical radius $R$. With this new interpretation, the Slim Disk Equations are a valid description of the radial variations of both slim, cooling-dominated flows and quasi-spherical, ADAFs. This result provides the technical underpinning for much of the work on ADAFs, nearly all of which is based on height-integrated equations.

It should be emphasized that although an ADAF has a quasi-spherical morphology, it is nevertheless very different from pure spherical accretion (Bondi 1952). The ADAF solutions discussed here describe rotating flows where the accretion is completely controlled by angular momentum transport via viscosity. Bondi spherical accretion, on the other hand, ignores rotation altogether and the flow involves merely a competition between gravity and pressure. Since in nearly all applications we expect the accreting gas to have significant angular momentum, the ADAF solutions are more realistic than the classical spherical solutions.

### *2.3. 1D Solution in R — Global Radial Solutions of ADAFs*

Recently, two groups have independently studied the global radial structure of ADAFs around black holes (Narayan, Kato & Honma 1996, Chen, Abramowicz & Lasota 1996). Using height-integrated equations (which are valid, see section 2.2) they have obtained numerical solutions over a range of radius $R$ with consistent boundary conditions at both ends. On the inside, the gas flows through a sonic point and falls supersonically into the black hole. Early work on a similar problem was reported by Matsumoto, Kato & Fukue (1985).

These studies show that physically self-consistent, stable, transonic, ADAF solutions are available for all values of $\alpha$ in the range $0 < \alpha \lesssim 0.3$. Therefore, there is no need for radial shocks in ADAFs. Standing shocks in viscous accretion disks have been discussed by Chakrabarti & Titarchuk (1995, and several other papers by Chakrabarti). It is claimed that low-$\alpha$ flows have shocks, while high-$\alpha$ flows do not, and that the former have sub-Keplerian rotation at large radii while the latter are Keplerian at all radii except very close to the black hole. These features are not confirmed by the global solutions of ADAFs obtained by Narayan et al. (1996) and Chen et al. (1996).

Another interesting result is that the 0D self-similar solution of Narayan & Yi (1994) is found to be quite a good representation of the global flow away from the boundaries. Near the sonic point, however, the radial velocity increases above the self-similar value, as does the angular velocity.

Global ADAF solutions with low values of $\alpha$ are found to be quite different from those with large $\alpha$. The former have sonic radii $R_s$ close to the marginally bound orbit, $R_s \sim 2R_g$ where $R_g = 2GM/c^2$ is the gravitational radius, while the latter have sonic radii close to or even outside the marginally stable orbit, $R_s \gtrsim 3R_g$.

Hydrostatic rotating thick tori with $R_s \sim 2R_g$ have been studied for many years as models of active galactic nuclei (Fishbone & Moncrief 1976; Abramowicz, Jaroszyński & Sikora 1978; Kozlowski, Jaroszyński & Abramowicz 1978; Paczyński & Wiita 1980; see Frank et al. 1992 for a simple discussion of the physics of these models). The models have (i) a region of super-Keplerian rotation near the inner edge, (ii) a radial pressure maximum within the disk which is the origin of the toroidal morphology, and (iii) twin empty funnels along the rotation axis. The funnels have been considered promising for the initiation of relativistic jets.

Global ADAF solutions with low values of $\alpha \lesssim 0.01$ are similar to thick tori in many respects. Because of the low viscosity, the radial velocity is small and there is a near hydrostatic balance between gravity, pressure and rotation. These flows have super-Keplerian rotation over a range of R and radial pressure maxima. Also, they have nearly empty funnels extending some distance along the rotation axis, though at large $R$ the funnels disappear and the flow becomes quasi-spherical as described in section 2.2. These models are thus improved and self-consistent versions of the earlier tori, where now angular momentum and energy are conserved at each $R$.

Global ADAF solutions with large $\alpha$, however, appear to be completely different. They have sub-Keplerian rotation at all radii, have no pressure maxima, and show no sign of an axial funnel. These flows represent truly dynamical structures where, in contrast to low-$\alpha$ flows, the radial velocity is substantial. Viscosity strongly influences the dynamics when $\alpha$ is large, as Matsumoto et al. (1984) recognized even in the case of thin accretion disks.

Narayan, Kato & Honma (1996) extended the global height-integrated 1D solutions to approximate 2D solutions in $R\theta$ by means of a $(1+1)$D approach, where they used the local $\theta$ structure described in section 2.2 to approximate the "vertical" structure at each $R$. The resulting isodensity contours confirm the quasi-spherical morphology and absence of funnels in large-$\alpha$ ADAFs.

### 2.4. In Defence of the $\alpha$ Viscosity Prescription

All the work on ADAFs so far has been based on the Shakura-Sunyaev $\alpha$ viscosity prescription. In this approach, the kinematic viscosity coefficient is written as $\nu = \alpha c_s^2/\Omega_K$ with $\alpha$ taken to be independent of $R$. A simpler approach is to set the shear stress equal to $\alpha P$, where $P$ is the pressure. In the opinion of this reviewer, the $\alpha$ prescription is well-motivated for an ADAF.

ADAFs have at least two linear instabilities, namely the Balbus-Hawley MHD instability (see articles by Balbus, Hawley and Brandenburg in this volume) and a convective instability (Narayan & Yi 1994, 1995a; Igumenshchev et al. 1995). One of these instabilities (or perhaps both) doubtless produces the shear stress which causes angular momentum transport. The effective $\nu$ must, therefore, depend on the nonlinear saturation of the chosen instability. Now, an ADAF has only one length scale, namely the radius $R$; in contrast to thin disks, where $H$ is a second scale, here we have $H \sim R$. Further, there is only one velocity scale in the problem, namely $v_{\rm ff}$; the bulk flow of the gas has a speed $\sim v_{\rm ff}$, the sound speed is $c_s \sim v_{\rm ff}$, and if we have equipartition magnetic fields as seems reasonable with the Balbus-Hawley instability, then the Alfvén speed too is $v_A \sim v_{\rm ff}$. This collapse of scales leads to an enormous simplification of the physics. Purely from dimensional analysis, it seems obvious that all linear instabilities must necessarily saturate with velocities $\sim v_{\rm ff}$ and coherence scales $\sim R$, so that the effective viscosity coefficient must necessarily scale as $\nu \propto v_{\rm ff} R \sim c_s^2/\Omega_K$. Further, the self-similar nature of the flow guarantees that the coefficient $\alpha$ in this relation will be independent of $R$. Therefore, the $\alpha$ prescription is particularly appropriate for ADAFs.

## 3. Optically Thin ADAFs

The rest of the article is devoted specifically to the optically thin branch of solutions. This section deals with general properties of the solutions, and the next section describes applications to a number of astrophysical objects.

### 3.1 The Critical Mass Accretion Rate

The optically thin branch is advection-dominated because of the poor radiative efficiency of the accreting gas. This requires the gas density to be sufficiently low that ion-electron coupling via Coulomb collisions (in a two-temperature plasma) becomes weak and cooling via synchrotron and bremsstrahlung radiation is unimportant. In addition, the optical depth has to be sufficiently low for Compton cooling to be modest. For these reasons, the optically thin branch exists only at low mass accretion rates.

In the following we scale radii in gravitational units, $r = R/R_g$, masses in solar units, $m = M/M_\odot$, and accretion rates in Eddington units, $\dot{m} =$

$\dot{M}/\dot{M}_{\rm Edd} = \dot{M}/1.39 \times 10^{18} m \ {\rm g\,s^{-1}}$, where we have assumed a standard efficiency factor of 10% in defining the Eddington rate.

Various investigators (Abramowicz et al. 1995; Narayan & Yi 1995b; Chen et al. 1995) have estimated, under different assumptions, the maximum or critical accretion rate $\dot{m}_{\rm crit}$ up to which optically-thin ADAFs exist. The value of $\dot{m}_{\rm crit}$ depends moderately on the radius ($\dot{m}_{\rm crit} \propto r^{-1/2}$ for large $r$), and quite strongly on the viscosity parameter ($\dot{m}_{\rm crit} \propto \alpha^2$). However, $\dot{m}_{\rm crit}$ is independent of $m$, and is insensitive to the magnetic field strength.

The most detailed calculations of $\dot{m}_{\rm crit}$ are due to Narayan & Yi (1995b). These calculations are based on a two-temperature plasma (Shapiro, Lightman & Eardley 1976), where all the energy from viscous dissipation first goes into the ions and then part of the energy is transferred to the electrons via Coulomb collisions. Cooling by bremsstrahlung and synchrotron radiation is included, along with Comptonization of these emissions. Pair processes are not included, but it has been checked a posteriori that the pair density is low (see Kusunose, this volume). All the gas parameters, including the ion temperature $T_i$ and the electron temperature $T_e$, are calculated self-consistently at each radius using the self-similar solution (section 2.1), and the advection parameter $f$ is calculated at each radius self-consistently via the local cooling.

These calculations show that $\dot{m}_{\rm crit} \sim 0.3\alpha^2$ for $r \lesssim 10^3$ and $\dot{m}_{\rm crit} \sim 0.3\alpha^2(r/10^3)^{-1/2}$ for $r \gtrsim 10^3$. This means that unless $\alpha$ is fairly large, say $\alpha > 0.1$, optically-thin ADAFs are restricted to extremely low $\dot{m}$ (see section 4.6).

### *3.2 Properties of Optically Thin ADAFs*

1. The accreting gas is extremely hot. The ion temperature is given by $T_i \sim 10^{12}{\rm K}/r$, which is nearly virial. The electron temperature at radii $r \lesssim 10^3$ is $T_e \sim 10^9 - 10^{10}$ K in the case of accreting black holes and $T_e \sim 10^{8.5} - 10^9$ K for accreting neutron stars (Narayan & Yi 1995b).

2. The accreting gas is thermally and viscously stable to long wavelength perturbations (Abramowicz et al. 1995; Narayan & Yi 1995b) and is only weakly unstable even to short wavelength perturbations (Kato, Abramowicz & Chen 1996; see Chen and Manmoto, this volume). Global stability is, therefore, assured.

3. The high electron temperature leads to a hard non-blackbody spectrum extending up to a few hundred keV. This, coupled with the thermal stability of the gas, makes these solutions extremely attractive for modeling accreting black holes (section 4), most of which are known to have hard spectra in X-rays and soft $\gamma$-rays (Tanaka & Lewin 1995; Maisack et al. 1993). Previously, the only hot accretion solution known was the two-temperature cooling-dominated solution of Shapiro et al. (1976), which is thermally very unstable (Piran 1978) and is, therefore, unsuitable for realistic models.

4. Because these flows are advection-dominated, they are by definition inefficient radiators. Therefore, the flows are underluminous for their accretion rate. Whereas a cooling-dominated accretion flow around a black hole has a luminosity $L \sim \dot{m} L_{\rm Edd}$, optically-thin ADAFs have $L \sim (\dot{m}^2/\dot{m}_{\rm crit}) L_{\rm Edd}$. Thus, especially for $\dot{m} \ll \dot{m}_{\rm crit}$, the radiative efficiency is extremely low.

## 4. Astrophysical Applications of Optically Thin ADAFs

Because (i) optically-thin ADAFs exist only for low values of $\dot{m}$, and (ii) the radiative efficiencies of these flows are low by construction, the most obvious applications are to low-luminosity systems. The majority of successful applications belong to this category, as we discuss in sections 4.3 and 4.4 (see also articles by Lasota and Yi, this volume). Recently, there has been an attempt to explain luminous sources as well using these solutions and this is discussed in section 4.5. So far, all applications have been limited to accreting black holes, but in principle the solutions could apply equally well to neutron stars (Narayan & Yi 1995b), and even to white dwarfs [for example, the hot inner flow in the siphon model of Meyer & Meyer-Hofmeister (1994) is likely to be advection-dominated.]

### *4.1 Spectral States of Accreting Black Holes*

It is well-known that black hole X-ray binaries (XRB) exhibit several distinct spectral states (Tanaka & Lewin 1995), and it is possible that active galactic nuclei (AGN) too have similar states. Before discussing specific applications, it is useful to describe a plausible scenario (Narayan 1995, 1996) in which we identify the role of advection-dominated flows in the various spectral states.

At the lowest luminosities, say $L < 10^{-4} L_{\rm Edd}$, we have the so-called Quiescent state or Off state. Soft X-ray transients (SXTs) exhibit this state in between outbursts, and very low luminosity AGN probably also correspond to this state. As we show in sections 4.3 and 4.4, there is very good evidence that all systems in the Quiescent state have optically-thin ADAFs, at least close to the black hole. The nature of the flow far from the black hole depends on the outer boundary condition. If the gas is introduced in a fairly hot or quasi-spherical state, then the flow would be advection-dominated at all $r$, as seems to be the case in low-luminosity AGN (e.g., Sagittarius A$^*$, section 4.3). However, if the incoming gas is cold and has a lot of angular momentum, then the flow would initially form a thin disk at large $R$. This is the case in SXTs. The observational evidence (section 4.4) suggests that the accreting gas switches from a thin disk to an ADAF at some large transition radius $r = r_{\rm tr} \gg 1$. Two physical mechanisms have been suggested for the transition: the coronal siphon mechanism of Meyer & Meyer-Hofmeister (1994), and evaporation via diffusive heating as described by Honma (1996) (see articles by both authors in this volume).

With increasing $\dot{m}$, the flow configuration is likely to remain essentially the same as in the Quiescent state until $\dot{m}$ crosses $\dot{m}_{\rm crit}$. However, the luminosity will increase quite dramatically ($L \propto \dot{m}^2$, section 3.2) and the system will become significantly brighter than in the Quiescent state. Narayan (1996) has suggested that systems with values of $\dot{m}$ approaching $\dot{m}_{\rm crit}$ correspond to the so-called Low State of accreting black holes. Observations indicate that sources in the Low state have luminosities $\sim (10^{-2} - 10^{-1})L_{\rm Edd}$, very hard spectra with photon indices $\sim 1.5 - 2$, and no hint of any soft component in the spectrum that might correspond to a thin accretion disk. These properties are consistent with optically-thin ADAFs with $\dot{m} \lesssim \dot{m}_{\rm crit}$ and $\alpha \sim 1$ (Narayan 1996, section 4.5).

Once $\dot{m}$ crosses $\dot{m}_{\rm crit}$, optically-thin ADAFs are no longer allowed, and the accreting gas has to switch to a different flow configuration; presumably, it becomes a pure thin accretion disk. Such a system would be quite luminous ($L \sim \dot{m}L_{\rm Edd}$), but with a soft multi-color blackbody spectrum. The High state of accreting black holes is known to match these properties. The change from the Low State to the High State happens quite suddenly according to the observations, and this is consistent with the ADAF model, where the transformation occurs when $\dot{m}$ crosses $\dot{m}_{\rm crit}$. The details of the transformation are probably as follows. When $\dot{m}$ approaches $\dot{m}_{\rm crit}$, we expect that the transition radius $r_{\rm tr}$ shrinks rapidly so that the outer thin disk progressively encroaches into the inner advection-dominated zone (Narayan 1996). During this stage, the spectrum would continue to be hard (though with a steepening power law index), and at the same time a soft component due to the thin disk would begin to increase in importance. Presumably, once $\dot{m}$ crosses $\dot{m}_{\rm crit}$, the thin disk extends all the way down to the black hole, and we have a full-fledged High state with a pure soft spectrum.

Finally, black hole systems also exhibit a so-called Very High state where the spectrum has both a luminous soft component and a substantial hard tail. It is claimed that this state corresponds to a higher $\dot{m}$ than the High state. The precise nature of the Very High state is not clear at this time. Perhaps, the flow consists of a standard thin disk and a corona, with the latter having a fairly substantial luminosity. The corona could be an ADAF in its own right (Narayan & Yi 1995b) or it may be efficiently cooled by Comptonization of disk photons (Haardt & Maraschi 1991).

### *4.2 Modeling Procedures*

The models of individual objects published in the literature involve a few approximations. The most serious simplification is that the flow is assumed to have a locally self-similar form (section 2.1), though the advection parameter $f$ is allowed to vary self-consistently with $r$ according to the local radiative efficiency (Narayan, McClintock & Yi 1996; Narayan 1996). Very recently, fully self-consistent models using the global flow described in section

2.3 have been constructed (Narayan, Kato & Honma 1996; Chen, Abramowicz & Lasota 1996; Nakamura, this volume) and it has been confirmed that the self-similar approximation is quite good for calculating the spectrum.

In addition to the black hole mass $m$ and mass accretion rate $\dot{m}$, each model requires a choice of the viscosity parameter $\alpha$, the strength of the magnetic field (e.g., the parameter $\beta$ in Narayan & Yi 1995b), and the transition radius $r_{tr}$ where the outer thin disk (if there is one) transforms to the inner advection-dominated flow. The plasma is assumed to be two-temperature (Shapiro et al. 1976), and thermal balance is enforced at each radius independently for the ions (viscous heating of ions balanced by Coulomb energy transfer to electrons) and the electrons (Coulomb heating balanced by radiative cooling). The radiative cooling is calculated with a radial Boltzmann-like radiative transfer code (rather than a full Monte Carlo calculation), with small corrections for non-spherical geometry and finite optical depth. The radiative transfer involves propagating the spectrum from one radial shell to the next, including emission via synchrotron and bremsstrahlung processes, Comptonization, and gravitational redshift. Electron positron pairs are included approximately, but their contribution is small.

The spectral calculations show that it is important to include synchrotron emission in the model since the Comptonization of synchrotron photons provides the bulk of the power in the X-ray band.

In the advection-dominated zone, the luminosity and spectrum depend only on two parameters, viz. $m$ and $\dot{m}/\alpha$, while in the outer thin disk, the relevant parameters are $m$, $\dot{m}$ and $r_{tr}$. The results are insensitive to the assumed strength of the magnetic field (provided it is not vanishingly small).

### *4.3 Sagittarius A* and Other Quiet AGN*

Sagittarius A* (Sgr A*) at the center of the Galaxy is an unusual source which has remained a longstanding mystery. Dynamical evidence suggests that it has a mass $\sim 10^6 M_\odot$ which, combined with its unusual spectrum, suggests that the object is a supermassive black hole. Observations of gas flows in the vicinity of Sgr A* indicate a mass accretion rate onto the central object $\sim 10^{-4} M_\odot \mathrm{yr}^{-1}$ (Melia 1992; Genzel, Hollenbach & Townes 1994). In a standard disk with an efficiency of $\sim 10\%$, this accretion rate would correspond to a luminosity of $\sim 10^{42}$ $\mathrm{ergs\,s^{-1}}$. In contrast, the actual luminosity observed is $\sim 10^{37}$ $\mathrm{ergs\,s^{-1}}$. Further, the spectrum is essentially flat in $\nu L_\nu$ from radio to X-rays, with a few bumps. This is very different from the spectrum expected with a standard thin disk. Both the dimness of Sgr A* and its peculiar spectrum were a puzzle until recently.

Narayan, Yi & Mahadevan (1995) showed that an optically-thin ADAF model with a $7 \times 10^5 M_\odot$ black hole accreting at $\dot{M}/\alpha \sim 10^{-5} M_\odot \mathrm{yr}^{-1}$ fits the spectrum of Sgr A* (including all upper limits) quite well (see also Rees 1982). The predicted spectrum extends from radio to hard X-rays and even fits the observed spectral bumps reasonably well. Although $\dot{M}$ in the model is not quite as large as the direct estimate of $10^{-4} M_{\odot} \mathrm{yr}^{-1}$ mentioned above, it is nevertheless quite an improvement over the $\dot{M} \sim 10^{-9} M_{\odot} \mathrm{yr}^{-1}$ which one obtains with a standard thin disk.

The key feature of this model of Sgr A* is that the flow is advection-dominated, which explains the low luminosity of the source despite a large $\dot{M}$. Fabian & Rees (1995) have extended the idea to a number of nearby elliptical galaxies which are believed to have large nuclear black holes ($M \sim 10^8 - 10^9 M_{\odot}$). The black holes in these galaxies must accrete at reasonable mass accretion rates (Fabian & Canizares 1988) and yet they are unusually dim. Once again, an ADAF model provides a natural explanation.

Recently, Lasota et al. (1996, see Lasota, this volume) have developed a model of the nucleus of NGC 4258 involving an optically-thin ADAF.

### *4.4 Quiescent Soft X-ray Transients*

The Quiescent state of SXTs, in between outbursts, has been very difficult to understand in models involving thin accretion disks. The spectrum of the prototypical SXT, A0620-00, in quiescence consists of two components (McClintock, Horne & Remillard 1995): (i) an optical/UV component which is nearly blackbody in shape, and (ii) a weak X-ray tail. The optical/UV component is consistent with a thin accretion disk except that the inner edge of the disk has to be at $r > 10^3$ instead of at $r = 3$ as one expects for a standard thin disk. This is very unusual. Further, the X-rays cannot be produced by any standard thin disk model; if one attempts to fit the observed temperature then the luminosity becomes too large, while if one fits the X-ray luminosity then the temperature is too low. In addition, if one assumes that the X-rays are produced close to the black hole, then it is hard to understand why the X-ray luminosity is much less than the optical luminosity when the viscous dissipation is expected to have just the opposite behavior.

Narayan, McClintock & Yi (1996) developed a model for quiescent SXTs which resolves these puzzles and explains the observations. The flow consists of a thin disk on the outside which extends down to $r = r_{\mathrm{tr}} \sim \mathrm{few} \times 10^3$, and an optically-thin ADAF on the inside from $r = r_{\mathrm{tr}}$ down to the black hole horizon at $r = 1$. The outer disk produces the optical and UV, while the ADAF produces the X-ray emission. The low luminosity of the X-rays is of course the result of advection-domination — nearly all the energy liberated by viscosity in the ADAF is advected into the black hole. Narayan et al. showed that they could fit the observed quiescent spectra of A 0620-00, V404 Cyg and Nova Muscae 1991 quite well. Mineshige (1996) and Lasota, Narayan & Yi (1996) have shown that the thermal disk instability model of SXT outbursts (Mineshige & Wheeler 1989) is in much better agreement with observations if the thin disk is truncated at a large $r_{\mathrm{tr}}$. This provides independent support for an ADAF at small radii in these systems.

### *4.5 Luminous Sources*

On the face of it, an advection-dominated model is not an obvious choice for modeling a luminous black hole system since the high luminosity probably implies fairly efficient radiation. The reason for attempting it is that many luminous systems have substantial fluxes in hard X-rays and $\gamma$-rays of up to $100 - 200$ keV (Narayan 1995), and one requires a hot model with electron temperatures $T_e > 10^9$ K. The optically-thin ADAF solution is currently the only known, self-consistent, dynamical model which is both hot and thermally stable. (Very recently, Esin et al. 1996 have discovered a cooling-dominated, hot, stable, one-temperature solution; applications of this solution have not yet been considered.)

Narayan (1996) has shown that advection-dominated models provide quite a good description of luminous black hole XRB and AGN, provided $\alpha$ is chosen to have a large value, $\alpha \sim 1$. The models correspond to $\dot{m}$ close to $\dot{m}_{\rm crit}$, where the radiative efficiency is moderately large. Spectral calculations give a good fit to observations in the Low State and provide a natural explanation for the transition from the Low State to the High State. The models also are consistent with the spectral evolution observed in SXTs during outburst.

The models indicate that $T_e$ is essentially independent of the black hole mass (cf. Narayan & Yi 1995b). This means that the X-ray/$\gamma$-ray spectra of black hole XRB and AGN should be qualitatively similar when they are in an advection-dominated state. Observations do suggest a close similarity. In contrast, a thin disk model predicts $T_{\rm eff} \propto m^{-1/4}$.

### *4.6 Constraint on $\alpha$*

The application to luminous black hole systems discussed above requires a large value of $\alpha \sim 1$. The reason is that $\dot{m}_{\rm crit}$ is proportional to $\alpha^2$ (section 3.1). If $\alpha$ is not large enough, then $\dot{m}_{\rm crit}$ becomes very small, and this means that even a system with the maximum allowed accretion rate, $\dot{m} = \dot{m}_{\rm crit}$, is just not luminous enough to match observations. Interestingly, even applications to low-luminosity systems seem to suggest large values of $\alpha$. In Sgr A*, the best agreement between the model $\dot{M}$ and the directly estimated $\dot{M}$ is obtained with $\alpha \sim 1$. Similarly, in quiescent SXTs, the models work best with $\alpha \sim 0.1 - 0.3$, and there are analogous indications in NGC 4258 as well.

All of these results have been obtained using the self-similar solution. It is possible that when the spectral calculations are done with a fully global model (section 2.3) the actual value of $\alpha$ needed may not quite be unity but may be somewhat smaller. However, it is not likely to reduce below $\sim 0.1 - 0.3$. Simulations of MHD instabilities in accretion flows generally do not give large values of $\alpha$ but rather give $\alpha \sim 0.01$ (see articles by Hawley and Brandenburg, this volume). The simulations, however, correspond to standard thin disks. If $\alpha$ increases with increasing $H/R$, then it is conceivable that $\alpha \gtrsim 0.1$ may be allowed in ADAFs which have $H/R \sim 1$.

## 5. Conclusion

To conclude, the optically-thin ADAF solution provides for the first time a thermally stable and dynamically self-consistent flow model which has the extremely high electron temperatures ($T_e > 10^9$ K) needed to explain X-ray/$\gamma$-ray observations of accreting black holes (and some neutron stars). Spectra calculated with these models resemble observations quite well. The models work convincingly in low-$\dot{m}$ systems (sections 4.3 and 4.4) and show promise even for high-$\dot{m}$ systems, provided $\alpha$ is large (section 4.5).

The detailed models developed so far are based on a two-temperature plasma. One may interpret the success of the models as empirical support for the two-temperature idea (Narayan, McClintock & Yi 1996).

Perhaps the most exciting aspect of these models is that advection is not merely a minor perturbation but is in some sense the whole story. Especially for the applications discussed in sections 4.3 and 4.4, it is only because of massive advection of viscous energy into the black hole that the models work at all. The existence of a horizon is critical in order to ensure that whatever energy falls into the central object disappears without being re-radiated. It is not enough just to have a relativistic central star — we specifically need a *black hole.* The success of the models in the various applications discussed in this article could, therefore, be considered as "proof" that these particular systems do have central black holes.

## Acknowledgements

The author thanks J.-P. Lasota, T. Manmoto and I. Yi for comments on the manuscript. This work was supported in part by NASA grant NAG 5-2837 (to the Harvard-Smithsonian Center for Astrophysics) and NSF grant PHY 9407194 (to the Institute for Theoretical Physics).

## References

Abramowicz M, Chen X., Kato S., Lasota J. P., Regev O. 1995, ApJL 438, L37
Abramowicz M., Czerny B., Lasota J. P., Szuszkiewicz E. 1988, ApJ 332, 646
Abramowicz M., Jaroszyński M., Sikora M. 1978, A&A 63, 221
Begelman M. C. 1978, MNRAS 184, 53
Begelman M. C., Meier D. L. 1982, ApJ 253, 873
Bondi H. 1952, MNRAS 112, 195
Chakrabarti S. K., Titarchuk L. G. 1995, ApJ 455, 623
Chen X. 1995, MNRAS 275, 641
Chen X., Abramowicz M., Lasota J. P. 1996, ApJ submitted
Chen X., Abramowicz M., Lasota J. P., Narayan R., Yi I. 1995, ApJL 443, L61
Esin A. A., Narayan R., Ostriker E., Yi I. 1996, ApJ in press

Fabian A. C., Canizares C. R. 1988, Nature 333, 829
Fabian A. C., Rees M. J. 1995, MNRAS 277, L55
Fishbone L. G., Moncrief V. 1976, ApJ 207, 962
Frank J., King A., Raine D. 1992, Accretion Power in Astrophysics (Cambridge Univ. Press, Cambridge)
Genzel R., Hollenbach D., Townes C. H. 1994, Rep. Prog. Phys. 57, 417
Haardt F., Maraschi L. 1991, ApJL 380, L51
Honma F. 1995, PASJ 48 in press
Igumenshchev I. V., Abramowicz M., Chen X. 1995, MNRAS in press
Kato S., Abramowicz M., Chen X. 1996, PASJ 48 in press (Feb. 25)
Kozlowski M., Jaroszyński M., Abramowicz M. 1978, A&A 63, 209
Lasota J. P., Abramowicz M. A., Chen X., Krolik J., Narayan R., Yi I. 1996, ApJ 462 in press
Lasota J. P., Narayan R., Yi I. 1996, A&A submitted
Maisack M., et al. 1993, ApJL 407, L61
Matsumoto R., Kato S., Fukue J. 1985, in Theoretical Aspects on Structure, Activity, and Evolution of Galaxies III, eds S. Aoki, M. Iye, Y. Yoshii (Tokyo Astr. Obs. Publ., Tokyo), p102
Matsumoto R., Kato S., Fukue J., Okazaki A. T. 1984, PASJ 36, 71
McClintock J. E., Horne K., Remillard R. A. 1995, ApJ 442, 358
Melia F. 1992, ApJL 387, L25
Meyer F., Meyer-Hofmeister E. 1994, A&A 132, 184
Mineshige S. 1996, PASJ in press
Mineshige S., Wheeler J. C. 1989, ApJ 343, 241
Narayan R. 1995, in Some Unsolved Problems in Astrophysics, ed J. N. Bahcall, J. P. Ostriker (Princeton Univ. Press, Princeton)
Narayan R. 1996, ApJ in press (May 1996)
Narayan R., Kato S., Honma F. 1996, ApJ submitted
Narayan R., McClintock J. E., Yi I. 1996, ApJ 457 in press
Narayan R., Popham R. G. 1993, Nature 362, 820
Narayan R., Yi I. 1994, ApJL 428, L13
Narayan R., Yi I. 1995a, ApJ 444, 231
Narayan R., Yi I. 1995b, ApJ 452, 710
Narayan R., Yi I., Mahadevan R. 1995, Nature 374, 623
Paczyński B., Wiita P. J. 1980, A&A 88, 23
Piran T. 1978, ApJ 221, 652
Rees M. J. 1982, in The Galactic Center, ed Riegler, R. D. Blandford
Rees M. J., Begelman M. C., Blandford R. D., Phinney E. S. 1982, Nature 295, 17
Shakura N. I., Sunyaev R. A. 1973, A&A 24, 337
Shapiro S. L., Lightman A. P., Eardley D. M. 1976, ApJ 204, 187
Shaviv G., Wehrse, R. 1991, A&A 251, 117
Spruit H. C., Matsuda T., Inoue M., Sawada K. 1987, MNRAS 229, 517
Tanaka Y., Lewin W. H. G. 1995, in X-Ray Binaries, eds W. H. G. Lewin, J. van Paradijs, E. P. J. van den Heuvel (Cambridge Univ. Press, Cambridge)

Photographs at the banquet.

Photographs at the banquet (upper) and Dr and Mrs Kato receiving bouquets (lower).

# Global Structure and Phase Transition of Advection-Dominated Accretion Disks

Fumio HONMA
*Department of Astronomy, Faculty of Science, Kyoto University, Sakyo-ku, Kyoto 606-01, Japan*

**Abstract**

A global accretion disk model is constructed taking the radial energy transport by means of the turbulent diffusion into account. Solutions in which the outer part of the disk is the standard optically thick disk and the inner part is the optically-thin advection-dominated disk are obtained. It is shown that if initially the hot advection-dominated flow forms in the innermost part of the cold disk, this kind of global solution automatically realizes within the accretion time scale of the outer cold disk. Comparison with the black hole candidates and AGN's are briefly discussed.

## 1. Introduction

The optically thin advection-dominated accretion disk is important as a model of various kinds of objects which have the non-thermal emission spectra, such as the black hole candidates and the active galactic nuclei.

In some situations, the advection-dominated disk is considered to realize at the innermost region of the standard optically-thick disk. As the transition mechanism from the standard disk to the advection-dominated disk, the thermal instability (Shapiro et al. 1976) and the vertical evapolation (Meyer & Meyer-Hofmeister 1994) are proposed so far (Narayan & Yi 1995). However, self-consistent global disk models have not yet constructed.

We constructed a global disk model by taking the energy transport in the radial direction by means of the turbulent diffusion into account. Basic idea of the modelling is simple. In the advection-dominated disk, the viscous heating rate per unit radius is $\propto \dot{M}/r^2$. On the other hand, the radiative cooling rate by means of the free-free emission is $\propto \alpha^{-2}\dot{M}^2 r^{-3/2}$. Here, $\dot{M}$ is the mass accretion rate and $\alpha$ the viscosity parameter. As a consequence,

*S. Kato et al. (eds.), Physics of Accretion Disks, 31–36.*
© 1996 OPA (Overseas Publishers Association) Amsterdam B.V.

the radiative cooling dominates the viscous heating outside a certain radius, which is $\propto \alpha^4 \dot{M}^{-2}$. If the energy is transported from the inside of this radius to the outside by means of the diffusion, the steady state realizes.

## 2. Global Structure

In this section we show the structure of the steady disk model. More detailed discussion is given in Honma (1996).

We consider accretion disks around a black hole. The general relativistic effects are simulated by using a pseudo-Newtonian potential $\psi = -GM/(r - r_g)$ introduced by Paczyński and Wiita (1980). We assume that the disk is geometrically thin.

Basic equations are as follows. Firstly, we use three components of the equation of motion:

$$v\frac{dv}{dr} + \frac{1}{\rho}\frac{dp}{dr} - \frac{l^2}{r^3} + \frac{GM}{(r - r_g)^2} = 0, \tag{1}$$

$$\dot{M}(l - l_{in}) = 4\pi r^2 H \tau_{r\varphi}, \tag{2}$$

$$\frac{p}{\rho} = (\Omega_K H)^2, \tag{3}$$

where $v$, $\rho$, $p$, and $l$ are the inward radial velocity, the density, the pressure, and the specific angular momentum of the gas, respectively. Quantity $H$ is the half-thickness of the disk, $\Omega_K$ the Keplerian angular velocity, and $\tau_{r\varphi}$ the radial-azimuthal component of the viscous stress tensor, and $l_{in}$ an integration constant. The energy equation of the gas is

$$\frac{\dot{M}}{4\pi}T\frac{ds}{dr} - \frac{d}{dr}(rHF_d) + rH\rho\kappa_P c(U - aT^4) - rH\tau_{r\varphi}r\frac{d\Omega}{dr} = 0, \tag{4}$$

where $s$ and $T$ are the specific entropy and the temperature of the gas, respectively, $F_d$ the radial diffusive energy flux, $\kappa_P$ the Planck mean opacity for the free-free absorption, and $U$ the energy density of the radiation field. The radiation field is determined by the transfer equation:

$$\frac{\dot{M}}{4\pi}\frac{U}{\rho}\left(\frac{d\ln U}{dr} - \frac{4}{3}\frac{d\ln\rho}{dr}\right) + \frac{d}{dr}\left(\frac{rHc}{3\rho\kappa_R}\frac{dU}{dr}\right) - \frac{rcU}{1 + 3\rho\kappa_R H} - rH\rho\kappa_P c(U - aT^4) = 0, \tag{5}$$

where $\kappa_R$ is the Rosseland mean opacity for the free-free absorption and the electron scattering. Finally, the equation of state is $p = \rho kT/\mu m_H + U/3$.

For the model of $\tau_{r\varphi}$ and $F_d$, we employ the $\alpha$-prescription (Shakura & Sunyaev 1973), i.e.,

$$\tau_{r\varphi} = -\alpha\rho c_s H r\frac{d\Omega}{dr} \quad \text{and} \quad F_d = -\alpha\rho c_s H T\frac{ds}{dr}, \tag{6}$$

where $c_s = \sqrt{p/\rho}$.

We obtain solutions of these set of equations which satisfy the condition that the inner part of the disk is the optically-thin advection-dominated disk and the outer part is the standard optically-thick disk. Figure 1 shows the radial temperature distribution of a typical solution for $M = 10M_\odot$, $\dot{M} = 0.01\dot{M}_{\rm Edd}$, and $\alpha = 0.4$, where, $\dot{M}_{\rm Edd} = 16L_{\rm Edd}/c^2$. It is notable that the radiative cooling-dominated region, which is expected to exist from the naive consideration given in section 1, is reduced to a thin boundary layer between the inner hot disk and the outer cold disk. In this region, the radiative cooling balances with the negative divergence of the diffusive energy flux.

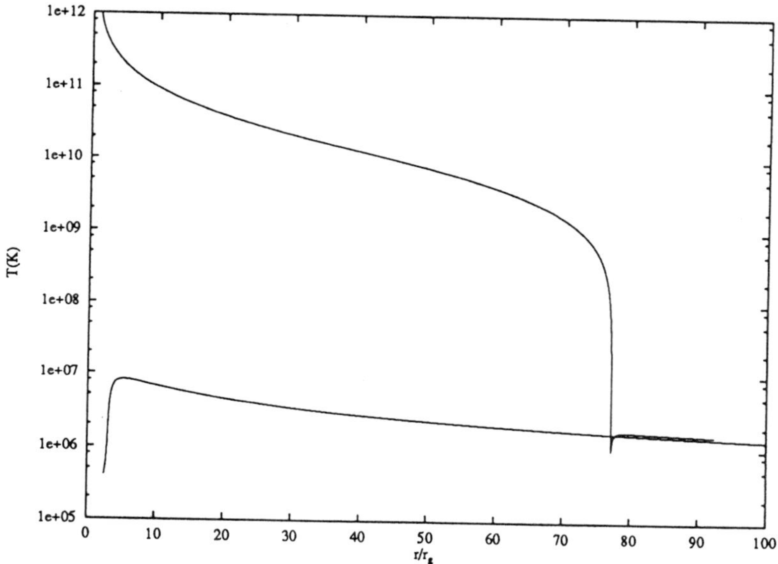


Fig. 1. Upper curve shows the radial temperature distributions of a solution for $M = 10M_\odot$, $\alpha = 0.4$, and $\dot{M} = 0.01\dot{M}_{\rm Edd}$. Lower curve shows that of the standard optically thick disk.

Figure 2 shows the radial distribution of the specific angular momentum. The rotation law of the inner hot disk and that of the outer cold disk are different: In the inner disk, $l \propto r$. On the other hand, $l \propto r^{1/2}$ in the outer disk. These two curves join continuously in the boundary layer.

Figure 3 shows the radii where the transition begins, $r_{\rm tr}$, as functions of the accretion rate for several values of $\alpha$. Two solutions exist for a given value of $\dot{M}$ smaller than a certain critical value. For the outer solution, the behavior of $r_{\rm tr}$ is well expressed as

$$r_{\rm tr} \sim 0.3\alpha^4 \left(\frac{\dot{M}}{\dot{M}_{\rm Edd}}\right)^{-2} r_{\rm g}. \tag{7}$$

Note that the parameter dependence of $r_{tr}$ coincides with that expected from the consideration in section 1. In the vicinity of the inner edge of the disk, $r_{tr}$ deviates from equation (7) due to the general relativistic effects.

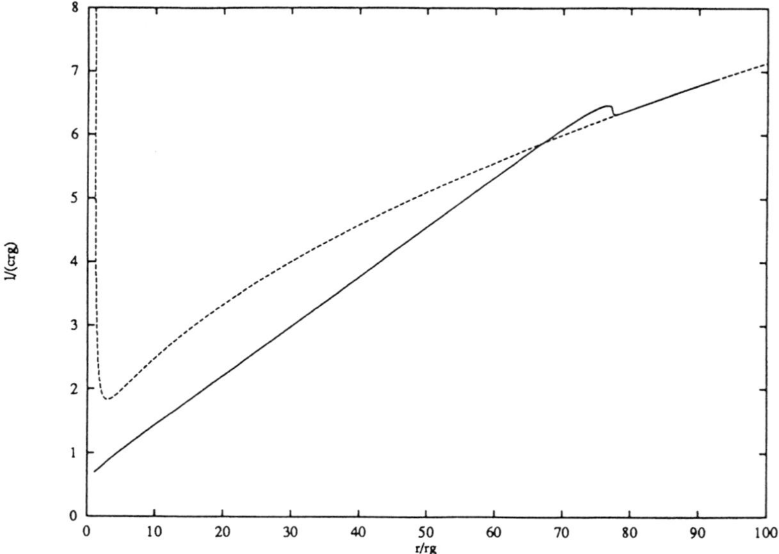


Fig. 2. Radial distribution of the specific angular momentum. Values of the parameters are the same as those of figure 1. Broken curve is the Keplerian angular momentum, $l_K = r^2\Omega_K$.

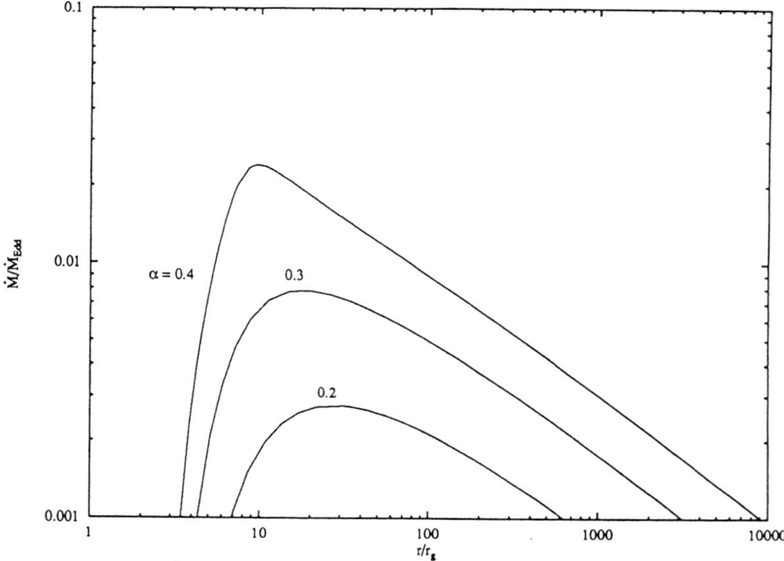


Fig. 3. Radii where the transition begins for $M = 10M_\odot$ and $\alpha = 0.2, 0.3,$ and 0.4.

## 3. Phase Transition

In order to consider the time dependent behavior, we divide the disk into the optically-thin hot inner part and the optically-thick cold outer part. Since the time scale of the disk evolution is the accretion time scale of the outer cold disk, the inner hot disk is considered to be steady at each time. The steady state is characterized by $r_{\rm tr}$. For each value of $r_{\rm tr}$, the accretion rate of the inner region, $\dot{M}_{\rm in}(r_{\rm tr})$, is determined by the conditions that $T = 0$, $l = l_{\rm K}$, and $F_{\rm d} = 0$ at $r = r_{\rm tr}$.

The behavior of the outer cold disk is described by the well-known set of equations for the optically thick Keplerian disk. That is, the mass conservation equation

$$\frac{\partial}{\partial t}(\rho H) = \frac{1}{r}\frac{\partial}{\partial r}\left[\frac{1}{dl_{\rm K}/dr}\frac{\partial}{\partial r}(r^2 H\tau_{r\varphi})\right], \tag{8}$$

and the energy equation

$$H\rho T\left(\frac{\partial s_{\rm tot}}{\partial t} - v\frac{\partial s_{\rm tot}}{\partial r}\right) = -\frac{caT^4}{3\rho\kappa_{\rm R}H} - H\tau_{r\varphi}r\frac{d\Omega_{\rm K}}{dr}, \tag{9}$$

where $s_{\rm tot}$ is the total specific entropy of the gas and the radiation field.

These two regions are joined at $r = r_{\rm tr}$ by the mass conservation condition:

$$4\pi r H\rho\frac{dr_{\rm tr}}{dt} = \dot{M}_{\rm in}(r_{\rm tr}) - \dot{M}. \tag{10}$$

In addition, we require the continuity of the viscous torque, $4\pi r^2 H\tau_{r\varphi}$, at $r = r_{\rm tr}$, in order that the angular momentum is conserved.

From the functional form of $\dot{M}_{\rm in}$ shown in figure 3 and equation (10), we can expect that if the initial value of $r_{\rm tr}$ is larger than that of the inner solution which corresponds to the mass supply rate from the outer boundary, the disk settles down to the outer steady state within the accretion time scale of the outer cold disk.

Figure 4 shows an example of the evolution of the disk with $M = 10M_\odot$ and $\alpha = 0.3$. Initially, the outer disk is the steady state with $\dot{M} = 5 \times 10^{-3}\dot{M}_{\rm Edd}$ and $\dot{M}_{\rm in} = 5.05 \times 10^{-3}\dot{M}_{\rm Edd}$.

## 4. Discussion

From the point of view of the comparison with the observation, the present model will be interesting in three points:

(1) The non-thermal spectra observed in the black hole candidates and AGN's are considered to be produced by means of the unsaturated Comptonization of the soft photon in the hot disk. In the present model, the soft photon with luminosity $\sim GM\dot{M}/r_{\rm tr}$ is generated in the boundary layer, which can be Comptonized in the inner disk.

(2) The rapid time variability is observed in the hard state of the black hole candidates. In the present model, the negative gradient of $l$, the strong shear and the entropy gradient in the boundary layer is expected to induce many kinds of instabilities. The fluctuation of the boundary layer due to these instabilities can be an origin of the time variability.

(3) The present model shows that if a sufficiently wide hot region is generated initially, the cold disk automatically transits to the steady state with the hot inner region. This can be a clue to understand the mechanism of the soft-hard state transition of the black hole candidates.

Various problems remains. Firstly, detailed calculation of the emission spectra and the comparison with the observations are needed. Secondly, mechanisms which generate the hot inner region and initiate the soft-hard transition is not known. Improvements of the model, such as inclusion of the high energy process and two dimensional treatment are also necessary.

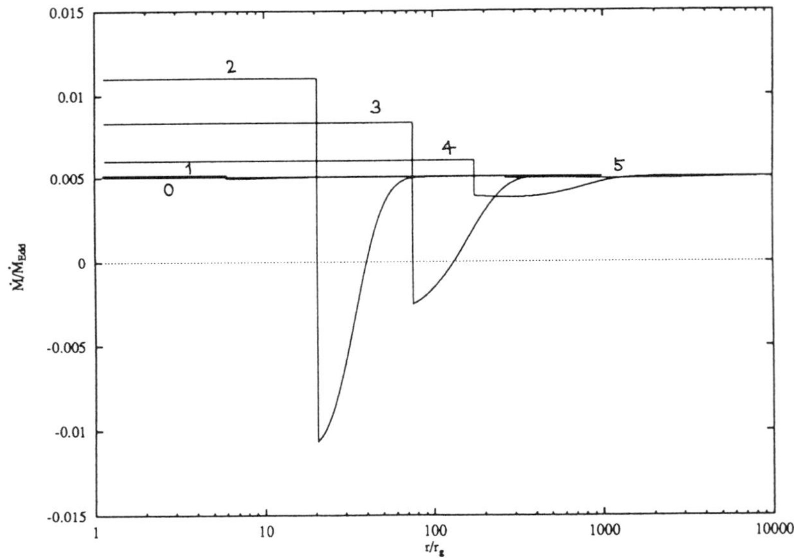


Fig. 4. Time evolution of the radial distribution of the accretion rate for $M = 10M_\odot$ and $\alpha = 0.3$. Curves labeled 0, 1, 2, 3, 4, and 5 correspond to the time 0, $10^5$, $10^6$, $10^7$, $10^8$, and $10^9$ in units of $r_g/c$, respectively.

## References

Honma F. 1996, PASJ in press

Meyer F., Meyer-Hofmeister E. 1994, A&A 288, 175

Narayan R., Yi I. 1995b, ApJ 452, 710

Paczyński B., Wiita P. J. 1980, A&A 88, 23

Shakura N. I., Sunyaev R. A. 1973, A&A 24, 337

Shapiro S. L., Lightman A. P., Eardley D. M. 1976, ApJ 204, 187

# The Radial Structure of Advection-Dominated Accretion Disks

Thomas BECKERT[1], Wolfgang J. DUSCHL[1,2], and Andrew R. KING[3]
*1. Institut für Theoretische Astrophysik der Universität Heidelberg, Tiergartenstr. 15, 69121 Heidelberg, Germany*
*2. Max-Planck-Institut für Radioastronomie, Auf dem Hügel 69, 53121 Bonn, Germany*
*3. Astronomy Group, University of Leicester, Leicester LE1 7RH, England*

**Abstract**

We give scaled forms of the equations governing the radial structure of advection-dominated flows. These can be used to show that the viscosity in such flows always takes a value close to the maximum which is physically reasonable ($\nu \sim r v_K$). Numerical results are presented and it is shown that the influence of the inner boundary condition can persist at large radii, calling into question the applicability of self-similar solutions.

## 1. Introduction

Since the appearance of the 'Slim Disk' models for the inner parts of accretion disks (Abramowicz 1988) it has been recognized that advection plays an important role for the structure and stability of accretion disks surrounding black holes. Recently optically thin advection-dominated flows have been investigated and applied to Sgr A*, the radio source at the dynamical center of our galaxy (Narayan et al. 1995). Here we present a scaled version of the vertically integrated equations which allows considerable insight into the nature of advection-dominated flows.

*S. Kato et al. (eds.), Physics of Accretion Disks, 37–40.*
© 1996 OPA (Overseas Publishers Association) Amsterdam B.V.

## 2. Equations and Scaling

We use the steady-state vertically-averaged equations given by Narayan & Yi (1994). In particular the transport of angular momentum is described by

$$\frac{\dot{M}}{2\pi}\frac{\partial}{\partial r}\left(r^2\Omega\right) = -\frac{\partial}{\partial r}\left(\nu\Sigma r^3\frac{\partial\Omega}{\partial r}\right), \tag{1}$$

and we allow for a radial entropy flow:

$$-\Sigma T v\frac{\partial s}{\partial r} = \nu\Sigma\left(r\frac{\partial\Omega}{\partial r}\right)^2 - 2F^-, \tag{2}$$

which is balanced by viscous heating $Q^+$ and radiative cooling. We assume that the cooling is proportional to the local heating, which can be expressed as $2F^- = (1-f)Q^+$.

To derive dimensionless equations we parametrize the radius in terms of the Schwarzschild radius, $r = xr_S$, and the radial velocity in terms of the local Keplerian azimuthal velocity, $v_r = -v = -\beta(x)v_K(x)$. We scale the angular velocity arbitrarily to the value at the outer boundary $\Omega = g(x)\Omega_0$, so $g(x)$ is normalized to $g(x_{\text{out}}) = 1$ at the outer radius. The angular momentum equation can be integrated directly, giving a constant of integration which can be used to satisfy the inner boundary condition. Here we consider solutions in which the constant vanishes; these are in any case asymptotically valid far from the inner edge of the disk. As a result we can write the viscosity as

$$\nu = -v\frac{g}{g'}, \tag{3}$$

which appears in the entropy equation (2). In terms of the scaled quantities the full set of equations now reduces to two equations for the radial disk structure:

$$g'' = a\,f\,\frac{x^4}{x_0^3}\frac{(g')^3}{\beta} - b\,\frac{g'}{x} + d\,\frac{\beta' g'}{\beta} + \frac{g'^2}{g}, \tag{4}$$

$$\beta' = \frac{\beta}{2x} - \frac{1}{\beta x} + \frac{(\Omega_0 g x)^2}{\Omega_K^2(x_{\text{out}})\beta x_0^3} - \frac{c_s^2}{\beta v_K^2}\frac{1}{\rho}\rho'. \tag{5}$$

Here a prime denotes differentiation with respect to the radial coordinate $x$ and $a, b, d$ are constants. From the scaled equations it is easy to show that the viscosity in all advection-dominated solutions must be close to its maximum conceivable value, i.e., $\nu \sim rv_K$. We shall present a full derivation of this result in a forthcoming paper; it holds quite independently of the $\alpha$ prescription (Shakura & Sunyaev 1973), which we adopt for the solutions given below.

## 3. Special Solutions

### *3.1. The self-similiar solution*

The equation for the radial velocity is reduced to an algebraic relation if $\beta'$ is assumed to vanish for all $x$. This means that the radial velocity is proportional to the Keplerian velocity $v = \beta(\alpha, f)v_K(r)$. Assuming a power-law for $g(x)$ the exponent is uniquely determined and there is no freedom for choosing $\Omega_0$;

$$\Omega(x) = \frac{c}{r_S}\sqrt{\frac{1}{2x_0^3}}\left(1 - \frac{\beta}{\alpha} - \frac{\beta^2}{2}\right)\left(\frac{x_0}{x}\right)^{\frac{3}{2}}. \tag{6}$$

As a consequence, the sound speed is also proportional to the Keplerian velocity $c_s^2 = c^2\beta/3\alpha x$ and the Mach number is a constant for all radii $\mathcal{M}^2 = 1.5\alpha\beta$. This is the self-similar solution found by Narayan & Yi (1994).

### *3.2. Solid-body rotation*

The other extreme solution can be obtained by recognizing that equation (2) has a trivial solution if $g'(x) \to 0$. This solution can only be reached asymptotically, because it has $H/r \to \infty$ and $\nu \to \infty$. The corresponding flow is either hyperthick ($H/r \gg 1$) or the temperature exceeds the virial temperature.

### *3.3. Numerical solutions*

The differential equations (4) and (5) can be integrated using a shooting method. Figure 1 shows the results for a fixed value of $f = 0.9$ of the ratio of advective cooling to heating, and $\Omega_0 = 0.2\Omega_K(r_{\rm out})$. The boundary value at the inner edge for $\Omega(x = 2)$ affects the structure of the disk for all radii. The flow stays subsonic in our Newtonian treatment, and the disks are slim ($H/r \sim 1$).

## 4. Discussion

We have given a scaled form of the equations governing the radial structure of advection-dominated accretion disks. These can be used to show that the viscosity always takes a value close to the maximum which is physically reasonable ($\nu \sim r v_K$). The self-similar solution can only be realized in a region where the boundary conditions have essentially no influence. The numerical solutions presented here show that such a region may not necessarily exist.

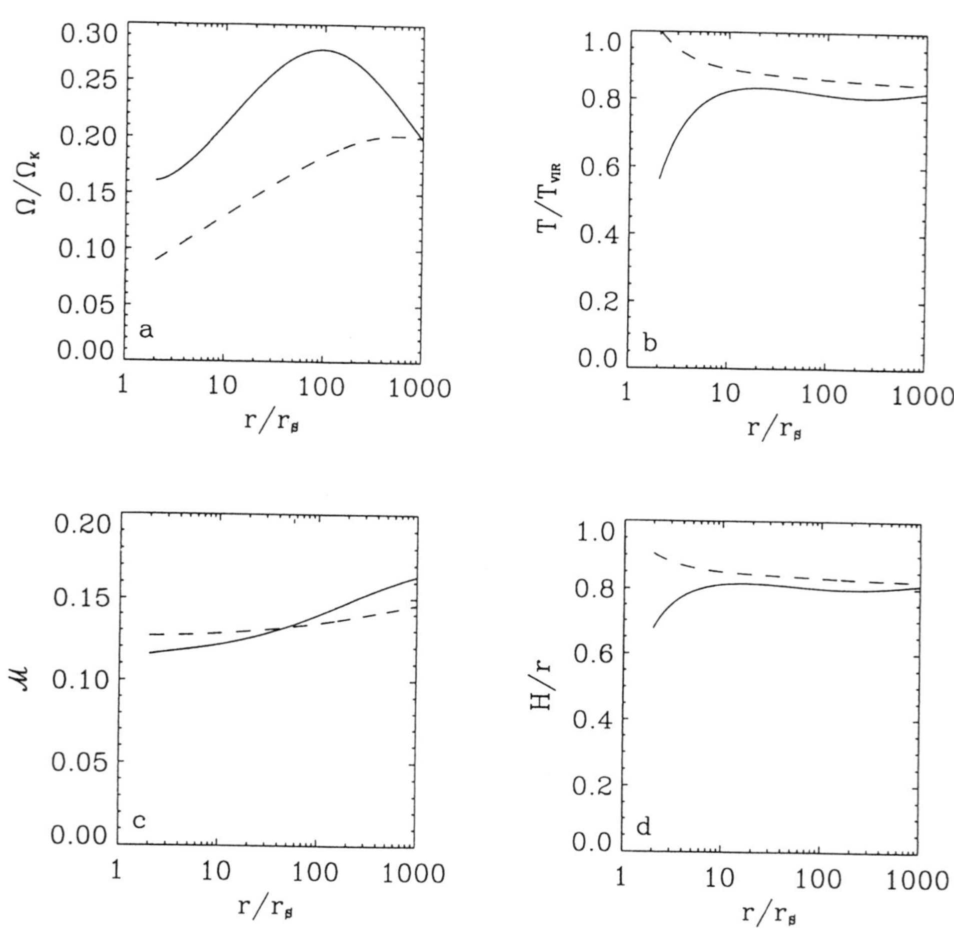


Fig. 1. Two numerical solutions for the structure of advection-dominated disks with $\Omega_0 = 0.2\Omega_K(10^3 r_S)$. The solid curves denote a solution defined by $\beta(x = 2) = 0.08$ , and the dashed curves a solution defined by $\beta(2) = 0.11$ . Panel *a* shows the angular velocity divided by the local Kepler velocity. Panels *b*, *c* and *d* show the temperature in units of the virial temperature, the Mach number, and the height divided by the radius respectively.

## References

Abramowicz M., Czerny B., Lasota J. P., Szuszkiewicz E. 1988, ApJ 332, 646
Narayan R., Yi I. 1994, ApJL 428, L13
Narayan R., Yi I., Mahadevan R. 1995, Nature 374, 623
Shakura N. I., Sunyaev R. A. 1973, A&A 24, 337

# Global Structures of Advection-Dominated Two-Temperature Accretion Disks

Kenji E. NAKAMURA[1], Ryoji MATSUMOTO[2], Masaaki KUSUNOSE[3], and Shoji KATO[1]
*1. Department of Astronomy, Faculty of Science, Kyoto University, Sakyo-ku, Kyoto 606-01, Japan*
*2. Department of Physics, Faculty of Science, Chiba University, Inage-ku, Chiba, 263, Japan*
*3. Department of Astronomy, University of Texas at Austin, Austin, TX, 78712, USA*

**Abstract**

Steady global models for an optically-thin, advection-dominated, two-temperature accretion disk surrounding a black hole are presented. Disks are assumed to be geometrically thin and the physical quantities are integrated in the vertical direction. The global structure which satisfies inner and outer boundary conditions is obtained by a shooting method.

## 1. Introduction

Recently, advection-dominated accretion disks are extensively studied theoretically by Narayan & Yi (1994), Abramowicz et al. (1995) and many other researchers (see also reviews by both Narayan and Abramowicz in this volume). These disks are important to explain the observed X-ray spectra of active galactic nuclei or X-ray binaries. Advection-dominated disk models obtained so far, however, are local or under some simplification on the basic set of differential equations (e.g., Chen 1995). A first step to solving the global structure of advection-dominated, optically-thin accretion disks was done by Matsumoto et al. (1985). A difficulty in obtaining such disk models lies in the transonic nature of flows. To simplify the calculations, they neglected the cooling term in the energy equation of ions; that is, viscous heating was balanced with advective cooling. With this approximation they decoupled the equation determining electron temperature from that for ion

*S. Kato et al. (eds.), Physics of Accretion Disks, 41–44.*
© 1996 OPA (Overseas Publishers Association) Amsterdam B.V.

temperature. We extend their transonic optically-thin advection-dominated disk models so that we no longer decouple equations. That is, the energy equation of electrons is solved simultaneously, coupled with other equations. The standard $\alpha$-viscosity is adopted.

## 2. Models

We consider a steady, optically thin, geometrically thin disk around a black hole. To describe axially symmetric disks, cylindrical coordinates $(r, \varphi, z)$ are employed. We use vertically integrated quantities. General relativistic effects are simulated by using the pseudo-Newtonian potential $\phi = -GM/(r - r_g)$, where $M$ is the mass of the black hole and $r_g$ is the Schwarzschild radius. The basic equations are

$$\dot{M} = -2\pi r \Sigma u_r, \qquad \dot{M}(l - l_{\rm in}) = 2\pi r^2 \alpha W, \tag{1}$$

$$u_r \frac{du_r}{dr} + \frac{1}{\Sigma}\frac{dW}{dr} = \frac{l^2 - l_{\rm k}^2}{r^3} - \frac{W}{\Sigma}\frac{d{\rm ln}\Omega_{\rm k}}{dr}, \tag{2}$$

$$\frac{k}{m}\dot{M}T_{\rm e}\left[\frac{3+\Gamma_{\rm e}}{2}\frac{d{\rm ln}T_{\rm e}}{dr} + \frac{\Gamma_{\rm i}}{2}\frac{d{\rm ln}T_{\rm i}}{dr} - \frac{d{\rm ln}\Sigma}{dr} - \frac{d{\rm ln}\Omega_{\rm k}}{dr}\right] = 2\pi r(-\Lambda_{\rm ie} + Q_{\rm brems}), \tag{3}$$

and

$$\frac{k}{m}\dot{M}T_{\rm i}\left[\frac{3+\Gamma_{\rm i}}{2}\frac{d{\rm ln}T_{\rm i}}{dr} + \frac{\Gamma_{\rm e}}{2}\frac{d{\rm ln}T_{\rm e}}{dr} - \frac{d{\rm ln}\Sigma}{dr} - \frac{d{\rm ln}\Omega_{\rm k}}{dr}\right] = 2\pi r\left(\alpha W r\frac{d\Omega}{dr} + \Lambda_{\rm ie}\right), \tag{4}$$

where $u_r$ is the radial component of the matter velocity, $\alpha$ is the viscosity parameter, $\dot{M}$ is the accretion rate, $\Sigma$ is the surface density, $W$ is the vertically integrated pressure, $l_{\rm k}$ is the Keplerian angular momentum, $l_{\rm in}$ is the angular momentum swallowed by the black hole, $T_{\rm e}$ is the electron temperature, $T_{\rm i}$ is the ion temperature, $\Lambda_{\rm ie}$ is the energy transferred by Coulomb collision from ions to electrons (Stepney & Guilbert 1983), $Q_{\rm brems}$ is the electron-electron and electron-ion bremsstrahlung amplified by Compton (Svensson 1984), $\Gamma_{\rm e} = T_{\rm e}/(T_{\rm i} + T_{\rm e})$ and $\Gamma_{\rm i} = T_{\rm i}/(T_{\rm e} + T_{\rm i})$. The synchrotron cooling is not considered.

By introducing $\eta = u_r^2\Sigma/W$, we can combine the above set of equations to give

$$\frac{d\eta}{dr} = \frac{N_1}{D}, \quad \frac{dT_{\rm e}}{dr} = \frac{N_2}{D} \quad \text{and} \quad \frac{dT_{\rm i}}{dr} = \frac{N_3}{D}. \tag{5}$$

Smooth transonic solutions must satisfy regularity conditions, $D = N_1 = N_2 = N_3 = 0$ at the sonic point. Outer boundary conditions adopted are $l = 0.45 l_{\rm k}$, $T_{\rm i} = 0.2 T_{\rm virial}$ and $\Lambda_{\rm ie} = Q_{\rm brems}$ at $r = 1000 r_g$. Solutions which satisfy the outer boundary conditions and the inner sonic conditions are obtained by a shooting method adjusting $l_{\rm in}$.

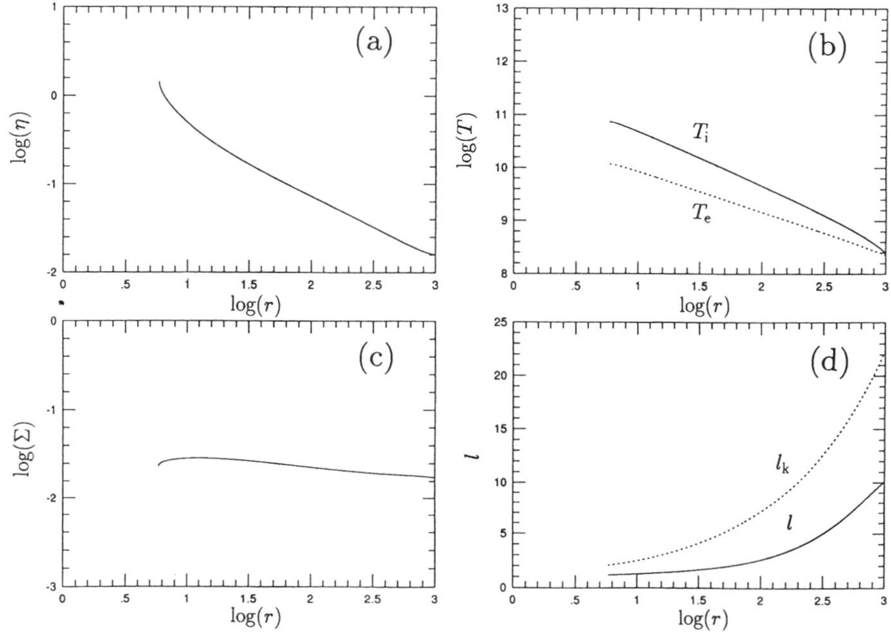


Fig. 1. Radial distributions of physical quantities. From (a) to (d), $\eta$, $T_e$, $T_i$, $\Sigma$, $l$ and $l_k$ are depicted as functions of $r$. Here $\dot{M} = 10^{-2}$ in unit of $2\pi c r_g/\kappa_{es}$ and $\alpha = 0.1$ are adopted.

## 3. Results

Radial variations of physical quantities in the case of $\alpha = 0.1$ and $\dot{M} = 10^{-2}$ are shown in figure 1. Here, the units of $\dot{M}$, $\Sigma$, $T$ and $l$ are $2\pi c r_g/\kappa_{es}$, g/cm$^2$, $K$ and $cr_g$, respectively ($\kappa_{es} = 0.40$). When $\alpha$ is as large as 0.1, the flow is viscosity-driven , not pressure-driven (Matsumoto et al. 1984). That is, the flow loses angular momentum so efficiently by viscosity that it falls towards the central object with a sub-Keplerian rotation (see panel (d)). This situation is clear in panel (a), where the Mach number of the accretion flow gradually increases inwards from a far outside. [This is in contrast to the case of a pressure-driven accretion flow, where the Mach number increases sharply near the sonic point.] The sonic point is $\sim 3.4 r_g$. The panel (a) shows that $\eta$ increases inwards in proportion to $\sim 1/r$, which implies $|u_r| \sim 1/r$, since the virial temperature increases inwards as $r^{-1}$. The mass conservation leads then to $\Sigma \sim$ const (panel (c)). A low density in the inner region makes the electron cooling process less efficient, causing a relatively high electron temperature (panel (b)). The above results are consistent with the heat balance between viscous heating $Q^+ \equiv -\alpha W r d\Omega/dr$ and advective cooling $Q^-_{ad} \equiv -(\dot{M}/2\pi r^2)(W/\Sigma)\xi$, where $\xi = -d\ln S/d\ln r$. If $\Omega \sim r^{-3/2}$ and $\xi \sim$ const, the energy balance leads to $\Sigma \sim r^{-1/2}$, inconsistent with $\Sigma \sim$ const in panel (c). In the inner region of the present disks, however,

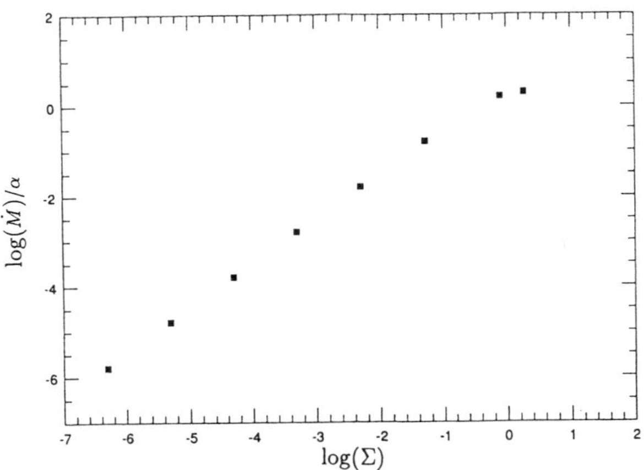


Fig. 2. $\dot{M} - \Sigma$ relation at $r = 10r_g$. We adopted $\alpha = 0.1$ and $\zeta = 0.5$.

$\xi$ decreases inwards by a rapid accretion flow and $\Omega$ increases inwards faster than $r^{-3/2}$. Hence the balance can lead to $\Sigma \sim$ const. In this way, a high electron temperature in disks will be a common characteristics of an advection-dominated transonic flow with a large $\alpha$, unless Compton cooling due to ample soft photons is efficient. Figure 2. shows the $\dot{M} - \Sigma$ relation at $r = 10r_g$ in the case of $\alpha = 0.1$ and $\zeta = 0.5 = l/l_k$ at outer boundary. The results agree well with that of Chen (1995). The advective cooling is more than 98 % of the total cooling at $r = 10r_g$. An outstanding issue is to extend our calculations to other parameter ranges, as well as to cases with synchrotron cooling in order to understand more generally the inner structure of optically-thin two-temperature disks.

**Acknowledgements**

The authors would like to express their sincere thanks to Drs. F. Honma and S. Mineshige for their valuable advice.

**References**

Abramowicz M.A., Chen X., Kato S., Lasota J.P., Regev O. 1995, ApJL 438, L37

Abramowicz M.A. 1996, in this volume

Chen X. 1995, MNRAS 275, 641

Matsumoto R., Kato S., Fukue J. 1985, in Theoretical Aspects on Structure, Activity and Evolution of Galaxies III, eds. S. Aoki, M. Iye, and Y. Yoshii (Tokyo Astr. Obs.;Tokyo), p102

Matsumoto R., Kato S., Fukue J., Okazaki A. 1984, PASJ 36, 71

Narayan R., Yi I. 1994, ApJL 428, L13

Narayan R. 1996, in this volume

Stepney S., Guilbert P.W. 1983, MNRAS 204, 1269

Svensson R. 1984, MNRAS 209, 175

# Electron-Positron Pairs in Advection-Dominated Disks

Masaaki KUSUNOSE
*Department of Astronomy, University of Texas, TX 78712, USA*

## Abstract

We study disk structure of steady-state advection-dominated disks, including electron-positron pairs. When pair production is possible, pairs contribute to radiative cooling. Hence thermal equilibrium curves are changed by pairs. Pair production reduces the value of the critical accretion rate when the value of $\alpha$ parameter is smaller than a critical value. The value of the critical $\alpha$, on the other hand, increases. However, these effects are minor and the structure of advection-dominated disks does not change much when pairs are included.

## 1. Electron-Positron Pairs in Hot Accretion Disks

Effects of pair production have been investigated for optically-thin accretion disks and nonthermal plasmas near black holes. In this paper, we review the effects of thermal pairs in no-advection disks and present new results on advection-dominated disks with pairs. We note that results shown here are only for two-temperature disks consisting of fully ionized hydrogen, $e^{\pm}$ pairs, and radiation.

First we review the effects of $e^{\pm}$ in optically-thin, hot accretion disks without advection (see Björnsson & Svensson 1992 for a detailed review). Since the work by Shapiro et al. (1976), optically-thin accretion disks have been thought to be promising to explain hard X-rays from galactic black holes and active galactic nuclei. They already noticed that $e^{\pm}$ pairs might play an important role in this kind of disks, because electron temperature $T_e$ is close to $6 \times 10^9$ K (the rest mass energy of an electron).

For a steady state we need to include pair balance equation:

$$(\dot{n}_+)_{\rm prod} - (\dot{n}_+)_{\rm ann} = 0\,, \tag{1}$$

*S. Kato et al. (eds.), Physics of Accretion Disks, 45–50.*
© 1996 OPA (Overseas Publishers Association) Amsterdam B.V.

where $(\dot{n}_+)_{\rm prod}$ and $(\dot{n}_+)_{\rm ann}$ are production and annihilation rates of pairs, respectively. This equation must be solved with disk equations simultaneously. Pair production processes are $pe^\pm \rightarrow pe^\pm e^+ e^-$, $e^\pm e^\pm \rightarrow e^\pm e^\pm e^+ e^-$, $p\gamma \rightarrow pe^+e^-$, $e^\pm\gamma \rightarrow e^\pm e^+ e^-$, and $\gamma\gamma \rightarrow e^+e^-$. Pair annihilation is by $e^+e^- \rightarrow \gamma\gamma$. Assuming that photon energy extends above 511 keV, $e^\pm\gamma$ is efficient, when $\tau_{\rm T} \lesssim 1$, because photon density becomes larger due to scattering. Here $\tau_{\rm T} \equiv \sigma_{\rm T} n_e H$ is the Thomson scattering depth, where $\sigma_{\rm T}$ is the Thomson cross section, $n_e = n_+ + n_-$, and $H$ is the scale height of a disk, where $n_+$ and $n_-$ are the number densities of positrons and electrons, respectively. When $\tau_{\rm T} > 1$, $\gamma\gamma$ contributes to pair production most efficiently. The pair annihilation rate is a function of $n_+$, $n_-$, and $T_e$. (For detailed properties of thermal pair plasmas, see Svensson 1984.)

Liang (1979) assumed that pairs are produced by $\gamma\gamma$ and that $\mu_+ + \mu_- = 2\mu_\gamma$ to solve equation (1), where $\mu_\pm$ and $\mu_\gamma$ are chemical potentials of $e^\pm$ and photons, respectively. These assumptions hold when photons are in Wien distribution. However, this is not always true for optically thin disks and we need to solve equation (1) for more general situations.

Kusunose & Takahara (1988) included pair production without assuming $\mu_+ + \mu_- = 2\mu_\gamma$; it was needed to obtain photon spectra to calculate the pair production rate. They found that there is a critical accretion rate above which there is no pair balance condition is satisfied (this critical rate corresponds to $\dot{m}^U_{\rm cr}$ below). They did not include Compton scattering, which is important when $\tau_{\rm T}$ is greater than unity. White & Lightman (1989) included Compton scattering and obtained qualitatively the same solutions as Kusunose & Takahara.

Later, Björnsson & Svensson (1992) studied optically thin disks in more detail. They found that there is another critical accretion rate, which is related with radiation pressure. Their results were confirmed by Kusunose & Mineshige (1992). Results for $\alpha = 0.5$, 1 and 2 are shown in figure 1 in $(\tau_p, \dot{m})$-plane (dashed curves). To obtain the curves in figure 1, we used the pseudo-Newtonian potential $\Phi = -GM_{\rm bh}/(R - R_{\rm S})$ to compare the results with the advection-dominated disks obtained by Abramowicz et al. (1995). Here $R_{\rm S} = 2GM_{\rm bh}/c^2$ is the Schwarzschild radius. Because of this pseudo-Newtonian potential, results in figure 1 are different from those of Björnsson & Svensson (1992) and Kusunose & Mineshige (1992). The mass accretion rate is normalized by the Eddington rate, i.e., $\dot{m} = \dot{M}/\dot{M}_{\rm Edd}$, where $\dot{M}_{\rm Edd} = L_{\rm Edd}/(\eta_{\rm eff}c^2) = 4\pi GM_{\rm bh}m_p/(\eta_{\rm eff}\sigma_{\rm T}c)$. Here $\eta_{\rm eff}$ is the efficiency of mass to energy conversion and we assume $\eta_{\rm eff} = 0.1$. We use $\tau_p \equiv \sigma_{\rm T} n_p H$ to measure the *pair-free* Thomson scattering depth of a disk in the vertical direction, where $n_p$ is the proton number density. Note that the surface density is given by $\Sigma = 2m_p n_p H = 2(m_p/\sigma_{\rm T})\tau_p \approx 5\tau_p$ (g/cm$^2$).

The radial distributions of $z \equiv n_+/n_p$, $T_e$, $T_p$, and $\tau_{\rm T}$ show different characteristics for three different accretion rates; namely, $\dot{m} < \dot{m}^L_{\rm cr}$, $\dot{m}^L_{\rm cr} < \dot{m} <$

$\dot{m}_{\rm cr}^U$, and $\dot{m} > \dot{m}_{\rm cr}^U$. When $\dot{m} < \dot{m}_{\rm cr}^L$, the solution is continuous and the value of $z$ is small ($z \ll 1$). When $\dot{m}_{\rm cr}^L < \dot{m} < \dot{m}_{\rm cr}^U$, the solution is continuous but solution is three-valued for a certain range of $R$. At the transition points from one-valued to three-valued solutions, the curve in $(R, z)$-plain has infinite radial gradients. And when $\dot{m} > \dot{m}_{\rm cr}^U$, there is an island solution at around $R = (16/3)R_{\rm S}$ in addition to a solution which extends from the outer boundary to the inner boundary.

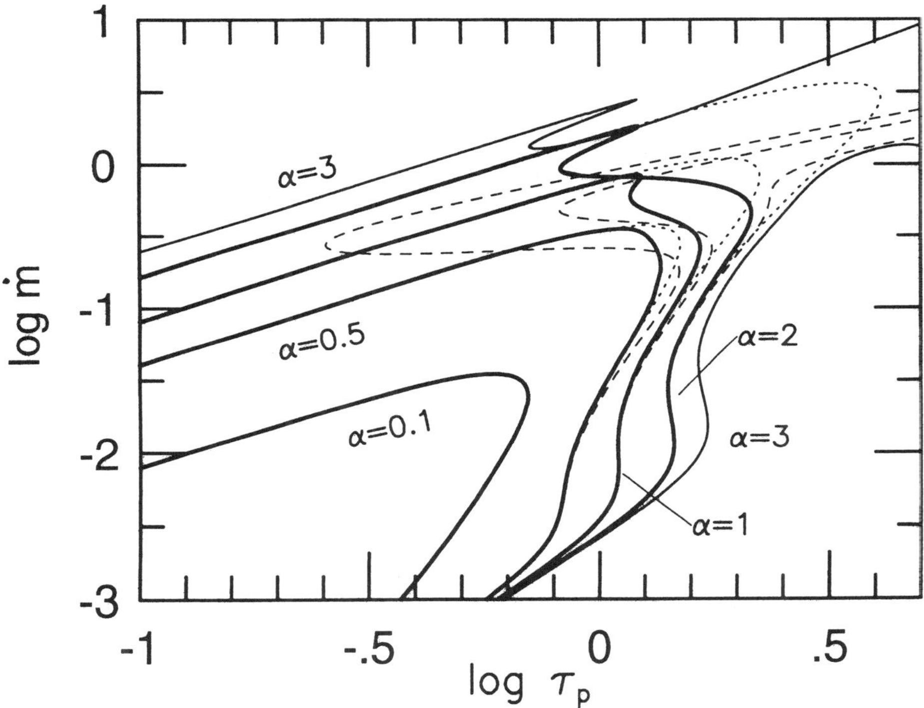


Fig. 1. Thermal equilibrium curves with pairs. Parameters are $M_{\rm bh} = 10M_\odot$ and $r = 5$. For solid and dotted curves, $\xi = 1$. For dashed curves, $\xi = 0$. $\zeta = 0$ for solid and dashed curves.

## 2. Electron-Positron Pairs in Advection-Dominated Hot Disks

We consider only steady states here. Then continuity equation of pairs is given by

$$\frac{1}{RH}\frac{d}{dR}(R v_r n_+ H) = (\dot{n}_+)_{\rm prod} - (\dot{n}_+)_{\rm ann}\,, \tag{2}$$

where $v_r$ is radial velocity of fluid. The left-hand side is the pair advection term. Here we neglect relativistic effects and assume that $n_+$, $(\dot{n}_+)_{\rm prod}$, and

$(\dot{n}_+)_{\rm ann}$ are averaged over the scale height. Using $\dot{M} = -2\pi R\Sigma v_r$, equation (2) is rewritten as

$$-\frac{40}{3}\frac{\dot{m}}{r^2}\frac{H}{R_{\rm S}}\frac{z}{\tau_p^2}\frac{d\ln z}{d\ln r} = \frac{8}{3}\frac{1}{\sigma_{\rm T}\, c\, n_p^2}[(\dot{n}_+)_{\rm prod} - (\dot{n}_+)_{\rm ann}]\,, \tag{3}$$

where $r = R/R_{\rm S}$. We include all the pair production processes mentioned in section 1; we assume photons are produced by Comptonized bremsstrahlung.

Equation (3) should be solved with disk equations simultaneously. Tritz & Tsuruta (1989) solved this equation with disk equations. The pair advection term is effectively an additional annihilation term and reduces pair density. However, they neglected advective cooling in energy balance equation and their solution was qualitatively the same as those without advection, except that they did not find the critical accretion rates $\dot{m}_{\rm cr}^L$ and $\dot{m}_{\rm cr}^U$.

Radial structure should be calculated to obtain solutions of advection-dominated disks, because equation (3) and the advection cooling rate (below) include radial gradients. The advection cooling rate per unit area is given by

$$Q_{\rm adv} = -\frac{\dot{M}}{2\pi R}T\frac{dS}{dR} = \frac{\dot{M}}{2\pi R^2}\frac{P}{\rho}\xi\,, \tag{4}$$

where

$$\xi = -\frac{4-3\beta}{\Gamma_3 - 1}\frac{d\ln T}{d\ln R} + (4-3\beta)\frac{d\ln\Sigma}{d\ln R} - (4-3\beta)\frac{d\ln H}{d\ln R}\,, \tag{5}$$

with $\beta = P_{\rm gas}/P$, $\Gamma_3 = 1 + (4-3\beta)(\gamma-1)/[\beta + 12(\gamma-1)(\beta-1)]$, and $\gamma$ being the ratio of specific heats (Abramowicz et al. 1995; Chen et al. 1995). To avoid the complication of solving for global solution, Abramowicz et al. (1995) assumed $\xi$ = constant. On the other hand, Narayan & Yi (1994) used self-similar solutions. Here we present how advection-dominated disks are modified by pair production. As we show below, the effects of pair production is not large. But the effects depend on which advection model is used and the final result is still open to exact calculations of advection.

### *2.1. $\xi$ = constant*

First we assume $\xi$ is constant as in Abramowicz et al. (1995). We also need to assume $\zeta \equiv d\ln z/d\ln r$ = const. to avoid solving for global solution.

In figure 1, we show equilibrium curves for $\xi = 1$ (solid curves); solutions without advection but with pairs are shown by dashed curves, those with advection but without pairs are shown by dotted curves. Here we assume $\zeta = 0$. As seen in figure 1, the effects of pairs are almost negligible except where the accretion rate approaches the maximum values for a given $\alpha$. There solution switches from the advection-dominated branch to the radiative-cooling-dominated branch where pairs cool the disk.

We note that because of pair balance the critical value of $\alpha$, $\alpha_{\rm cr}$ found by Chen *et al.* (1995) increases. If there is no pair production, $\alpha_{\rm cr} \sim 2.2$ at $r = 5$. But because of pairs, $\alpha_{\rm cr} \sim 2.4$.

### *2.2. Self-similar solution*

Here we use the method by Narayan & Yi (1994) to obtain pair balance solutions (we assume that gas is fully ionized hydrogens and pairs). Since $v_r \propto r^{-1/2}$ and $n_p \propto r^{-3/2}$ for self-similar solution, equation (3) becomes

$$- 0.43\,\alpha^2\, c_1^2\, c_3^{1/2} \dot{m}^{-1}\, z\,\zeta = \frac{8}{3} \frac{1}{\sigma_{\rm T}\, c n_p^2} [(\dot{n}_+)_{\rm prod} - (\dot{n}_+)_{\rm ann}] \qquad (6)$$

where $c_1$ and $c_3$ are given in Narayan & Yi (1995). Note that we use Newtonian potential in this section whereas we used pseudo-Newtonian potential in a previous section. To solve equation (6), we assume $\zeta$ is constant.

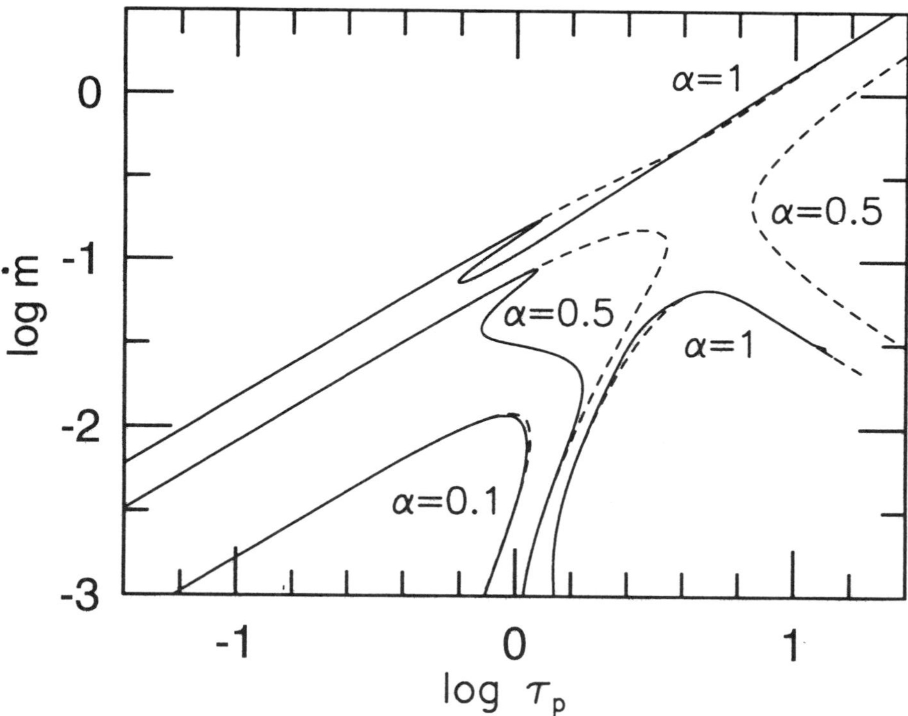


Fig. 2. Thermal equilibrium curves of self-similar solutions with pairs. Parameters are $M_{\rm bh} = 10M_\odot$, $r = 5$, and $\zeta = -1$. Dashed curves are solutions without pairs.

In figure 2, we show solutions in ($\tau_p$, $\dot{m}$)-plane (*solid*: with pairs, *dashed*: without pairs); parameters are $M_{\rm bh} = 10M_\odot$ and $R = 5R_{\rm S}$. The solution is similar to those shown for $\xi$ = const., though there are quantitative differences. For example, the critical value of $\alpha$ is about 0.66 at $r = 5$ (when pairs are not included, $\alpha_{\rm cr} = 0.56$).

## 3. Summary

Advection dominated solutions of optically-thin, hot disks are modified by $e^\pm$ pairs; (1) the critical value of $\alpha$ increases, (2) the maximum accretion rate for $\alpha < \alpha_{\rm cr}$ decreases. These effects appear for $\tau_p \sim 1$ and are not significant quantitatively. Then when $Q_{\rm adv}/Q_{\rm vis} \sim 1$, internal energy is carried into the gravitational horizon without converting a large fraction of energy to pairs.

Although we have not carried out pair instability analysis, pairs might trigger some instabilities in non-steady disks.

Finally we note that Björnsson et al. (1995) obtained similar results for local solutions (i.e., $\xi$ = const.) (see the paper by Abramowicz in these proceedings).

## Acknowledgements

The author is grateful to the organizing committee of this workshop for inviting him to this valuable meeting.

## References

Abramowicz M. A., Chen X., Kato S., Lasota J.-P., Regev O. 1995, ApJL 438, L37

Björnsson G., Abramowicz M. A., Chen X., Lasota J.-P. 1995, in preparation

Björnsson G., Svensson R. 1992, ApJ 394, 500

Chen X., Abramowicz M. A., Lasota J.-P., Narayan R., Yi, I. 1995, ApJL 443, L61

Kusunose M., Mineshige S. 1992, ApJ 392, 653

Kusunose M., Takahara F. 1988, PASJ 40, 709

Liang E. P. 1979, ApJ 234, 1105

Narayan R., Yi I. 1994, ApJL 428, L13

Narayan R., Yi I. 1995, ApJ 452, 710

Shapiro S. L., Lightman A. P., Eardley D. N. 1976, ApJ 204, 187

Svensson R. 1982, ApJ 258, 335

Svensson R. 1984, MNRAS 209, 175

Tritz B. G., Tsuruta S. 1989, ApJ 340, 203

White T. R., Lightman A. P. 1989, ApJ 340, 1024

# Instabilities of Advection-Dominated Accretion Flows

Xingming CHEN
*Department of Astronomy and Astrophysics, Göteborg University, and Chalmers University of Technology, 412 96 Göteborg, Sweden*

## Abstract

Accretion disk instabilities are briefly reviewed. Some details are given to the short-wavelength thermal instabilities and the convective instabilities. Time-dependent calculations of two-dimensional advection-dominated accretion flows are presented.

## 1. Introduction

Accretion disk is central to our understanding of many diverse astronomical objects and phenomena. In particular, it is believed that disk instabilities are fundamentally responsible for some of the luminosity variations observed in these systems.

Various kinds of instabilities may arise in a rotating, shearing gas flow. For example, the MHD instability in a shearing flow (Balbus & Hawley 1991) is found to be able to survive and develop into nonlinear turbulence. In fact, an external magnetic field is not required to sustain the instability since magnetic field can be generated and maintained self-consistently through dynamo-effect within the flow (Brandenburg et al. 1995). The Balbus-Hawley instability is not considered here, instead, I model the small-scale turbulence by assuming that it introduces viscosity, which can be described by the $\alpha$ prescription (Shakura & Sunyaev 1973),

$$\nu = \frac{2}{3}\alpha c_s H, \tag{1}$$

where $c_s$ is the local sound speed and $H$ is the scale-height of the disk. The disk is also assumed to be axisymmetric and nonself-gravitating, therefore the global non-axisymmetric Papaloizou-Pringle instability (Papaloizou & Pringle 1984) and all the self-gravitating-related instabilities are not present here.

*S. Kato et al. (eds.), Physics of Accretion Disks, 51–56.*
© 1996 OPA (Overseas Publishers Association) Amsterdam B.V.

An $\alpha$-model accretion disk may experience the thermal and viscous instabilities (Piran 1978). They could be coherent global instabilities and are often applied to explain the dwarf nova outbursts and the X-ray transients (see related discussion in this volume and references within). This type of instability can be analysed conveniently in terms of the $\dot{M}(\Sigma)$ relation (where $\dot{M}$ is the mass accretion rate and $\Sigma$ is the surface density of the disk). Here I consider the case of accretion disks around black holes. The steady state solutions or the thermal equilibria of accretion disks can be described in a unified scheme (see Chen et al. 1995). Depending on the magnitude of the viscosity parameter $\alpha$, the cooling mechanisms, and the equation of states, there exist several different branches (see figure 1). On the $\dot{M}(\Sigma)$ plane, it can be shown that, if the slope is negative, then the disk is viscously unstable (Lynden-Bell & Pringle 1974). The thermal instability can be examined by comparing the cooling and heating rates near each equilibrium curve. One may consider a perturbation of $\dot{M}$ (increase $\dot{M}$ but keep $\Sigma$ unchanged), or effectively an increase of temperature, if the cooling rate exceeds the heating rate, then the gas cools down and the original thermal equilibrium is restored. The disk is therefore thermally stable. On the other hand, if the heating rate exceeds the cooling rate, then the gas heats up further and an instability occurs.

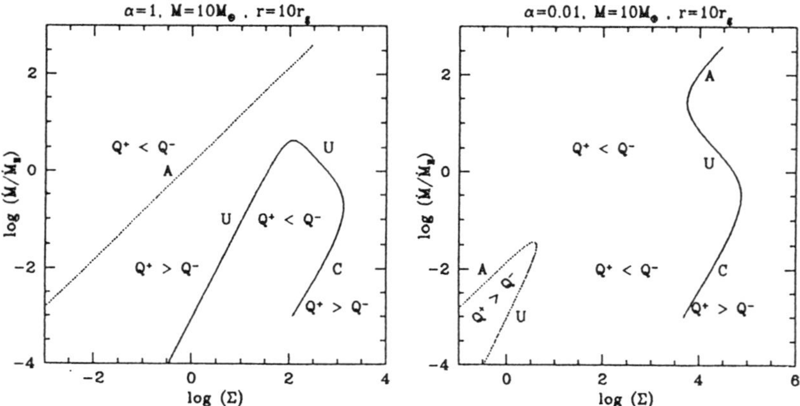


Fig. 1. The thermal equilibria of accretion disks. Here $\dot{M}_{\rm E} = 4\pi GM/\kappa c$ is the Eddington limit ($\kappa = 0.34$), $Q^+$ is the heating rate, and $Q^-$ is the cooling rate including both local radiative cooling and radial advective cooling. A: 'advection' and stable, U: 'unstable', and C: 'local cooling' and stable.

Thus, it is easily understood that for optically thick disks (large surface density), the thermal and viscous instabilities occur at the same time on the middle-branch of the **S**-shaped curve (or on the upper-branch of the upside down **U**-curve) when the disk is radiation-pressure-dominated and the radial advection is not important. In this case a limit-cycle type disk evolution may be resulted (Taam & Lin 1984) or a transition to an optically-thin disk state may occur (Lightman & Eardley 1974). For optically-thin local cooling dominated disks, on the other hand, the viscous mode is stable but the thermal mode is always unstable. Therefore, a stable optically-thin local cooling dominated accretion disk may not exist, it may either collapse to an optically thick disk or heat up catastrophically without bound. Finally but most importantly, the advection-dominated disks are both thermally and viscously stable independent on the optical depth.

The thermal and viscous instabilities are probably the most violent global instabilities an accretion disk around black hole may suffer. Under such instabilities, a disk will most probably be destroyed entirely, especially in the case of optically thin disks. Accretion disk models which do not have such type of instabilities include advection-dominated disks and gas-pressure-dominated optically-thick disks. In the following, I discuss the less severe instabilities which may exist in advection-dominated accretion disks.

## 2. Short-Wavelength Thermal Instability

The above analysis for thermal instability, however, can only be applied to the long-wavelength perturbations. To examine the short-wavelength instability, a dispersion relation equation needs to be solved. The dispersion relation can be derived from the time-dependent disk equations by assuming the perturbation (of the steady state solution), $\delta X$, of each physical quantity, $X$, has the following functional form with respect to time ($t$) and space ($r$):

$$\delta X/X \propto e^{\xi t - ikr}, \tag{2}$$

where $k$ is the wavenumber of the modes, $k = 2\pi/\lambda$, and $\xi$ is, in general, complex. Specifically, $\tau = \mathrm{Re}(\xi)$ represents the growth or damping rate depending upon whether $\tau$ is positive or negative, and $\omega = \mathrm{Im}(\xi)$ represents the frequency of the mode.

I here consider only the thermal mode instability. In a recent work, Kato et al. (1996) demonstrated, analytically, that short-wavelength thermal mode instability is still present in advection dominated accretion disks. The unstable short-wavelength thermal mode has been obtained numerically in Chen & Taam (1993). However, detailed discussion has not been made there.

By solving the dispersion relation of a transonic black hole accretion disk model (see Chen & Taam 1993), it is seen that the short-wavelength ($\lambda \lesssim 6H$) thermal instability is a traveling mode with frequencies in the order of the local dynamical time-scale. Figure 2 shows the growth rate and the oscillation frequency of the thermal mode instability for two cases with $\dot{M} = 16\dot{M}_{\mathrm{E}}$ and $\dot{M} = 160\dot{M}_{\mathrm{E}}$, where $\dot{M}_{\mathrm{E}} = 4\pi GM/\kappa c$ is the Eddington limit and $\kappa = 0.34$. The disk is optically thick with the latter one fully advection-dominated. Because of the nature of the propagation, the instability is expected to be weak globally and exist only in the inner regions near the black hole. Local properties of the short-wavelength thermal mode instability suggest that its non-linear behavior may be similar to that of the inertial-acoustic mode instability discussed in Chen & Taam (1995). Preliminary results of Manmoto et al. (1995, this volume) shows that a perturbation indeed grows due the short-wavelength thermal mode instability, but then it is advected into the black hole.

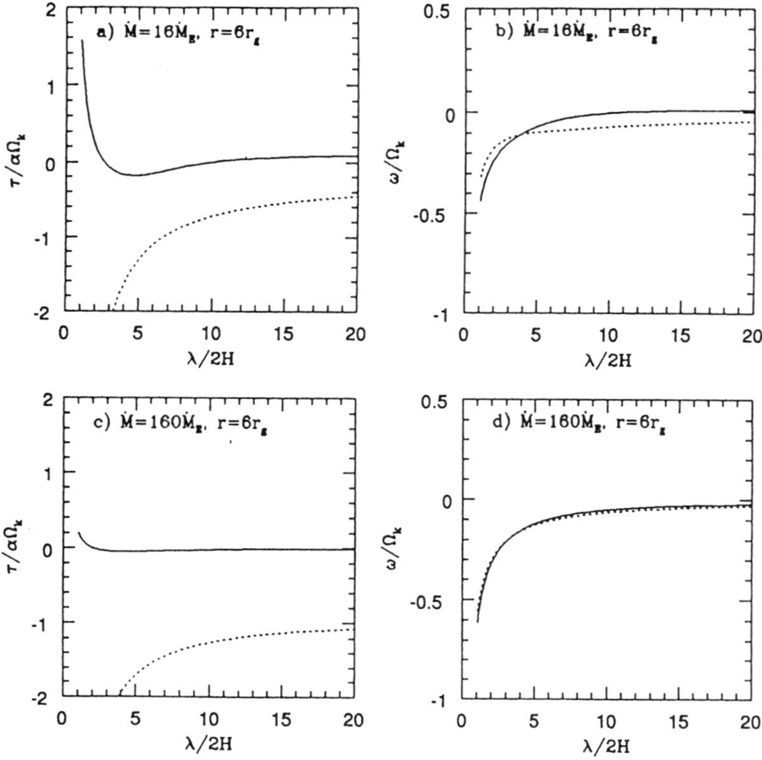


Fig. 2. The growth rate and the oscillation frequency of the thermal mode (solid line, the dotted line represents the viscous mode). Note the increase of the thermal mode growth rate for short-wavelength ($\lambda \lesssim 6H$), and also the corresponding high frequencies ($\omega \sim \Omega_K$).

## 3. Convective Instability: Time-Dependent Simulations

In an advection-dominated disk, the specific entropy of the flow increases inwards, therefore, it is likely that the convective instability is present. For a static rotating fluid, the convective stability condition is the Høiland criterion which states that the fluid is stable against local, axisymmetric, adiabatic perturbations if and only if the following two conditions are satisfied (Tassoul 1978):

$$\frac{1}{r^3}\frac{\partial \ell^2}{\partial r} - \left(\frac{\partial T}{\partial p}\right)_S \nabla p \cdot \nabla S > 0, \tag{3a}$$

$$-\frac{1}{\rho}\frac{\partial p}{\partial z}\left(\frac{\partial \ell^2}{\partial r}\frac{\partial S}{\partial z}-\frac{\partial \ell^2}{\partial z}\frac{\partial S}{\partial r}\right)>0. \qquad (3b)$$

Here $(r, z)$ is the radial and vertical coordinates in a cylindrical coordinates, $\ell$ is the specific angular momentum, $p$ is the pressure, and $S$ is the specific entropy. Note that the first terms in criterion (3a) is the epicyclic frequency $\kappa^2$ and this criterion reduces to that of Narayan & Yi (1994) for one-dimensional gas pressure dominated disks.

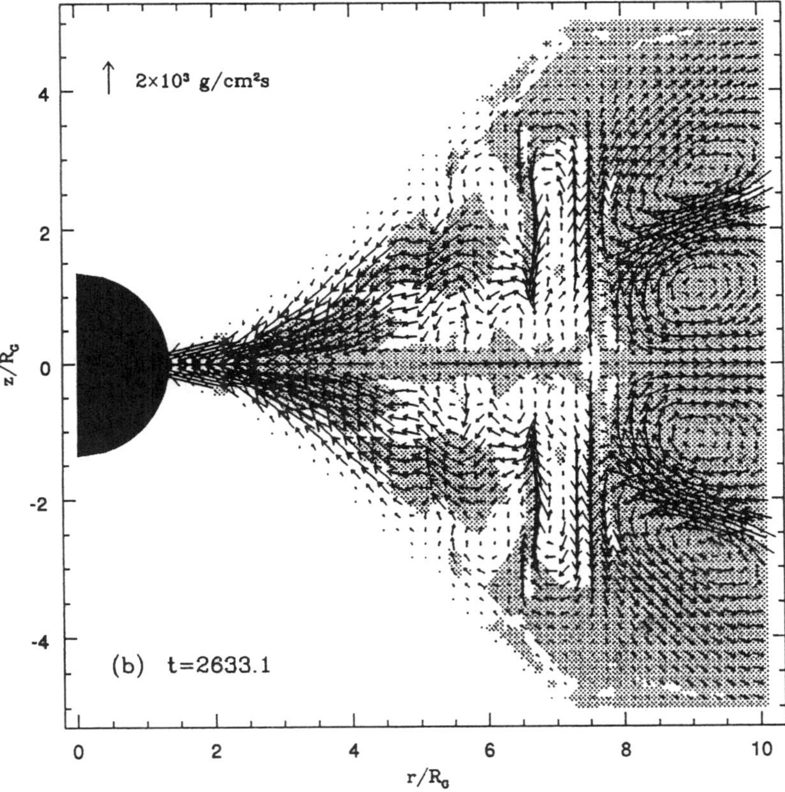


Fig. 3. The momentum vector field of advection-dominated flows. The regions which are convectively unstable according to Høiland criterion (3ab) are indicated as grey areas. The time is in units of $r_g/c$. This model has $\alpha = 0.001$ and the initial condition is a static torus.

We have performed two-dimensional hydrodynamical simulations for axisymmetrical advection-dominated accretion flows. 2D numerical models of accretion disks around black holes have been studied in different situations. For example, Hawley et al. (1984) simulated the formation of gas pressure supported tori. Eggum et al. (1987) calculated radiative hydrodynamical models of optically-thick accretion disks. Our detailed results have been reported elsewhere (Igumenshchev et al. 1996). A typical flow pattern (momentum vector field) is shown in figure 3. Here, we have used a viscosity prescription $\nu = (2/3)\alpha c_s^2/\Omega_K$, and $\alpha = 0.001$. The grey area represents the convective unstable regions according to the Høiland criterion (3ab). We see a very efficient mixing of matter. Further analysis shows that this process of meridional circulations and vortices transports angular momentum outwards and thus the effective $\alpha$ becomes larger.

## 4. Summary

Advection-dominated flows subject to convective instability. By time dependent simulations, we showed that this process transports angular momentum outwards. However, how important does it effect the dynamics of the flow is yet to be examined. There were some numerical simulations which showed that convection transports angular momentum inwards (e.g., Stones & Balbus 1995). The difference may be caused by the different physical conditions considered. For example, in our simulation, an $\alpha$-viscosity is assumed whereas in Stones & Balbus (1995) no viscosity is included.

In advection-dominated flows, there is short-wavelength thermal instability. This type of instability is expected to be weak and localized and it may be termed as "flicker-type" instability (Kato et al. 1996). High resolution time-dependent calculation should be the future investigation.

I wish to thank Dr. Shoji Kato for support.

### References

Balbus S. A., Hawley J. F. 1991, ApJ 376, 214
Brandenburg A. et al. 1995, ApJ 446, 741
Chen X. et al. 1995, ApJL 443, L61
Chen X., Taam R. E. 1993, ApJ 412, 254
Chen X., Taam R. E. 1995, ApJ 441, 354
Eggum G. E., Coroniti F. V., Katz J. I. 1987, ApJ 323, 634
Hawley J. F., Smarr L. L., Wilson J. R. 1984, ApJ 277, 296
Igumenshchev I. V., Chen X., Abramowicz M. A. 1996, MNRAS in press
Lightman A. P., Eardley D. M. 1974, ApJL 187, L1
Lynden-Bell D., Pringle J. 1974, MNRAS 168, 603
Kato S., Abramowicz M. A., Chen X. 1996, PASJ in press
Manmoto T. et al. 1995, this volume
Narayan R., Yi I. 1994, ApJL 428, L13
Papaloizou J., Pringle J. E. 1984, MNRAS 208, 721
Piran T. 1978, ApJ 221, 652
Shakura N. I., Sunyaev R. A. 1973, A&A 24, 337
Stones J. M., Balbus S. A. 1995, preprint
Taam R. E., Lin D. N. C. 1984, ApJ 287, 761
Tassoul J.-L. 1978, Theory of Rotating Stars (Princeton University Press, Princeton)

# Stability of Advection-Dominated Disks

Tadahiro MANMOTO[1], Mitsuru TAKEUCHI[1], Shin MINESHIGE[1]
Shoji KATO[1], and Ryoji MATSUMOTO[2]
*1. Department of Astronomy, Faculty of Science, Kyoto University, Sakyo-ku, Kyoto 606-01, Japan*
*2. Department of Physics, Faculty of Science, Chiba University, Inage-ku, Chiba 263, Japan*

**Abstract**

The stability of an optically-thin, advection-dominated accretion flow (ADAF) is examined by means of time-dependent numerical simulations. The results indicate that the optically-thin ADAFs are unstable against local short-wavelength perturbations but are still globally stable. These properties of optically-thin ADAFs may be relevant to time variations of X-rays from active galactic nuclei and stellar black hole candidates in the hard state.

## 1. Introduction

There have been various arguments indicating that advective cooling is likely to stabilize the optically-thin accretion flows, but these arguments are based on the criteria derived under the approximation of fixed surface density during the growth of instability. This approximation is related to the assumption that the disks are geometrically thin. For this reason the arguments adopting such criteria are not appropriate for the instability of ADAFs whose vertical thickness is comparable to the radius.

Kato et al. (1996) examined instability of ADAFs more rigorously than ever by local analysis and showed the presence of thermally unstable modes. They concluded that optically-thin ADAFs are locally unstable against short-wavelength perturbations as long as the standard alpha-viscosity is assumed and the effects of thermal diffusion are neglected.

We have performed, under the same situations mentioned above, numerical simulations on time evolution of optically-thin ADAFs around black holes to confirm the presence of unstable modes and to obtain some insight into the global stability of optically-thin ADAFs. The possible astrophysical applications are briefly discussed in the last section.

*S. Kato et al. (eds.), Physics of Accretion Disks, 57–60.*
© 1996 OPA (Overseas Publishers Association) Amsterdam B.V.

## 2. Basic Equations, Method, and Results

### 2.1. Steady Solution

As initial disks on which perturbations are superposed, we have made global steady disks by solving the full set of vertically integrated equations numerically. General relativistic effect is included by adopting the pseudo-Newtonian potential, $\psi = -GM_{\rm bh}/(r-r_{\rm g})$ (Paczyński & Wiita 1980), where $M_{\rm bh}$ is the mass of the black hole, and $r_{\rm g}$ the Schwarzshild radius. The basic equations are (Matsumoto et al. 1984):

$$2\pi r\Sigma v = -\dot{M}, \tag{1}$$

$$v\frac{dv}{dr} + \frac{1}{\Sigma}\frac{dW}{dr} = r(\Omega^2 - \Omega_{\rm K}^2) - \frac{W}{\Sigma}\frac{d\ln\Omega_{\rm K}}{dr}, \tag{2}$$

$$\dot{M}(l - l_{\rm in}) = -2\pi r^2 \alpha W, \tag{3}$$

$$B_1\dot{M}T\frac{ds}{dr} = \dot{M}(l - l_{\rm in})\frac{d\Omega}{dr} - 2\pi r Q_{\rm rad}^-, \tag{4}$$

where $\dot{M}$ is the accretion rate, $W$ the vertically integrated pressure, $\Sigma$ the surface density, $l_{\rm in}$ the angular momentum swallowed by the black hole, $\Omega_{\rm K}$ the Keplerian angular velocity, $\alpha$ the viscosity parameter, $T$ and $s$ are the temperature and the specific entropy of the gas on the equatorial plane, respectively. We assume the local radiative cooling is provided by thermal bremstrahlung with a cooling rate per unit surface area defined by

$$Q_{\rm rad}^- = 6.22 \times 10^{20} B_2 \rho^2 T^{1/2} H, \tag{5}$$

where $H$ is the half thickness of the disk, $\rho$ the density of the equatorial plane. The hydrostatic balance in the vertical direction and the equation of state are, respectively,

$$B_3\frac{W}{\Sigma} = \Omega_{\rm K}^2 H^2 \quad \text{and} \quad B_4\frac{W}{\Sigma} = \frac{\Re}{\mu}T, \tag{6}$$

where $\Re$ is the gas constant, $\mu(=0.62)$ is the mean molecular weight, $B_1$, $B_2$, $B_3$, and $B_4$ are constants which depend on the vertical disk structure and $B_1 = 8/9, B_2 = 858\pi/4096, B_3 = 9$, and $B_4 = 9/8$. Here, polytropic relation with a polytropic index $n = 3$ in the vertical direction is assumed (Hōshi 1977).

Outer boundary conditions are adopted at $r = 1000r_{\rm g}$. They are $\Omega = \Omega_{\rm K}$ and that the viscous heating is balanced with the radiative cooling. Basic equations are integrated numerically toward the inner boundary which is set at $r = 2r_g$. In the present study, we assign $\alpha = 0.3$, $M_{\rm bh} = 10M_\odot$, $\dot{M} = 10^{-5}\dot{M}_{\rm crit}$. Here $\dot{M}_{\rm crit} = L_{\rm Edd}/16c^2$, where $L_{\rm Edd}$ is the Eddington luminosity. The remaining parameter $l_{\rm in}$ is determined in such a way that the flow passes the sonic point smoothly.

### 2.2. *Time-Dependent Simulation*

For time-dependent calculations, the four conservational equations (1)–(4) must be replaced with those of time-dependent versions;

$$\frac{\partial}{\partial t}(r\Sigma) + \frac{\partial}{\partial r}(r\Sigma v) = 0, \tag{7}$$

$$\frac{\partial}{\partial t}(r\Sigma v) + \frac{\partial}{\partial r}(r\Sigma v^2) = -r\frac{\partial W}{\partial r} + r^2\Sigma(\Omega^2 - \Omega_K^2) - W\frac{d\ln\Omega_K}{d\ln r}, \tag{8}$$

$$\frac{\partial}{\partial t}(r^3\Sigma\Omega) + \frac{\partial}{\partial r}(r^3\Sigma v\Omega) = -\frac{\partial}{\partial r}(r^2\alpha W), \tag{9}$$

$$\frac{\partial}{\partial t}(r\Sigma\varepsilon) + \frac{\partial}{\partial r}(r\Sigma\varepsilon v) = -\frac{\partial}{\partial r}(rWv) - \frac{\partial}{\partial r}(r^2\alpha W\Omega) - rQ^-_{\rm rad}, \tag{10}$$

where $\varepsilon$ is the internal energy of gas per unit mass defined by

$$\varepsilon = \left(\frac{1}{\gamma - 1} + \frac{1}{2}\right)\frac{W}{\Sigma} + \frac{1}{2}\left(v^2 + r^2\Omega^2\right) + \psi, \tag{11}$$

with $\gamma$ $(= 5/3)$ being the specific heat ratio.

We set a fixed outer boundary at $r = 20r_{\rm g}$ and a free inner boundary at $r = 2r_{\rm g}$ for time-dependent calculations. In the region between the outer and the inner boundaries, the cooling is dominated by advection and the vertical thickness is nearly equal to the radius.

On the steady disk obtained previously, we put a small perturbation with the functional form of

$$\frac{\delta\Sigma}{\Sigma} = 0.01 \times \exp\left[-\left(\frac{r - 18r_{\rm g}}{0.1r_{\rm g}}\right)^2\right] \times \sin(kr), \tag{12}$$

where $k$ is a wave number. Small perturbations are added to all other physical variables for consistency. Their order-of-magnitude relations are

$$(\delta\Sigma/\Sigma) \approx kr(\delta W/W) \approx \sqrt{kr}(\delta v/v) \approx \sqrt{kr}(\delta\Omega/\Omega), \tag{13}$$

according to the linear analysis (Kato et al. 1996). The results are shown in figure 1.

For the case of $k = 30$, small local perturbations grow rapidly as is predicted by the local analysis. They drift inwards on a comparable timescale with that of growth of perturbations without largely affecting the steady-state structure. For the case of $k = 6$, the growth of perturbation is not so rapid. These results are consistent with the growth rate derived analytically, which is of the order of $\alpha\Omega(kr)^{1/2}$.

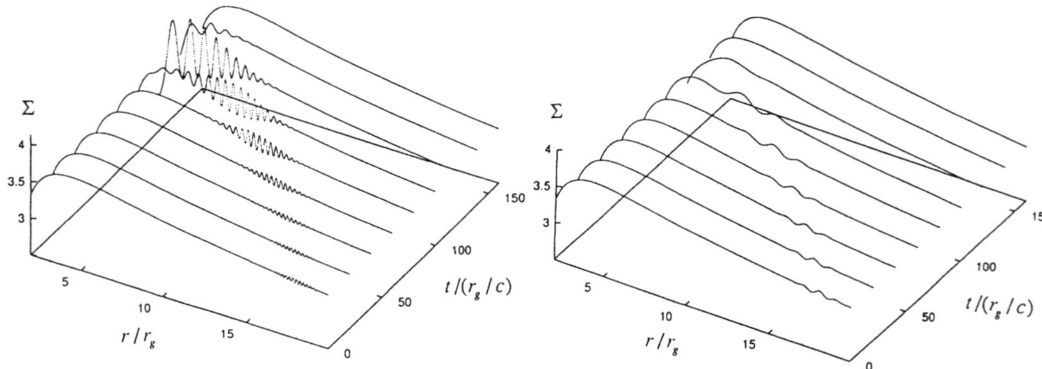


Fig. 1. Growth of short wavelength perturbation in the radial Σ profile for $k = 30$ (left) and $k = 6$ (right), respectively.

## 3. Discussion

The results indicate there *is* a local instability, but that the disk is globally stable. These features of the optically-thin ADAFs are quite appreciable for explaining the rapid X-ray fluctuations which are observed in active galactic nuclei and stellar black hole candidates during the hard state. Our results suggest that once the disk is perturbed locally, the perturbations will be drifted inward without diffusing, and hard X-ray emission with substantial variations is expected. The global stability, however, assures a persistent X-ray emission, although perturbations may always be present.

Lastly, we have to mention that the adoption of vertically integrated equations under the assumption of vertical hydrostatic balance may not be a good approximation to the study of time evolution of local perturbations (see Kato et al. 1995). For this reason, a more detailed two-dimensional study, as well as a consideration of effects of turbulent thermal diffusion, are needed for the quantitative discussion concerning physical applications.

## References

Hōshi R. 1977, PTP 58, 1191
Kato S., Abramowicz M. A., Chen X. 1996, PASJ 48 in press
Matsumoto R., Kato S., Fukue J., Okazaki A. T. 1984, PASJ 36, 71
Paczyński B., Wiita P. J. 1980, A&A 88, 23

# Structure and Dynamics of an Accretion Disk around a Black Hole

Masayoshi YOKOSAWA
*Department of Physics, Ibaraki University, Mito, Ibaraki 310, Japan*

**Abstract**

The dynamical evolution of accreting flow onto a black hole is investigated using a general-relativistic, hydrodynamic code which contains a viscosity based on the $\alpha$-model. We find three types of flow's pattern, depending on the thickness of the accretion disk. In a case of a thin disk, the accreting flow through a surface of the marginally stable orbit becomes thinner due to additional cooling caused by a general-relativistic Roche-lobe overflow and horizontal advection of heat. An accretion disk with a middle thickness, $2r_{\mathrm{h}} \leq h \leq 3r_{\mathrm{h}}$, divides into two flows: the upper region of the accreting flow expands into the atmosphere of the black hole, and the inner region of the flow becomes thinner, smoothly accreting onto the black hole. The expansion of the flow generates a dynamically violent structure around the event horizon. The kinetic energy of the violent motion becomes equivalent to the thermal energy of the accreting disk. The shock heating due to violent motion produces a thermally driven wind. A very thick disk, $4r_{\mathrm{h}} \leq h$, forms a shock front near to the midplane of the accreting flow. The accretion flowing through the thick disk, $h \geq 2r_{\mathrm{h}}$, cannot only form a single, laminar flow falling into the black hole, but also produces turbulent-like structure above the event horizon. The thick disk may possibly emit the X-ray radiation observed in active galactic nuclei.

## 1. Introduction

Studies of viscous accretion onto a black hole are astrophysically interesting in the context of active galactic nuclei (AGN). According to the standard model for an accretion disk, the effective temperature of emission from an

*S. Kato et al. (eds.), Physics of Accretion Disks, 61–64.*
© 1996 OPA (Overseas Publishers Association) Amsterdam B.V.

optically thick disk decreases with the mass of the central collapsed object, $T \propto M^{-1/4}$. The effective temperature of the disk near to a massive black hole with a mass of $M_8$ in units of $10^8 M_\odot$ is $T \approx 5 \times 10^5 M_8^{-1/4}$ K. Although the UV radiation of AGNs may be explained by the accretion disk, the standard model can not directly account for the X-ray radiation. The X-ray radiation from AGNs must be emitted in an optically-thin region around the black hole. A power-law spectrum may be produced by a stochastic process of photons or electrons in turbulent motion. The turbulent motion of a fluid might be produced by a magnetic field or radiation pressure with critical luminosity, $L = L_{\mathbf{Eddington}}$. We consider here the dynamics of the accretion disk around a black hole, and examine the condition which produces turbulent motion without a global magnetic field nor a large amount of radiation.

## 2. Hydrodynamics of Accreion Disk around a Black Hole

A stationary disk with a large viscosity gives a simple expression for the thickness of the disk, such as $h \approx \kappa_{\rm es}/c\dot{M}f(r) \approx 10 r_h \dot{m} f(r)$, at the inner region of the disk where the radiation pressure dominates over the gas pressure. Here, $\dot{M}$ is the accretion rate and $\dot{m}$ is the normarized one by the critical-mass flux of the Eddington limit. $f(r)$ is a function with a value of unity far from the hole. An approximation by using a thin disk has been successful far from the hole, but invalid near to the horizon when $\dot{m} = 0.1 - 1$. The angular momentum of the Keplerian orbit has a minimum at the marginally stable orbit ($r_{\rm ms}$). Even a small amount of extraction of the angular momentum may drive the fast inflow at $r \approx r_{\rm ms}$. The dynamical time of the innermorst part of the accretion disk could be smaller than the thermal time. The dynamical interaction of fluid near to a black hole may redistribute the energy among thermal, kinetic, turbulent, and gravitational energies of accreting fluid. The continuously accreting flow through $r_{\rm ms}$ is stabilized by additional cooling due to general-relativistic Roche lobe overflow and a horizontal advection of heat. This cooling might drive a steep configulation of accreting flow in the vicinity of $r_{\rm ms}$ when $\dot{m} = 0.1 - 1$. Then, the rapid heating due to compression of the flow would be accompanied by falling motion.

## 3. Numerical Calculation of an Accretion Disk in Curved Space-time

We consider the structure and evolution of the innermost parts of the accretion disk based on a numerical simulation. We used a numerical technique of flux-corrected transport, which was developed from numerical codes used for special relativistic hydrodynamics and for general-relativistically magnetohydrodynamic accretion. In order to examine a standard accretion disk, we include the viscosity in the transport of angular momentum and in the heat equation. The heat generated by viscosity, $-t_{\alpha\beta}\sigma^{\alpha\beta}$, is transported by the radiation. The transport of radiation was solved by the diffusion approximation.

In the case of a thin disk, $h_0 = (0.1\text{–}1)\ r_{\rm h}$, the accreting fluid is smoothly swallowed by a hole. Within the critical surface, $r = r_{\rm ms}$, the fluid rapidly falls, and then the gradient of the velocity becomes very large. This causes an additional cooling due to general-relativistic Roche lobe overflow and horizontal advection of heat. The rate of work due to the pressure force in the radial direction, $\dot{E}_{P\nabla V_r} = -P/\sqrt{\gamma}\ \partial_r(W\sqrt{\gamma}v^r)$, is negative there. On the other hand, its rate becomes positive in the vicinity of the horizon. Its sign is determined by the divergent $\dot{E}_{P\nabla V_r} \propto -\partial_r(W\sqrt{\gamma}v^r) = -\partial_r(\sqrt{g}u^r) \propto -\partial_r(r^2u^r)$. When $\partial_r(u^r)/u^r < -2/r$, then $\dot{E}_{P\nabla V_r} < 0$. If the fluid freely falls into a hole, $u^r = -\sqrt{2/r}$, then $\dot{E}_{P\nabla V_r} > 0$. The rapidly falling fluid through the critical surface enters a state of free fall near to the horizon.

In the case of the medium thickness of the disk, $h_0 = (2\text{–}3)\ r_{\rm h}$, some portion of the fluid in the disk can not be swallowed by a hole, and then begins to blow out into the atmosphere. The initial density at the vicinity of the critical surface, $r = r_{\rm ms}$, distributes along the lines of iso-density, $z(r)_\rho \propto r^n$, $n > 1$. These lines bend strongly at $r \approx r_{\rm ms}$. The fluid in the neighbourhood of the equatorial plane is attracted by the hole and cooled by the horizontal advection of heat. Although the upper part of the disk is also attracted by a hole, its direction of force is radial. Then, the advection of heat, $-P/\sqrt{\gamma}\partial_r(W\sqrt{\gamma}v^r)$, becomes positive. The work due to the movement of fluid heats the flow, and, thus, causes an expansion of the flow. The expanding fluid comes into collision at the central axis. A shock wave expands from the axis. The fluid heated by the shock wave blows up into a rarefied atmosphere. The configuration of the heated flow depends on the structure of the atmosphere surrounding the disk. If the atmosphere forms a rarefied region along the central axis, the heated fluid expands along the axis. Its flow becomes a beam. In this case the atmosphere is rare all over the disk. The heated fluid expands over the disk. The other calculation with a rarefied region along the axis shows that the flow initially expands along the axis.

In the case of a thick disk, $h_0 = (4\text{–}5)\ r_{\rm h}$, the initial profiles of iso-density are approximately radial, $z(r)_\rho \propto r$. At an early epoch, $t \leq 20$, the fluid smoothly accretes onto the hole. The horizontal advection of heat within the critical surface, $r = r_{\rm ms}$, cools the inner disk, such that the pressure of the disk can not suport the radialy distributed plofile. The flow becomes bent, and azimuthally collides at the equatorial plane in the critical surface. A shock front is formed. The kinetic energy of falling fluid is directly converted into heat energy at the shock front. The disk is no longer a viscously heated one, but a shock-heating one. The temperature of the disk is then given by a 'virial temperature'. The shock wave expands radially. The heated fluid blows up into rarefied region. A large amount of beaming flow is produced by the dense wall of the disk.

## 4. Summary and Discussion

We had calculated the accretion disk using a general-relativistic hydrodynamic code which included the $\alpha$-viscosity. The above numerical calculations show that the accretion flow with the trajectory $z(r) \propto r^n, n > 1$ is dynamically unstable in the vicinity of $r = r_{\rm ms}$. On the other hand, the radial flow with $n \approx 1$ is stable. The form of flow trajectory at $r \approx r_{\rm ms}$ depends strongly on the advection of heat and viscosity. The advection of heat rapidly cools the flow at $r \approx r_{\rm ms}$, and, thus, makes a thinner flow. In the case of a medium thickness, $h_0 = (2–3)\ r_{\rm h}$, the fluid near to the upper surface of the disk expands mainly due to compressive heating accompanied by a falling motion. The upper part of the accreting disk becomes dynamically unstable, and the inner part of the disk smoothly accretes onto a black hole. The above role of the advection is different from the local stabilization of accretiong flow indicated by Abramowicz (1981). The thickness of the stationary disk at the vicinity of $r \approx r_{\rm ms}$ is given by the accretion rate $\dot{m}$ with the assumption of $\alpha$-viscosity. This leads to the fact that the accretion flow with $\dot{m} > 0.1$ is dynamically unstable.

The dynamically unstable flow produces a complex structure around a black hole. A part of accreting fluid blows up into the atmosphere. The kinetic energy of turbulent-like motion becomes comparable to the thermal energy of an accreting disk in the vicinity of critical radius $r \approx r_{\rm ms}$. However, this kinetic energy can not exceed largely over the thermal energy in our case of $\alpha$-viscosity. The turbulent motion with a strong magnetic field would supply a large bulk of energetic particles. The stochastic process in the turbulent motion may produce a power-law spetrum of X-ray emission. This X-ray emission may be strong near to a black hole, and be diffusive at the atmosphere around the accretion disk. The pair creation of electrons and positrons may form the thick cloud of pairs around a black hole.

The very thick disk might produce a narrow beam whose energy is largely supplied from the hot region generated by the shock wave. Radio sources with a large luminosity are associated with an elliptical galaxy. In many cases these galaxies have a dust lane. A large amount of gas which may be suddenly supplied by an encounter with a nearby galaxy forms a thick disk at the central region of the elliptical galaxy. On the other hand, a UV hump and strong X-ray emission are observed in a Seyfert galaxy which has a large scale disk of stationary gas. The viscosity of this disk will not rapidly supply a large amount of gas around a massive black hole.

## References

Abramowicz M. A. 1981, Nature 294, 235
Yokosawa M. 1995, PASJ 47, 605

# Standing Shock Instability in General Relativistic Accretion Flows

Kunji NAKAYAMA
*Faculty of Education, Kochi University, Kochi 780, Japan*

**Abstract**

This paper deals with a stability analysis of an adiabatic accretion flow with a standing shock wave in the general relativistic gravitational field. One example of sufficient conditions for instability is presented.

## 1. Introduction

Recent theoretical studies have shown that transonic flows can be unstable by existence of a standing shock wave (Nakayama 1992, 1993, 1994). The instability condition is roughly stated as

(∗) If the postshock flow is accelerated, the flow is unstable.

As is well known, steady discontinuous accretion flows onto a black hole can take mutually exclusive two modes: i.e., an inner shock accretion flow and an outer shock one (Fukue 1987). The criterion (∗) implies that the former is unstable. This is supported also by numerical experiments (Nobuta & Hanawa 1994).

The above-mentioned studies of the instability are, however, limited to a nonrelativistic treatment. In the present paper we give a summary of Nakayama's (1995) theory, which deals with a general relativistic regime.

## 2. Basic Assumptions

We use $c = G = 1$ unit and $- + ++$ signature. Background spacetime is assumed to be stationary and axisymmetric (typically, the Kerr spacetime):

$$ds^2 = g_{tt}dt^2 + 2g_{t\phi}dtd\phi + g_{\phi\phi}d\phi^2 + g_{rr}dr^2 + g_{\theta\theta}d\theta^2. \tag{1}$$

*S. Kato et al. (eds.), Physics of Accretion Disks, 65–68.*
© 1996 OPA (Overseas Publishers Association) Amsterdam B.V.

Acceting gas is assume to be perfect and adiabatic. The particle number density, the internal energy density, the pressure, the temperture, the specific entropy, the contravariant velocity, and the covariant velocity are, respectively, denoted by $n$, $\rho$, $p$, $T$, $s$, $u^i$, and $u_i$.

We are interested in steady/non-steady and axisymmetric ($\partial/\partial\phi = 0$) accretion flows with a shock wave. For simplicity, we consider only equatorial $r$-direction as a spatial dimension and neglect 2-dimensional (meridional) structure of the flow by forcing $u^\theta = 0 = u_\theta$ and $\partial/\partial\theta = 0$.

A steady state ($\partial/\partial t = 0$) is accompanied by a standing shock wave positted at $r = R$. Since it is a black hole accretion flow, the postshock flow has a sonic point at $r = r_c$ outside the event horizon.

## 3. Linear Instability Problem

A linear instability problem is formulated under the following assumptions (Nakayama 1992, 1993, 1994): (i) perturbations are localized only on the postshock flow and in particular (ii) the angular momentum distribution is not disturbed. This is because effects of preshock perturbations and angular momentum perturbations are transient, and hence, do not influence stability properties. Notice, however, that in the relativistic regime the assumption (ii) must be interpreted as $\delta(hu_\phi) = 0$, where $h \equiv (\rho + p)/n$ and the prefix $\delta$ means an Eulerian perturbation.

Furthemore, we do not take into account the postshock supersonic region $r < r_c$ because any disturbance on the supersonic flow cannot propagate upstream beyond the sonic point; the *linear* instability problem is, therefore, defined on the interval $(r_c, R)$.

Thus, we obtain the basic equation:

$$\frac{d^2y}{dz^2} - \sigma^2 P(z)y = y_- \Delta(\sigma)\Theta(z;\sigma)\exp\left[-\sigma\int_0^z Q(\zeta)d\zeta\right], \tag{2}$$

the boundary condition at $r = R$ derived from linearized jump conditions:

$$\left(\frac{dy}{dz}\right)_- - (1 - l\Omega_R)\left\{(\nu+\sigma)v_-^2 + \lambda_-\left[\nu v_-^2 + \sigma(v_-^2 - \hat{a}^2)\right]\right\}y_- = 0, \tag{3}$$

and the boundary condition at $r = r_c$:

$$\lim_{z\to\infty}\left|y\exp\left\{-\sigma\int_0^z \frac{u^t v^2(1-a^2)}{u_t[u^t u^t(1-v^2)(1-a^2) - g^{tt}(a^2-v^2)]}dz\right\}\right| < \infty, \tag{4}$$

which means that the perturbations must be finitely bounded. Here, $\sigma$ is the growth rate,

$$z \equiv \int_R^r \frac{-u_t[u^t u^t(1-v^2)(1-a^2) - g^{tt}(a^2-v^2)]}{u^r(a^2-v^2)}dr, \tag{5}$$

$$y \equiv \delta(\sqrt{-g}nu^r)\exp\left\{\sigma\int_0^z \frac{u^t v^2(1-v^2)}{u_t[u^t u^t(1-v^2)(1-a^2)-g^{tt}(a^2-v^2)]}dz\right\}, \tag{6}$$

$$P \equiv \frac{u^r u^r[u_r u_r(1-v^2)(1-a^2)+g_{rr}(a^2-v^2)]}{u_t u_t[u^t u^t(1-v^2)(1-a^2)-g^{tt}(a^2-v^2)]}, \tag{7}$$

$$Q \equiv -\frac{u^t u^r[u_r u_r(1-v^2)(1-a^2)+g_{rr}(a^2-v^2)]}{u_t u_r[u^t u^t(1-v^2)(1-a^2)-g^{tt}(a^2-v^2)]}, \tag{8}$$

$$\Delta \equiv -(1-l\Omega_R)[\nu v_-^2-\sigma(\hat{a}^2-v_-^2)]\left(\frac{u^t}{T}\right)_-, \tag{9}$$

$$\Theta \equiv -\frac{d\theta_1}{dz}+\frac{\sigma(a^2-v^2)\theta_2}{-u_t[u^t u^t(1-v^2)(1-a^2)-g^{tt}(a^2-v^2)]}, \tag{10}$$

$$\lambda \equiv \frac{u^t u^t(1-v^2)}{u^t u^t[(1-v^2)(1-a^2)-g^{tt}(a^2-v^2)]}\frac{1}{T}\left(\frac{\partial h}{\partial s}\right)_n-1, \tag{11}$$

$$\nu \equiv \frac{1}{[u_r u^t(1-v^2)]_-}\left[\frac{d}{dR}\ln\left|g^{tt}-2lg^{t\phi}+l^2g^{\phi\phi}\right|+\hat{a}^2\frac{d}{dR}\ln|g^{rr}g|\right], \tag{12}$$

$$\theta_1 \equiv \frac{u^t(1-v^2)}{u^t u^t[(1-v^2)(1-a^2)-g^{tt}(a^2-v^2)]}\left(\frac{\partial h}{\partial s}\right)_n, \tag{13}$$

$$\theta_2 \equiv \frac{u^t u^t(1-v^2)+g^{tt}v^2}{u^t u^t[(1-v^2)(1-a^2)-g^{tt}(a^2-v^2)]}\left(\frac{\partial h}{\partial s}\right)_n-T, \tag{14}$$

the adiabatic sound speed $a \equiv (\partial p/\partial\rho)_s^{1/2}$, the radial velocity of the fluid observed by a corotating observer $v \equiv -[u_r u^r/(1+u_r u^r)]^{1/2}$, the specific angular momentum of the steady state $l \equiv -u_\phi/u_t$, the physical angular velocity of the fluid $\Omega \equiv u^\phi/u^t$, and finally, $\hat{a}^2 \equiv (p_+-p_-)/(\rho_+-\rho_-)$. [ The subscript + (resp. −) means a value at the *pre*shock (*post*shock) side of the shock wave in the steady state and $d/dR$ means $(d/dr)_{r=R}$. ]

## 4. Results and Remarks

Although we do not specify equations of state of gas, we need a (natural) prerequisite $(\partial p/\partial s)_n > 0$. Then, we obtain sufficient conditions for a *monotonically growing* mode without oscillations to exist by detailed analysis of the eigenvalue problem (2)-(4). We give the following example. [ For other theorems and proof, see Nakayama (1995). ]

**Theorem 1 (Sufficient condition for instability)** *Suppose that*

$$\nu < 0, \tag{15}$$

$$\lambda_-(v_-^2-\hat{a}^2)+v_-^2+\frac{\inf_{z\in(0,\infty)}[P(z)]^{1/2}}{1-l\Omega_R} > 0, \tag{16}$$

$$1 < \frac{\theta_{1c}}{\theta_{1-}} + \frac{2}{\pi} \left[ \frac{\inf_{z \in (0,\infty)} P(z)}{\sup_{z \in (0,\infty)} P(z)} \right]^{1/2}, \tag{17}$$

*and*

$$\frac{d\theta_1}{dz} < 0 \qquad \forall z > 0, \tag{18}$$

*then, there exists an unstable mode whose growth rate $\sigma$ is a positive real number.*

Since we must analyse the eigenvalue problem describing intricate interactions between sound waves and entropy waves, also the obtained theorem is somewhat complicated. Nevertheless, the postshock acceleration condition (*) can be perceived. In the limit of a weak standing shock, hence, that of $R \to r_c$, the condition $\nu < 0$ of Theorem 1 just means that the flow is accelerated behind the shock wave in the sense that $(dv^2/dR)_- < 0$. Other conditions except the inequality $d\theta_1/dz < 0$ become trivial. Therefore, if the shock is not extremely strong, the conditions are satisfied by the flow accelerated at the postshock side. Also in the relativistic case, thus, the statement (*) in section 1 still remains an available rule to predict the occurence of the standing shock instability.

## References

Fukue J. 1987, PASJ 39, 309
Nakayama K. 1992, MNRAS 259, 259
Nakayama K. 1993, PASJ 45, 167
Nakayama K. 1994, MNRAS 270, 871
Nakayama K. 1995, MNRAS submitted
Nobuta K., Hanawa T. 1994, PASJ 46, 257

# A Slim Accretion Disk with Hydrogen Burning

Kenzo ARAI[1], Masa-aki HASHIMOTO[2], and Shin-ichiro FUJIMOTO[1]
*1. Department of Physics, Kumamoto University, Kumamoto 860, Japan*
*2. Department of Physics, Kyushu University, Ropponmatsu, Fukuoka 810, Japan*

**Abstract**

We construct stationary models of a slim accretion disk around a balck hole of $10M_\odot$ with including hydrogen burning. Energy generation through the CNO cycle is explicitly incorporated into the equation of energy balance. When enough energy is generated by hydrogen burning, the disk expands vertically to the equatorial plane. More work is done by pressure than the net energy generated, causing temperature to decrease. Hydrogen is processed into helium as the gas flows towards the black hole. The most abundant is $^{15}$O among the CNO elements in the central region of the disk. It may emit the $\gamma$-ray line at energy 1.38 MeV because a part of the processed material is expelled as jets.

## 1. Introduction

The existence of an accretion disk is confirmed growingly by observations in low mass X-ray binaries and cataclysmic variables. In an accretion disk around a compact object, inflowing gas loses its angular momentum outwards due to viscous stress. Correspondingly its gravitational energy is efficiently converted into thermal energy and radiated away as X- and $\gamma$-ray photons.

As the accretion rate $\dot{M}$ increases, the liberated energy increases so that the disk becomes more and more thick. Abramowicz et al. (1988) have investigated stationary models of a slim disk and found that advective heat flux plays a crucial role for stabilizing the disk. Arai & Iimori (1992) have studied slim models supported exclusively by radiation pressure. When $\alpha$ becomes as small as $10^{-5}$, temperature rises more than $10^8$ K at the central part of the disk, causing hydrogen to burn into helium.

*S. Kato et al. (eds.), Physics of Accretion Disks, 69–72.*
© 1996 OPA (Overseas Publishers Association) Amsterdam B.V.

Equilibrium models of a thick disk have been constructed by many authors. Nucleosynthesis has been investigated by Chakrabarti (1988), Jin et al. (1989), and Arai & Hashimoto (1992). Although the nuclear energy release has been neglected, it becomes comparable to the gravitational energy release as the nuclear reactions proceed. Hence the nuclear energy generation might affect the structure of the disk.

Lamb et al. (1983) observed $\gamma$-ray lines at energies 1.2 and 1.5 MeV from SS 433 and interpreted these as blueshifted and redshifted components of the 1.37 MeV line of $^{24}$Mg. Boyd et al. (1984) proposed an alternative identification as the 1.38 MeV deexcitation of $^{15}$O. Kotani et al. (1994) have observed X-ray lines for various heavy elements, each of which has two different Doppler shifts. Since these elements might be produced inside the disk, it is worthwhile to study the accretion disk with nuclear reactions. It is also expected that the onset of nuclear burnings may change the equilibrium structure of the disk. To emphasize the difference between the models with nuclear reactions switched on and off, we follow the formalism given by Abramowicz et al. (1988). Details of the results given in the present paper have been discussed by Arai & Hashimoto (1995).

## 2. Equation of Energy Balance

In constructing a slim model of an accretion disk around a black hole of mass $M$, the disk is assumed to be stationary and axisymmetric. General relativistic effects on gas motion is described by a pseudo-Newtonian potential. Physical quantities are averaged vertically and represented in terms of the corresponding values at the equatorial plane. The mass accretion rate is measured by a dimensionless quantity:

$$\dot{m} = \dot{M}/\dot{M}_{\rm cr}\,, \quad \dot{M}_{\rm cr} = 64\pi G M m_p c/\sigma_{\rm T}\,. \tag{1}$$

The energy generation rate through the CNO cycle is written as

$$\begin{aligned}\epsilon_{\rm CNO} &= 7.94\times10^{27}\rho X_H X_{\rm CNO}\exp[-152.228/T_6^{1/3}] \\ &\quad\times T_6^{-2/3}\left(1+0.0027T_6^{1/3}-0.0078T_6^{2/3}\right)\ {\rm erg\ g^{-1}\ s^{-1}}, \end{aligned} \tag{2}$$

where $X_H$ and $X_{\rm CNO}$ are the abundances of hydrogen and CNO isotopes by weight and $T_6 = T/(10^6\,{\rm K})$. In the above expression we have used the rate of the reaction $^{14}{\rm N}(p,\gamma)^{15}{\rm O}$ updated by Caughlan & Fowler (1988), because the cycle time is nearly equal to the life time of $^{14}$N. Although $X_{\rm CNO}$ is fixed to be 0.014, $X_H$ is evaluated from the rate equation and consumed by the amount in accordance with the generated energy.

The equation of energy balance is written as

$$Q^+_{\rm vis} + Q^+_{\rm nuc} = Q^-_{\rm rad} + Q^-_{\rm adv}\,. \tag{3}$$

Here $Q^+_{\rm vis}$ is the energy generated by viscous stress, $Q^+_{\rm nuc}$ is the energy generated by the CNO cycle, $Q^-_{\rm rad}$ is the energy lost by radiative flux from the disk surface, and $Q^-_{\rm adv}$ is the energy transported by advection.

## 3. Results and Discussion

Sequences of stationary models are specified by the values of $M$, $\dot{m}$, and $\alpha$. In the present paper we discuss only the results for a model with $M = 10M_\odot$, $\dot{m} = 3$, and $\alpha = 10^{-6}$. For models with other sets of parameters, see Arai & Hashimoto (1995).

Figure 1 shows the temperature distribution. The solid line refers to the case with the hydrogen burning switched on and the dashed line is the case with the burning artificially switched off. The occurence of the burning lowers the maximum temperature. This is ascribed to the fact that the disk becomes thicker by about 10 % with the nuclear reactions. When energy is generated by hydrogen burning inside the disk, pressure increases and the disk expands vertically to the equatorial plane. More work is done than the net energy generated. Eventually, temperature inside the disk decreases in spite of the energy generation. This is a typical character of a gravitating system such as a main sequence star, which has negative heat capacity.

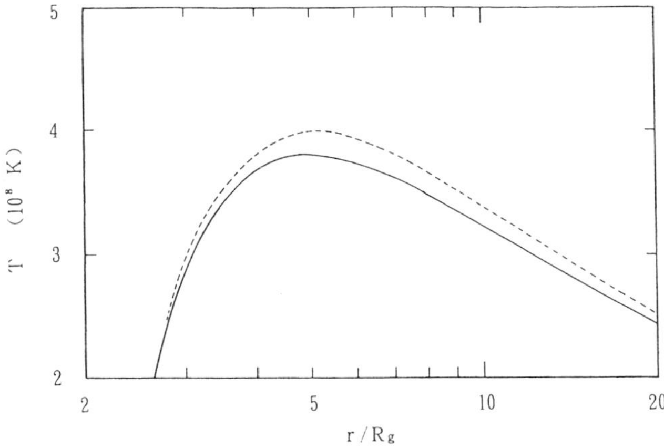


Fig. 1. Temperature distribution of the slim disk.

The distribution of chemical abundance is shown in figure 2. The initial composition of the CNO isotopes is set to be the population I abundance. Note that hydrogen is processed into helium as the gas flows towards the black hole. The hydrogen abundance is depleted from 0.7 to 0.041 at the inner edge of the disk. The burning occurs through the hot CNO cycle in the region $r = 30 - 8R_{\rm g}$, where $T = (2.0 - 3.5) \times 10^8$ K, $\rho = 3.3 - 5.9$ g cm$^{-3}$ and $t_{\rm dr} = 1.5 \times 10^5 - 5.2 \times 10^3$ s. CNO seeds are redistributed to approach equilibrium. About a half of the original CNO isotopes is transmuted to $^{15}$O before the gas reaches the inner edge of the disk. It may emit the 1.38 MeV $\gamma$-ray line because a part of the processed material is expelled as jets.

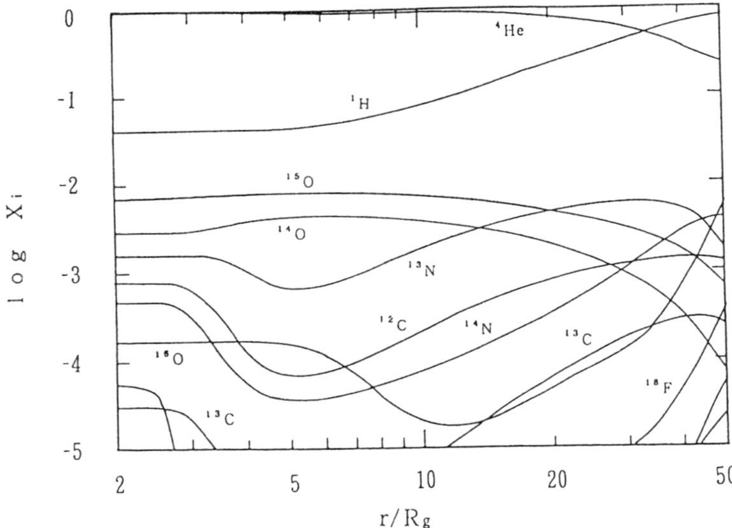


Fig. 2. Distribution of chemical elements in the slim disk.

According to nucleosynthesis in a thick disk studied by Arai & Hashimoto (1992), no significant hydrogen burning is recognized for a model with $M = 10M_\odot$, $\dot{m} = 100$ and $\alpha = 10^{-6}$. This model has $T = 4.3 \times 10^8$ K, $\rho = 10$ g cm$^{-3}$ and $t_{\rm dr} = 33$ s at $r \simeq 5R_{\rm g}$. On the other hand, our slim model with $\dot{m} = 3$ has similar values of temperature and density, but the drift time is $t_{\rm dr} = 1.1 \times 10^3$ s, which is longer by a factor of 33. This implies that the magnitude of the infall velocity varies as $|v_r| \sim \dot{m}$. Although a thick disk model has not enough time for hydrogen to burn, when $\alpha$ is as small as $10^{-10}$, temperature is high enough to proceed nuclear burning up to Fe.

Note that our slim disk supported by radiation pressure is hot enough to operate the CNO cycle, but the density is rather low. The total mass of the accretion disk is only $1.5 \times 10^{25}$ g. It justifies the neglect of the self-gravity.

## References

Abramowicz M.A., et al. 1988, ApJ 332, 646
Arai K., Hashimoto M. 1992, A&A 254, 191
Arai K., Hashimoto M. 1995, A&A 302, 99
Arai K., Iimori T. 1992, Phys. Rep. Kumamoto Univ. 9, 51
Boyd R. N. et al. 1984, ApJL 276, L9
Caughlan G. R., Fowler W. A. 1988, Atomic Nucl. Data Tables 40, 283
Chakrabarti S. K. 1988, ApJ 324, 391
Jin L., Arnett W. D., Chakrabarti S. K. 1989, ApJ 336, 572
Kotani T. et al. 1994 PASJ 46, L147
Lamb R. C. et al. 1983, Nature 305, 37

# Emission from Optically Thin Advection-Dominated Accretion Flows

Insu Yi
*Institute for Advanced Study, Olden Lane, Princeton, NJ 08540, USA*

**Abstract**

In accreting black hole systems with low mass accretion rates, most of dissipated accretion energy can be radially transported inward, resulting in very low radiative cooling efficiencies. Optically thin advection-dominated flows (ADFs) are hot and emit high energy X-rays with very low luminosities. We apply emission spectra from optically thin, magnetized plasma to astrophysical systems including some black hole soft X-ray transients (BSXTs), Sgr A*, and NGC 4258 for which we find remarkable agreements with observations. Further implications of ADFs are discussed.

## 1. Advection-Dominated Accretion Flows and Optically Thin Emission

When the cooling time scale becomes longer than the radial flow time scale in a plasma accreted to the black hole, the viscously dissipated energy is not efficiently radiated but stored as entropy and advected inward (Narayan & Yi 1994; Abramowicz et al. 1995; Chen et al. 1995; Abramowicz 1996; Narayan 1996). Such a flow is characterized by a low luminosity for a given mass accretion rate (Narayan & Yi 1995).

Since the temperature of the flow is very high, inclusion of magnetic fields in the plasma provides an efficient cooling mechanism through synchrotron emission which together with the bremsstrahlung cooling dominates the cooling of electrons. A fraction of these photons scatter off hot electrons and get Comptonized. These processes determine essential spectral characteristics of hot flows (figure 1). Depending on whether the accretion flow can exist in the form of a thin disk at some radii, the spectrum may also show thin disk emission. The emission from ADFs may be a crucial test for the nature of

*S. Kato et al. (eds.), Physics of Accretion Disks, 73–78.*
© 1996 OPA (Overseas Publishers Association) Amsterdam B.V.

the accretion flow when mass accretion rate is sufficiently low (Rees et al. 1982; Narayan & Yi 1995).

The optically thin ADFs exist for mass accretion rates below the critical accretion rate $\dot{M}_{\rm crit} \sim 0.3\alpha^2 \dot{M}_{\rm E}$ where $\dot{M}_{\rm E} = 1.39 \times 10^{18}(M/M_\odot)$ g s$^{-1}$ is the Eddington accretion rate, $M$ is the mass of a black hole, and $\alpha$ is the usual viscosity parameter (Narayan & Yi 1995). The critical accretion rate scaled by the Eddington rate is mass scale-free which suggests physical similarity between accretion flows around stellar mass black holes and those around massive black holes. Below this accretion rate, the radiative luminosity is determined to be much lower than the nominal efficiency of $\sim$ 10% (Narayan & Yi 1995).

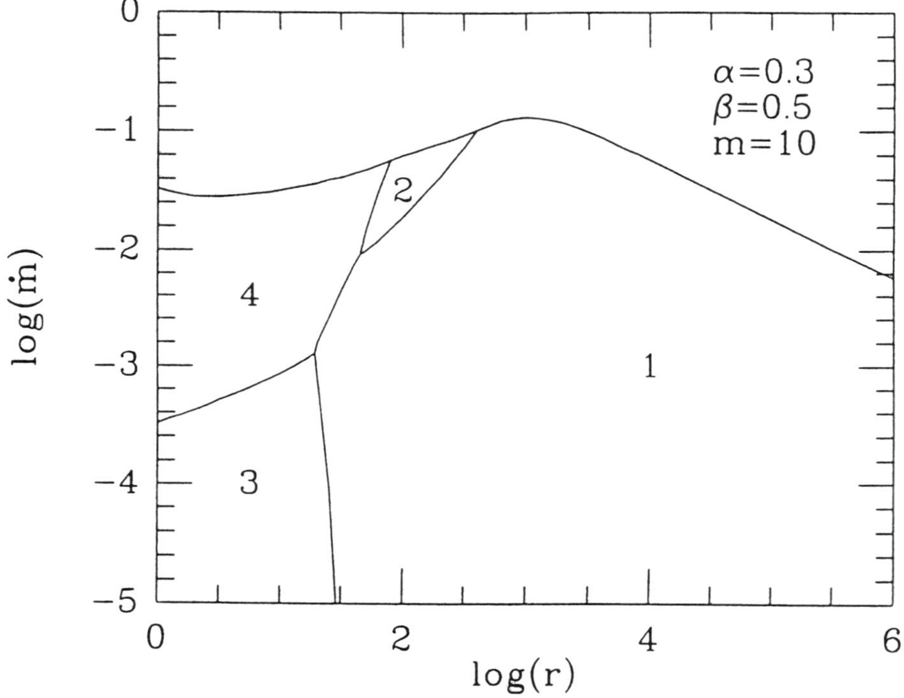


Fig. 1. Dominant cooling processes in the plane defined by $\dot{m} = \dot{M}/\dot{M}_{\rm E}$ and $r = R/R_{\rm s}$ where $R_{\rm s}$ is the Schwartzscild radius. The uppermost line gives the critical accretion rate as a function of radius. Each region marked by a number is dominated by a cooling process (1: bremsstrahlung, 2: Comptonized bremsstrahlung, 3: synchrotron, 4: Comptonized synchrotron). The emitted spectra are results of combinations of these cooling processes (Narayan & Yi 1995).

## 2. Black Hole Soft X-ray Transients

A0620-00 is a BSXT which occasionally undergoes outbursts during which luminosity dramatically increases. Between outbursts, the system resides in a low luminosity quiescent state with X-ray emission (McClintock et al. 1995). During quiescent, the optical luminosity suggests that the accretion rate $\dot{M} \sim 2 \times 10^{-10} M_{\odot} \mathrm{yr}^{-1}$ while the X-ray luminosity implies only $\dot{M} \sim 10^{-15} M_{\odot} \mathrm{yr}^{-1}$ if both emission components are from a thin disk. It is challenging for the thin disk model to explain the accretion rates in the two emission regions differing by more than 4 orders of magnitude (cf. Mineshige & Wheeler 1989; Huang & Wheeler 1989). In the thin disk model, it is unclear how high energy X-ray photons are produced with such a low mass accretion rate.

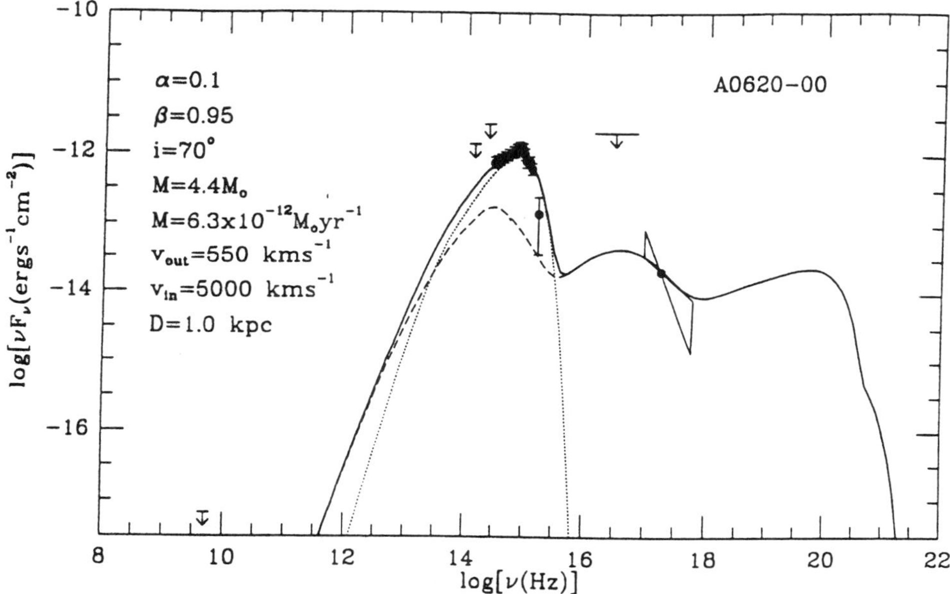


Fig. 2. The A0620 emission spectrum from an accretion flow which has an outer thin accretion disk truncated at a radius a few $\times 10^3 R_s$. For model parameters indicated in the figure, see Narayan et al. (1996).

This problem is naturally resolved by the ADF. The outer thin disk emits optical-UV photons with a mass accretion rate $\sim 10^{-11} M_{\odot} \mathrm{yr}^{-1}$. For the same mass accretion rate, below a radius of a few $\times 10^3 R_s$, the accretion flow becomes advection-dominated and emits X-ray photons with an accretion efficiency $\sim 10^{-4} - 10^{-3}$. The spectrum and luminosity are simultaneously determined (Narayan et al. 1996). The best fit spectrum is shown in figure 2. For details, see Narayan et al. (1996). The emission spectrum has four distinct peaks corresponding to optically thin disk emission (overlapping with a synchrotron peak), singly Comptonized synchrotron emission, bremsstrahlung emission (overlapping with doubly Comptonized synchrotron emission) in increasing frequencies. The thin disk emission alone is unable to fit the observed spectrum.

Other BSXTs such as V404 Cyg and Nova Mus 1991 may also harbor

ADAFs during quiescence (Narayan et al. 1996). Due to lack of data, these two systems are not sufficiently constrained. Nevertheless, spectra from ADFs have been shown to be compatible with observations although the thin disk models are not ruled out. The predicted spectra based on the ADFs indicate drastically different UV-X-ray emission which could be either readily testable (V404 Cyg) or testable by future observations (Nova Mus).

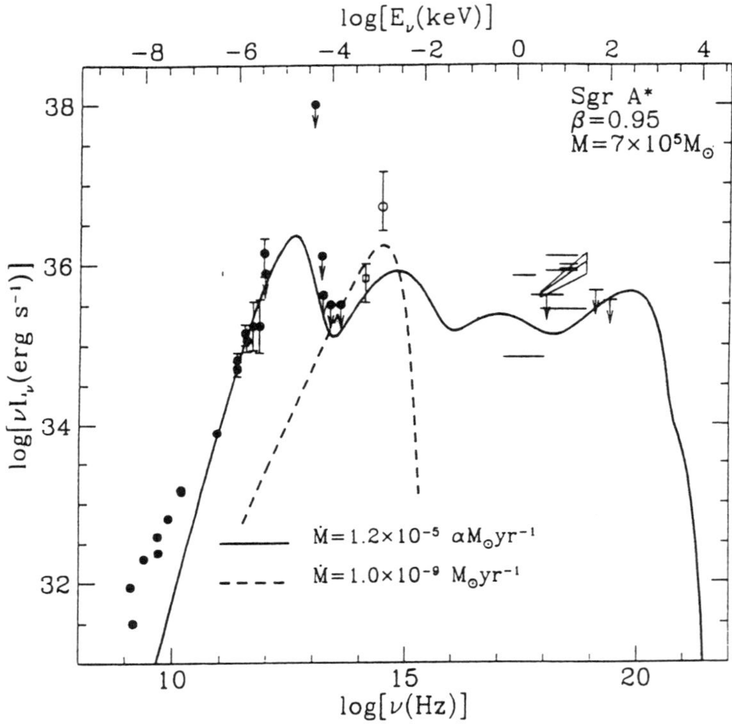


Fig. 3. The multi-frequency spectrum and spectral fit based on the ADF for Sgr A* (Narayan et al. 1995, see also Narayan 1996 in these proceedings).

## 3. Sgr A* and NGC 4258

The radio source Sgr A* is likely to be a black hole of dynamically determined mass $\sim 10^6 M_\odot$ (Genzel & Townes 1987). This source has been enigmatic as it shows a very low luminosity despite a relatively large estimated mass accretion rate. Moreover, its radio to gamma-ray spectrum is essentially flat, which is not explained by a thin disk with a mass accretion rate $\sim 10^{-9} M_\odot$ which satisfies the submm/IR constraint. An ADF provides a self-consistent resolution (Narayan et al. 1995).

Despite a large estimated mass accretion rate, the luminosity of the source can be as low as the observed value $\sim 10^{37}$ erg s$^{-1}$ due to the low accretion efficiency in the ADF (Narayan & Yi 1995). In figure 3, we show the spectrum from an ADF which fits various observed luminosities and upper limits

(Narayan et al. 1995). The dashed line corresponds to the thin disk emission with a mass accretion rate $\sim 10^{-9} M_{\odot} \mathrm{yr}^{-1}$. In the ADF model, the entire spectral range is explained by the emission processes essentially identical to those in the inner flows of BSXTs during quiescence. The estimated mass accretion rate is $\dot{M} \sim 10^{-5} \alpha M_{\odot} \mathrm{yr}^{-1}$ (Genzel & Townes 1987). In agreemnet with the dynamical estimate, the mass of the black hole is estimated to be $\sim 10^{6} M_{\odot}$.

NGC 4258 is an interesting galaxy with an accurately determined central point mass of $3.6 \times 10^{7} M_{\odot}$. The observed X-ray luminosity is $\sim 4 \times 10^{40}$ erg s$^{-1}$ while its optical-UV luminosity is $\leq 1.5 \times 10^{42}$ erg s$^{-1}$. The observed sub-Eddington luminosity suggests the possibility of an ADF. The observed X-ray spectrum and luminosity can be satisfactorily fit by the ADF model. If the observed emission is indeed from such a flow, then the estimated mass accretion rate is $\sim 10^{-2} \dot{M}_{\mathrm{E}}$ (Lasota et al. 1996; Lasota 1996 in these proceedings). This possibility is interesting since some extragalactic nuclei with $\dot{M} < \dot{M}_{\mathrm{E}}$ can be extremely dim despite substantial accretion rates. This may suggest why there are not many bright galactic nuclei at the present epoch as a result of QSO evolution (Fabian & Rees 1995).

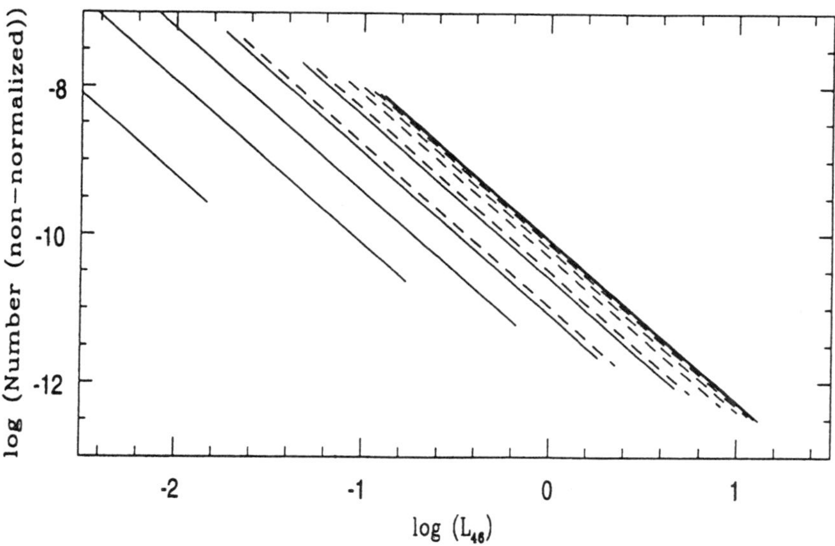


Fig. 4. The QSO luminosity evolution as a function of redshift. $L_{46}$ is the QSO luminosity in units of $10^{46}$ erg s$^{-1}$. The curves correspond luminosity functions for $z = 3$ (right) to 0 (left) in decreasing $z$ by 0.5 (Yi 1996).

## 4. Further Implications

ADFs may have interesting consequences on cosmological evolution of QSOs. The number of observed QSOs show little evolution between redshift $z \sim 2$ and $\sim 3$ with little change in their luminosity distribution (Hartwick &

Shade 1990). However, below $z \sim 2$ and certainly by $z \sim 1.5$, the bright QSOs disappear. This sudden disappearance of QSOs has remained largely unexplained. The transition of accretion flows from high efficiency flows to low-efficiency ADFs may be responsible for this enigmatic evolutionary behavior (Yi 1996). Even when the accretion continues with nearly constant accretion rate, the growth of black holes favor the arrival of ADFs toward low $z$. In figure 4, the evolution of QSO luminosity function is shown. The solid lines correspond to luminosity distributions (corresponding to a simple initial power-law mass function) for which evolution includes transition to ADFs. The evolution below $z \sim 2$ is rapid enough to explain the observed behavior of QSOs. The dotted line is the evolution without ADFs. In the latter case, the rapid disappearance of QSOs is not naturally explained.

The ADFs provide a unique way of distinguishing black holes from neutron stars (Narayan & Yi 1995). In the latter, advected energy is eventually radiated at or near the surface of neutron stars (Yi et al. 1996) whereas in the case of black holes the advected energy is lost beyond the horizon (Narayan et al. 1995).

**Acknowledgements**

The author would like to express his thanks to Jean-Pierre Lasota, Ramesh Narayan, and Craig Wheeler for helpful discussions.

**References**

Abramowicz M. 1996, this volume
Abramowicz M. et al. 1995, ApJL 438, L37
Chen X. et al. 1995, ApJL 443, L61
Fabian A. C., Rees M. J. 1995, MNRAS in press
Genzel R., Townes C. H. 1987, ARA&A 25, 377
Hartwick F. D. A., Shade D. 1990, ARA&A 28, 437
Huang M., Wheeler J. C. 1989, ApJ 343, 229
Lasota J.-P. 1996, this volume
Lasota J.-P. et al. 1996, ApJ in press
McClintock J. E., Horne K., Remillard R. A. 1995, ApJ 442, 358
Mineshige S., Wheeler J. C. 1989, ApJ 343, 241
Narayan R. 1996, this volume
Narayan R., McClintock J. E., Yi I. 1996, ApJ in press
Narayan R., Yi I. 1994, ApJL 428, L13
Narayan R., Yi I. 1995, ApJ 452, 710
Rees M. J. et al. 1982, Nature 295, 17
Yi I. 1996, ApJ submitted
Yi I. et al. 1996, A&A in press

# Optically Thick Accretion Disks: Their Structure, Spectra and Time Evolution

Ewa SZUSZKIEWICZ
*International School for Advanced Studies - SISSA, via Beirut 2-4, 34013 Trieste, Italy*

**Abstract**

We discuss here properties of the vertically integrated optically-thick accretion disks. The slim disk models with high accretion rates predict the soft X-ray excesses reported in some AGN. A sequence of slim disks with fixed mass and different accretion rates can explain the time evolution of the observed spectra in variable objects.

## 1. Introduction

Optically-thick accretion disk models are among the best studied accretion flows. In these models the advective cooling is important and it is explicitly included (slim disk models). Much of the attention has recently been shifted towards the optically thin disks with advection being dominant (Abramowicz, Kato, Lasota, Narayan, in this volume). However, optically thick disks are far from being understood and they are essential for finding a global solution for a realistic accretion flow (Szuszkiewicz et al. 1996).

The assumptions necessary for getting a simple picture of the disk accretion introduce a limitation of the applicability of such a model to a realistic situation. The standard thin disk is a local approximation of the global vertically-integrated slim accretion disk. The comparison between these solutions defines the region in the parameter space where the simplest thin model can be used and where the full slim solution must be adopted. The applicability of the vertical integration can be tested by a detailed comparison between one- and two-dimensional models (Papaloizou & Szuszkiewicz 1994).

*S. Kato et al. (eds.), Physics of Accretion Disks, 79–84.*
© 1996 OPA (Overseas Publishers Association) Amsterdam B.V.

## 2. The Properties of the Models

Slim disk models take advantage of the simplification due to vertical integration as in the case of the standard thin disk models, but at the same time they correctly describe important physical effects which dominate the flow for the accretion rates $\dot{m} > 0.2$, but which are omitted in the standard models ($\dot{m} = \dot{M}/\dot{M}_c$, where $\dot{M}_c = 2.6 \times 10^{18} m$ g s$^{-1}$ and $m$ is the mass of the central object in solar masses). The momentum equation for slim disks retains the inertial term $v_r(dv_r/dr)$, describing the dynamical importance of the accretion velocity $v_r$, and the horizontal pressure gradient $\rho^{-1}(dP/dr)$. The advective, horizontal heat flux, $v_r T(dS/dr)$, is included in the energy equation. The remaining equations are the same as in Shakura & Sunyaev (1973) except that we use the pseudo-Newtonian potential to describe the gravitational field of the central black hole. The inner boundary condition uses the fact that there is no viscous torque across the horizon of the black hole, while the outer boundary condition states that at large radii the model of the flow is similar to that of Shakura-Sunyaev. We consider disks of Population I chemical composition. The opacities are taken from the tables of Cox & Stewart (1970). The models use a family of different viscosity prescriptions in which the relevant viscous stress tensor component is proportional to the following combination of the total, $P$, and gas, $P_g$, pressures: $\tau_{r\varphi} = -\alpha P^{1-\mu/2} P_g^{\mu/2}$, where $\alpha$ is the viscosity parameter, $\mu$ is a measure of the relative importance of $P$ and $P_g$ in the process of viscous energy generation, and it ranges from 0 to 2.

We require that the mass of the disk is small in comparison with the mass of the central black hole, so the self-gravity is not important. This sets an upper limit on the accretion rate. The outer radius is determined by the condition that local self-gravitational instabilities are not present. These instabilities may develop when the disk density exceeds $M/r^3$ where $M$ and $r$ are the mass of the central object and the radius of the disk respectively. We use only those models which, checked a posteriori, are effectively optically thick. This means that the geometric mean of the total and free-free opacities multiplied by the surface density is greater than unity. The slim disk models discussed in this paper are determined by the above assumptions and by differential equations which describe mass, energy and momentum conservation (Abramowicz et al. 1988). The structure equations for a simple geometrically-thin, optically-thick accretion disk can be obtained from the slim disk equations by making appropriate approximations (Szuszkiewicz 1990).

Figure 1 shows the radial structure of the slim models with different luminosities, and the total luminosity - accretion rate relation. The luminosity is expressed in terms of the Eddington luminosity. For $\dot{m} < 1$ the inner edge is located very close to $r = 3r_G$ ($r_G$ is the Schwarzschild radius). At $\dot{m} = 1$ the location changes abruptly and for still higher accretion rate tends to $r = 2r_G$.

The outer radius of the disk, defined as the radius where its self-gravity is comparable to the gravity from the black hole, reaches a minimum for $\dot{m}$ around 0.1. A region of radiation-pressure dominance is situated above the short dashed line, and the region of electron-scattering dominance is above the long dashed line. In the case of $\mu = 0$ the models become optically thin in the very inner part of the disk for $1 < \dot{m} < 50$ (small region in the upper left corner). For higher accretion rates the surface density increases and the disk is again optically thick. The model with accretion rate $\dot{m} = 0.001$ is gas-pressure dominated. The main opacity source is free-free absorption. It is a pure example of the Shakura-Sunyaev "outer" region of the disk. Matter under the conditions described above will radiate as a blackbody. In the model with $\dot{m} = 0.01$, there are several different regions. For higher accretion rates the innermost region shrinks towards the central object and the outermost shifts further away. The whole disk becomes radiation-pressure dominated with electron scattering as the principal opacity source. This is the pure "inner" Shakura-Sunyaev region.

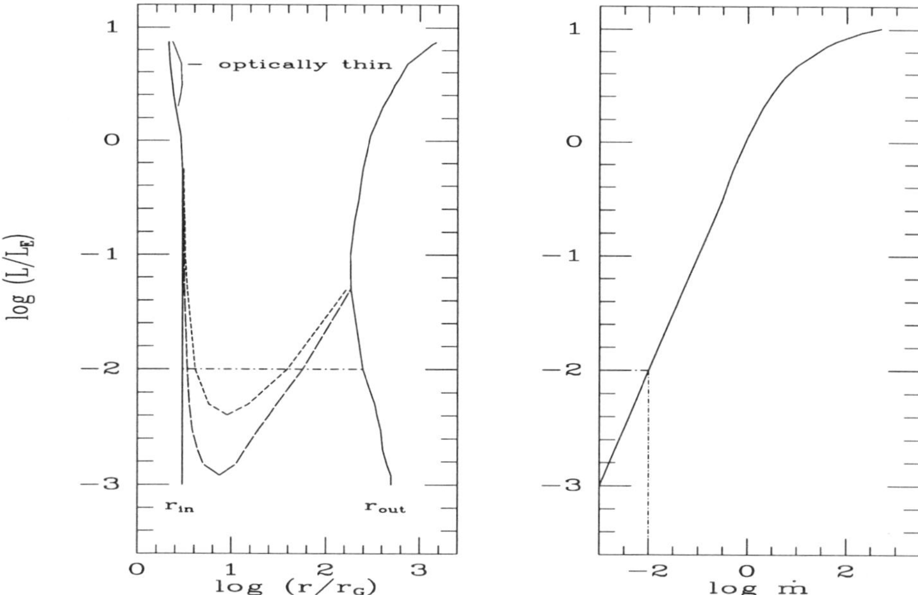


Fig. 1. Global structure of the slim disk models with $m = 10^8$, $\alpha = 0.001$, $\mu = 0$ and different luminosities (left panel) and the total luminosity - accretion rate relation for these models (right panel). For a given accretion rate (for example the $\dot{m} = 0.01$, dot-short dashed line), translated into a luminosity in the right panel, disk extends between two solid lines marked $r_{\rm in}$ and $r_{\rm out}$ respectively. Along the short dashed line the gas and radiation pressures are equal to each other. Along the long dashed line electron scattering and free-free absorption are of the same importance.

There is a satisfactory agreement between the structure and radiative flux of the slim models and the thin models for disks with $L < 0.2L_{\rm E}$. This justifies usage of thin disk approximation for small accretion rates to evaluate the gross properties of the flow, such as the density, temperature or local flux, everywhere apart from their innermost regions, where the thin model is inadequate for all accretion rates.

Accretion rate and efficiency determine total luminosity. The spectrum of the radiation emitted from the disk surface depends on its structure and surface temperature $T_S$. This temperature plays a similar role as effective temperature in stars. To calculate the spectra we need to know how and where the energy is released. The simplest assumption that can be made about the emergent spectrum is that it is emitted locally at the rate prescribed by viscous energy transport. For each disk annulus, we calculated the modified blackbody spectrum, which takes into account the dominance of electron scattering over absorption. The overall spectrum is computed by integrating the local spectra over the radial extent of the whole disk. The comparison of the slim and thin disk spectra can be summarized as follows: within the valid range of the thin disk approximation, the slim disk gives the same spectrum as the thin one. Moreover the small differences in local flux for sub-Eddington models do not appear in their total spectra at all. Only for super-Eddington models the difference is significant. It is particularly pronounced in the EUV. In the optical or UV the difference between the two spectra are so small that it is impossible to determined systematically how the disk parameters $m$ and $\dot{m}$ would change if we applied the slim disk models instead of the thin ones.

## 3. Comparison with Observations

We reexamine the hypothesis that the optical/UV/soft X-ray continuum of Active Galactic Nuclei is thermal emission from an accretion disk. Previous studies have shown that fitting the spectra with the simple blackbody spectra produced by the standard, optically-thick and geometrically-thin accretion disk models often led to luminosities which contradict the basic assumptions adopted in the standard model. The problem was removed by calculating spectra in a more sophisticated way. Our approach is different. We start from the more general disk model – slim accretion disk – and we calculate spectra in the simple modified blackbody approximation. We find that the spectra can be fitted not only by models with a high mass and a low accretion rate (as in the case of thin disk fitting) but also by models with a low mass and a high accretion rate. In the first case fitting the observed spectra in various redshift categories gives black hole masses around $10^9 M_\odot$ for a wide range of redshift, and for accretion rates ranging from 0.4 (low redshift) to $8 M_\odot$ yr$^{-1}$ (high redshift). In the second case the accretion rate is around $10^2 M_\odot$ yr$^{-1}$ for all AGN and the mass ranges from $3 \times 10^6$ (low redshift) to $10^8 M_\odot$ (high redshift). Moreover, the low mass - high accretion rate models extend further into the soft X-rays and can produce the observed excess component in this waveband, while the high mass-low accretion rate models can be responsible only for the optical and UV part of the spectrum.

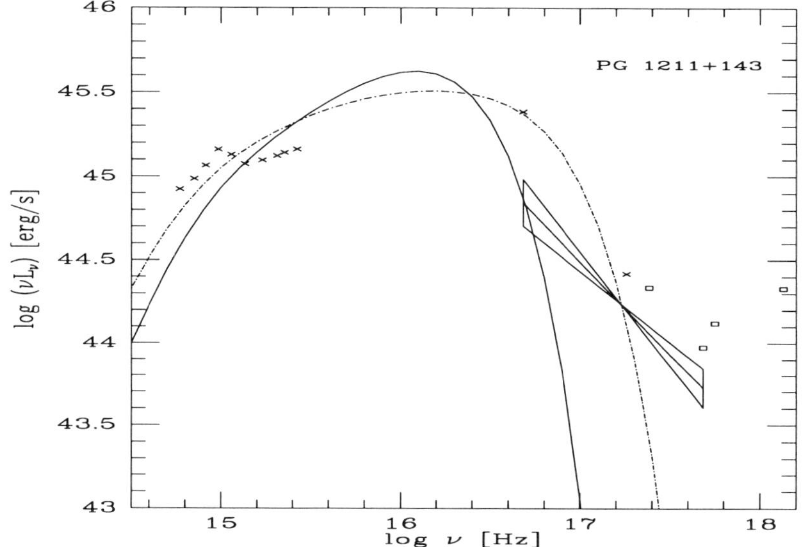


Fig. 2. Observed spectrum of PG 1211+143: crosses - Band & Malkan (1989), bow-tie - Walter & Fink (1993), open squares - Pounds et al. (1994). Computed spectra are obtained from the slim disk models: $\dot{m} = 1.5$, $m = 5 \times 10^7$ (solid line) and $\dot{m} = 100$, $m = 10^7$ (dot-short dashed line).

In figure 2 we show the spectrum of PG 1211+143, particularly well observed bright low-redshift AGN with a high soft X-ray flux relative to its optical emission. Other objects have less pronounced soft-X-ray excesses, while still others show no excess at all. However, the strong soft-X-ray-excess AGN are exactly the objects which present the greatest challenge to accretion models. The model spectra fall steeply (exponentially) in the soft X-ray regime, while the majority of observed spectra are less steep (see e.g. Laor et al. 1994). This may indicate that the flux from the optically thick accretion disk extends no further than 0.5 keV.

Abramowicz et al. (1988) suggested that a limit cycle instability can explain the variability of X-ray sources. If this mechanism is present in AGN, one can expect to see not only intensity variations, but also spectral variability. In figure 3 we illustrate how this mechanism might work. The difference in the UV flux (for example at 1544 Å) between low and high states is about a factor 20, as was observed in Fairall 9. We have plotted the three characteristic energy distributions for this high-luminosity Seyfert 1 nucleus, corresponding to maximum, minimum and average brightness levels. The instability operates in the inner part of the disk. Predicted thermal instability timescales for the variability is of the order of 1000 days. This picture is in a rough agreement with the proposed mechanism for variability. However, our investigations presented here are too approximate to make strong conclusion about the detailed spectral variability properties. This will require calculation of the full time evolution for such models (Miller & Szuszkiewicz 1996).

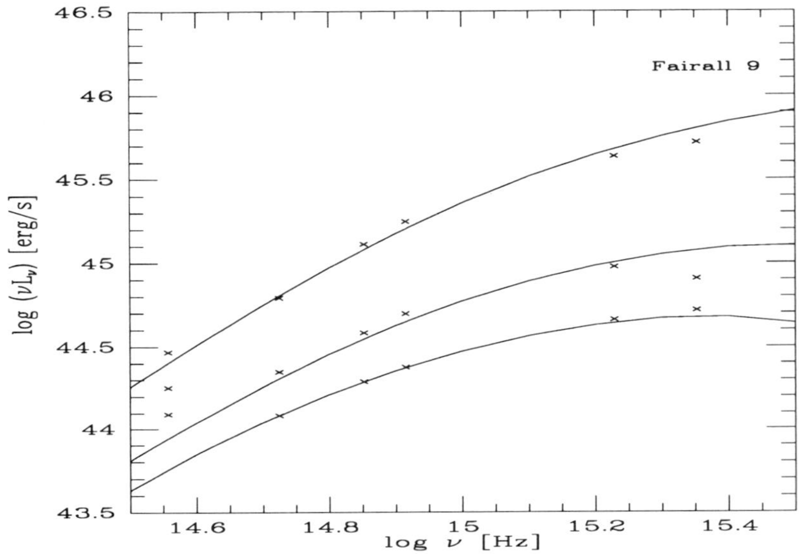


Fig. 3. Spectral evolution in Fairall 9: (Clavel et al. 1989) and the best fit obtained from a sequence of slim disk models with $m = 2 \times 10^8$ and $\dot{m} = 0.03$, 0.08, and 0.6 for minimum, intermediate and high states, respectively.

## References

Abramowicz M. A. et al. 1988, ApJ 332, 646
Band D. L., Malkan M. A. 1989, ApJ 345, 122
Clavel J. C., Wamsteker W., Glass I. 1989, ApJ 337, 249
Cox A. N, Steward J. N. 1970, ApJ Suppl 19, 243
Laor A. et al. 1994, ApJ 435, 611
Miller J. C., Szuszkiewicz E. 1996, in preparation
Papaloizou J. C. B., Szuszkiewicz E. 1994, MNRAS 268, 29
Pounds K. A. et al. 1994, MNRAS 267, 193
Shakura N. I., Sunyaev R. A. 1973, A&A 24, 337
Szuszkiewicz E. 1990, MNRAS 244, 377
Szuszkiewicz E., Malkan M. A., Abramowicz M. A. 1996, ApJ in press
Walter R., Fink H. H. 1993, A&A 274, 105

# Advection-Dominated Galactic Nuclei

Jean-Pierre LASOTA[1,2]
*1. Belkin Visiting Professor, Department of Condensed Matter Physics, Weizmann Institute of Science, 76100 Rehovot, Israel*
*2. UPR 176 du CNRS; DARC, Observatoire de Paris, Section de Meudon, 92195 Meudon Cédex, France*

**Abstract**

Underluminous active galactic nuclei may be powered by advection-dominated accretion flows in which the radiative efficiency is very low. The presence of such flows could explain the properties of our Galactic Center, of some 'dwarf' active galactic nuclei, and of the nuclei of giant elliptical galaxies.

## 1. Introduction

It is a truth universally acknowledged that accretion on to compact objects is the most efficient way of transforming gravitational potential energy into radiation (see, e.g., Frank et al. 1992). Comparing the 'low' efficiency of hydrogen-burning thermonuclear reactions ($\sim 0.007$) with the high efficiency ($\sim 0.1$) of the accretion on to a neutron star or a black leads to the conclusion that the power emitted from the vicinity of these compact objects is due to accretion of matter.

There is no doubt that radiation from such bright systems as quasars, Active Galactic Nuclei, and X-ray binaries, as well as the luminosity of cataclysmic variables (where the compact object is a white dwarf) is due to accretion. The only exceptions are X-ray bursts on the surfaces of neutron stars, which are due to thermonuclear explosions, and the nova outbursts which also result from runaway thermonuclear reactions. The emission of the so-called super-soft source could be due to steady nuclear burning on the surface of compact objects. For black holes which have no 'hard' surface the only source of energy can be accretion.

Problems arise when one tries to apply the standard accretion picture to very weak sources which are known to contain black holes, such as quiescent

*S. Kato et al. (eds.), Physics of Accretion Disks, 85–90.*
© 1996 OPA (Overseas Publishers Association) Amsterdam B.V.

Soft X–ray Transients (SXTs) (see Lasota 1995), our Galactic Center (Genzel et al. 1994), nuclei of giant elliptic galaxies (Fabian & Canizares 1988) and some weak AGNs (see, e.g., Petre et al. 1993).

The problem is that if one derives the value of the accretion rate from the relation:

$$\dot{M} = \frac{LR}{GM} \sim \frac{10L}{c^2} \tag{1}$$

where $M$, $R$, $L$ are respectively the mass of the compact objects, its radius and luminosity, one obtains values of accretion rates that are much smaller than the values one gets from other, independent, observational estimates. For SXTs in quiescence, the estimate of the rate at which mass is transferred to the outer disk regions gives results several orders of magnitude higher than the value of the accretion rate on to the black hole given by equation (1).

Clearly a non-radiative energy loss is at work in many underluminous systems. Models of advection-dominated accretion flows (ADAFs), in which a very low radiative efficiency causes most of the heat to disappear into the black hole, have been successfully applied to the description of various objects.

## 2. Nuclei of Galaxies

In our Galaxy the source Sgr A* is supposed to host a supermassive (Genzel et al. 1994) black hole ($M \sim 7 \times 10^5 M_\odot$). The mass infall rate is estimated to be $\sim 10^{22}$g s$^{-1}$ (Melia 1992). The X-ray luminosity (see however Duschl in this volume) is $\lesssim 10^{37}$, which by equation (1) would correspond to $10^{17}$g s$^{-1}$ falling down the black hole.

The ADAF model of Narayan et al. (1995) reproduces the Sgr A* spectrum over nearly ten decades assuming a constant accretion rate of $\sim 6 \times 10^{20}$g s$^{-1}$.

Most giant elliptical galaxies should have formed supermassive black holes in their nuclei. Judging from the properties of the galactic hot gas available for accretion, such galaxies should emit $\sim 10^{43}$erg s$^{-1}$ whereas the observed luminosity usualy does not exceed $\sim 10^{42}$erg s$^{-1}$. In a recent article Fabian & Rees (1995) suggest that in elliptic galaxies the flow into the central black hole could be advection-dominated.

## 3. The Case of NGC 4258

The giant spiral (SABbc) galaxy NGC 4258 (M106) shows weak nuclear activity in radio (Miyoshi et al. 1995; Hummel et al. 1989) and X-ray (Makishima et al. 1994) frequencies. Recently a compact polarized nucleus was found in the optical band (Wilkes et al. 1995).

NGC 4258 has been classified (Stauffer 1982) as a LINER (i.e., a galaxy with a 'low ionization nuclear emission-line region'). Since LINERs are defined just by the properties of some line intensity ratios (e.g., Osterbrock

1989) they might form a heterogeneous class of objects. Some LINERs could be starburst galaxies. Those however that show nuclear activity should contain 'dwarf' AGN.

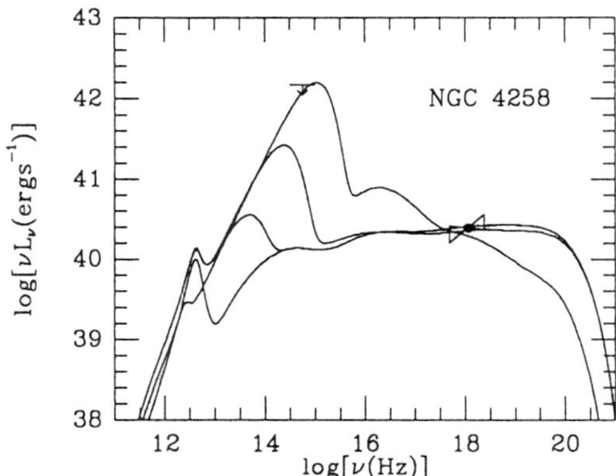


Fig. 1. Spectra of NGC 4258 calculated with an advection-dominated flow extending from $r = 3$ to $r = r_{tr}$, and a standard 'cold' disk extending from $r = r_{tr}$ to $r_{out} = 10^6$. The parameters are $m = 3.6 \times 10^7$, $\alpha = 0.1$, $\beta = 0.95$, and $\dot{m} = 0.016$. Four models are shown: (i) No outer cold disk ($r_{tr} = r_{out}$), (ii) $r_{tr} = 1000$, (iii) $r_{tr} = 100$, (iv) $r_{tr} = 10$. As $r_{tr}$ decreases the optical/UV emission increases. The upper limit of $1.5 \times 10^{42}$ erg s$^{-1}$ in the optical is indicated. The bow-tie represents the measured flux and the 1 $\sigma$ range of slopes (1.49 to 2.07) of the X-ray spectrum measured by Makishima et al.(1994).

The concept of 'weak activity', or 'dwarf' AGN makes sense only if it is defined with respect to some reference luminosity. The Eddington luminosity $L_E = 4\pi GMc/\kappa_{es}$, where $\kappa_{es}$ is the electron-scattering opacity, provides such a fundamental reference unit. Since masses of AGN are difficult to determine in most cases even the order of magnitude of $L_E$ is not known.

The first reliable measurement of $L/L_E$ in an AGN was published only recently by Miyoshi et al. (1995) for NGC 4258. In this source the motions of $H_2O$ maser spots permit a very accurate measurement of the central mass, which is $M = 3.6 \times 10^7 M_\odot$. The Eddington luminosity corresponding to this mass is $L_E = 4.5 \times 10^{45}$ erg s$^{-1}$. The X-ray luminosity of the nucleus of NGC 4258 between 2 – 10 keV is $(4 \pm 1) \times 10^{40}$ erg s$^{-1}$ (Makishima et al. 1994) while the optical/UV luminosity is less than $1.5 \times 10^{42}$ erg s$^{-1}$ (Rieke & Liebofsky 1978). The luminosity of NGC 4258 is therefore extremely sub-Eddington, $L \sim 10^{-5} L_E$ in X-rays and $L \sim 3 \times 10^{-4} L_E$ even if we take the maximum optical/UV luminosity. The uncertainty in the inferred $L/L_E$ in this source is due to the difficulties of estimating the bolometric luminosity from the data in infrared and UV.

The upper limit of $L/L_{\rm E} \lesssim 10^{-4}$ implied by the observations in NGC 4258 is surprisingly small. If it is typical of all AGN, luminous quasars with bolometric luminosities $\sim 10^{47}$ erg s$^{-1}$ would have masses $\sim 10^{13} M_\odot$, or $\sim 100$ times the mass of a typical galaxy. NGC 4258 must therefore have an unusually small value of $L/L_{\rm E}$ relative to the rest of the AGN population. A recent determination of the mass of the prototypical Seyfert 2 galactic nucleus in NGC 1068 confirms this conclusion. Also in this source observations of motions of $H_2O$ maser spots gave the value of the central mass: $4.4 \times 10^7 M_\odot$ (Gallimore et al. 1995). The intrinsic luminosity of this heavily obscured nucleus is likely to be $> 10^{45}$erg s$^{-1}$ (see Halpern 1992 and reference therein), so that $L/L_{\rm E} > 0.2$. We clearly have here two cases of 'hidden' galactic nuclei of similar masses but the luminosity of NGC 1068 is three orders of magnitude higher than that of NGC 4258.

According to the geometrically thin accretion disk model, luminosities of the order of $(10^{-4} - 10^{-5})L_{\rm E}$ should correspond to accretion rates $\dot{m} \equiv \dot{M}/\dot{M}_{\rm E} \sim 10^{-3} - 10^{-4}$, assuming a 10% efficiency of accretion, where $\dot{M}_{\rm E} = L_{\rm E}/c^2$. The standard disk model cannot however explain the hard X-rays observed in NGC 4258. The effective temperature at low accretion rates is (Frank et al. 1992)

$$T_{\rm eff} \approx 3.5 \times 10^7 m^{-1/4} \dot{m}^{1/4} r^{-3/4} {\rm K}, \tag{1}$$

and the optical depth is

$$\tau \approx 56 \alpha^{-4/5} \dot{m}^{1/5} m^{1/5}, \tag{2}$$

where $m = M/M_\odot$, $r$ is the radius in units of the Schwarzschild radius $r_S = 2GM/c^2$, and $\alpha \leq 1$ is the viscosity parameter. For parameters appropriate to NGC 4258 the standard inner disk would be cool ($T_{\rm eff} \sim 10^4$ K) and optically thick and the properties of such accretion disks are close to those of accretion disks in cataclysmic variables [equations (1) and (2)]. In non-magnetic cataclysmic variables however, if X-rays are observed at all, they originate in a boundary layer so that it is not clear how, in the AGN case, X-rays could be produced in the framework of the geometrically thin disk model.

The photon spectral index above 3 keV in NGC 4258 is $\Gamma = 1.78 \pm 0.29$, typical of AGNs (Nandra & Pounds 1994). The absorbing column is large, $1.5 \pm 0.2 \times 10^{23}$cm$^{-2}$, so there is no way directly to search for any optical/ultraviolet continuum from the nucleus of NGC 4258. If present, a strong continuum in this wavelength range could be seen when its luminosity is reradiated by the obscuring dust in the mid-infrared. As yet, a large aperture upper limit on the mid-infrared flux, $1.5 \times 10^{42}$ erg s$^{-1}$ (Rieke &

Lebofsky 1978), is the sum of our available data. Much of this infrared flux is likely to be reprocessed starlight, so this upper limit is probably considerably greater than the true nuclear infrared flux. Recently Wilkes et al. (1996) have observed the polarized, reflected nuclear continuum. Their results imply a luminosity $\sim 10^{40} - 10^{43}$erg s$^{-1}$, with a likely value of $\lesssim 10^{41}$ (Krolik, private communication) and a spectral shape of $F_\nu \propto \nu^{-1.1}$.

Lasota et al. (1996; hereafter LACKNY) used the two-temperature ADAF model of Narayan & Yi (1995) to model the observed properties of the continuum emitted by the nucleus of NGC 4258. The accretion flow in the LACKNY model consists of an inner ADAF and an outer 'standard' geometrically thin disk.

The properties of the output spectrum are a function of the black hole mass $m$, $\alpha$, the ratio of gas to total pressure $\beta$, the accretion rate $\dot{m}$, and the inner and outer radii of the ADAF and the thin disk. The results depend significantly on only two parameters. $\dot{m}/\alpha$ fixes the absolute luminosity and weakly influences the X-ray slope, in the sense that larger values of $\dot{m}/\alpha$ gradually lead to harder spectra (Narayan 1996). The inner radius of the cold thin disk $r_{\rm tr}$ determines the contribution of standard disk emission (in the optical/ultraviolet) to the total spectrum.

Since the optical upper limit is not very constraining, LACKNY fitted the spectrum of NGC 4258 with four models differing in the contribution of the standard disk to the total emission. The results are shown in figure 1.

In all four models, $m = 3.6 \times 10^7$, $\alpha = 0.1$, $\beta = 0.95$, and $\dot{m} = 0.016$. The value of $\dot{m}/\alpha$ has been adjusted so as to fit the observed X-ray luminosity of NGC 4258. The inner edge of the advection-dominated flow is fixed at $r = 3$, and the outer edge of the outer thin disk at $r_{\rm out} = 10^6$. In the first model there is no outer cold disk, while in the others $r_{\rm tr} = 1000$, $r_{\rm tr} = 100$ and $r_{\rm tr} = 10$ respectively. Decreasing $r_{\rm tr}$ gives increasing optical/UV emission. The model with $r_{\rm tr} = 10$ does not fit the X-ray spectrum since the advection-dominated flow is too efficiently cooled by the soft photons from the standard disk. The other three models all have slopes in the X-ray band consistent with the one observed. On the other hand, limits implied by the observations of Wilkes et al. (1996) would imply $r_{\rm tr}$ greater than a few hundred.

The spectrum of the outer thin disk is assumed to emerge from a stationary disk. One should therefore check that this disk, which could be subject to the instabilities caused by the partial ionization of hydrogen, is globally stable. These dwarf–nova–type instabilities can also operate in AGN disks (Lin & Shields 1986). Siemiginowska et al. (1996) give approximate expressions from which one can deduce the value of the critical radius above the disk is globally stable:

$$r_{\rm crit} \sim 400\ \dot{m}^{0.4} m_8^{-0.2} \alpha_{0.1}^{0.05}, \tag{2}$$

where $m_8 = (M/10^8 M_\odot)$. Since the acceptable spectral fits of NGC 4258 correspond to $r_{tr}$ larger than a few hundred, the assumption of a steady outer disk is justified.

Various other weak AGNs could contain ADAFs (see LACKNY) but reliable mass determinations are needed for model calculations and comparison with observations.

**Acknowledgements**

I am grateful to Julian Krolik, Ramesh Narayan, and Belinda Wilkes for enlightening discussions and useful information.

**References**

Fabian A. C., Canizares C. R. 1988, Nature 333, 829
Fabian A. C., Rees M. J. 1995 MNRAS 277, L55
Frank J., King A. R., Raine D. 1992, Accretion Power in Astrophysics, (Cambridge University Press, Cambridge)
Gallimore J., Baum S., O'Dea C., Brinks E., Pedlar A. 1995, preprint
Halpern J. P. 1992, in Testing the AGN Paradigm, AIP Conf. Proc. 254, ed S. S. Holt S. G. Neff C. M. Urry (AIP, New York) p524
Genzel R., Hollenbach D., Townes C. H. 1994, Rep. Prog. Phys. 57, 417
Hummel E., Kraus M., Lesch H. 1989, A&A 211, 266
Lasota J.-P., 1995, in Compact Stars in Binaries. Proceedings of the IAU Symposium 165, ed van den Heuvel E. P. J., van Paradijs J. (Kluwer, Dordrecht), in press
Lasota J. P., Abramowicz M. A., Chen X. M., Krolik J. H., Narayan R., Yi, I. 1996, ApJ in press
Lin D. N. C., Shields G. A. 1986, ApJ 305, 28
Melia F. 1992, ApJL 387, L25
Makishima K. et al. 1994, PASJ 46, L77
Miyoshi M., Moran J., Herrnstein J., Greenhill L., Nakai N., Diamond P., Inoue M. 1995, Nature 373, 127
Nandra K., Pounds K. 1994, MNRAS 268, 405
Narayan R., Yi I., Mahadavan R. 1995, Nature 374, 623
Narayan R., Yi I. 1995b, ApJ 452, 710
Narayan R. 1996, ApJ in press
Osterbrock D. E. 1989, Astrophysics of Gaseous Nebulae and Active Galactic Nuclei (University Science Books, Mill Valley)
Petre R., Mushotsky R. F., Serlemitsos P. J., Jahoda K., Marshall F. E. 1993, ApJ 418, 644
Rieke G. R., Lebofsky M. 1978, ApJ 220, L37
Siemiginowska A., Czerny B., Kostyunin. V. 1996, ApJ in press
Wilkes B. J., Schmidt G. D., Smith P. S., Mathur S., McLeod K. K. 1996, ApJ in press

# Free-Free Absorption in Sgr A* and Its Implications for the Interpretation of X-Ray Observations of the Galactic Center

Wolfgang J. Duschl[1,2], Thomas Beckert[1],
Peter G. Mezger[2], and Robert Zylka[1,2]
*1. Institut für Theoretische Astrophysik, Tiergartenstr. 15, D-69121 Heidelberg, Germany*
*2. Max-Planck-Institut für Radioastronomie, Auf dem Hügel 69, D-53121 Bonn, Germany*

## Abstract

The radio spectrum of Sgr A* has an upper cut-off and a lower turnover. Its radio/IR luminosity amounts to $L_{\rm radio,IR}[1-10^4\,{\rm GHz}] \sim 3 \times 10^2\,L_\odot$. We discuss free-free absorption due to the HII region Sgr A West and synchrotron self-absorption in Sgr A* as two possible physical processes which could explain the low-frequency ($\nu \lesssim 1.5\,{\rm GHz}$) turnover in the radio spectrum of Sgr A*. We show that both processes must occur. It is found that Sgr A* must be located close to the center of Sgr A West and in front of the Bar of the Minispiral as seen from the Sun. We discuss X-ray emission from Sgr A* based on recent ROSAT and ASCA observations of a weak soft X-ray source that coincides within 10″ with the radio position of Sgr A*. Since our radio observations exclude a large intrinsic X-ray absorption, Sgr A* is – if at all – only a low-luminosity X-ray emitter ($L_{\rm X}[0.5-10\,{\rm keV}] \lesssim 2 \times 10^2\,{\rm L}_\odot$).

## 1. Introduction

Sgr A* is the enigmatic radio source located at (or very close to) the dynamical center of the Milky Way. Due to absorbing material along the line of sight from the Sun to the Galactic Center (GC) corresponding to $A_{\rm V} \sim 30^{\rm m} - 31^{\rm m}$, Sgr A* is observable only in the IR and radio wavelength regime for $\lambda \gtrsim 1\,\mu{\rm m}$ and in the X-ray domain for energies $E \gtrsim 1\,{\rm keV}$.

*S. Kato et al. (eds.), Physics of Accretion Disks, 91–94.*
© 1996 OPA (Overseas Publishers Association) Amsterdam B.V.

In section 2 we give a short overview of the current status of the observation and understanding of the radio/IR spectrum of Sgr A*. We will show that the occurence of a low-frequency turn-over at $\nu \sim 1\,\mathrm{GHz}$ allows us to determine an upper limit of the column density of absorbing material in the immediate vicinity of Sgr A*. Based upon this determination, we conclude that the hard X-radiation ($E \gtrsim 10\,\mathrm{keV}$) claimed to come from Sgr A*, most likely is due to another source (section 3). From this we can give an upper limit for the X-radiation from Sgr A*: $L_\mathrm{X}[0.5 - 10\,\mathrm{keV}] \lesssim 2 \times 10^2\,L_\odot$.

## 2. The Radio Spectrum of Sgr A*

Radio measurements of the flux density $S_\nu$ from Sgr A* are available for the frequency range between $\sim 0.5$ and $\sim 600\,\mathrm{GHz}$. In the MIR domain significant upper limits are known. The time averaged radio – MIR spectrum of Sgr A* is characterized by three features: (i) Between $\sim 1$ and $\sim 600\,\mathrm{GHz}$ the spectrum is inverted and very close to $S_\nu \propto \nu^{1/3}$ (Duschl & Lesch 1994). (ii) At high frequencies $\nu \gtrsim 1\,000\,\mathrm{GHz}$ the flux density $S_\nu$ drops sharply; the maximum of $S_\nu$ is attained at $\nu_\mathrm{max} \sim 600\,\mathrm{GHz}$. (iii) In the low frequency domain, at $\nu \sim 1\,\mathrm{GHz}$ a turnover is observed with steeply decreasing flux densities towards lower frequencies. The total radio luminosity in the range between 1 and $10^4$ GHz amounts to $L_\mathrm{radio} \sim 3 \times 10^2\,L_\odot$

Duschl & Lesch (1994) explained the spectrum for $\nu \gtrsim 1\,\mathrm{GHz}$ as optically-thin synchrotron radiation from relativistic electrons with a monoenergetic electron distribution. Beckert et al. (1996) worked out this model in more detail and showed that a quasi-monoenergetic electron distribution can in addition account for the low-frequency turnover as being due to synchrotron self-absorption in Sgr A* itself. This allowed them to determine the size of the emitting region as $\sim 2.4 \times 10^{13}\,\mathrm{cm}$, in good agreement with recent mm-VLBI observations (e.g., Krichbaum et al. 1993). Other model parameters were a mean electron energy of $\sim 115\,\mathrm{MeV}$, a magnetic field strength of $\sim 11\,\mathrm{G}$ and a density of relativistic electrons of $\sim 3 \times 10^4\,\mathrm{cm}^{-3}$.

Alternatively, the low-frequency turnover could also be caused by free-free absorption in the immediate vicinity of Sgr A*. The turnover would be due to an absorbing plasma of electron temperature $T_\mathrm{e} \sim 6\,000\,\mathrm{K}$ with an emission measure of $E_\mathrm{t} \sim 10^6\,\mathrm{pc\,cm}^{-6}$.

Sgr A West, the HII region surrounding Sgr A*, consists of an extended component, and the so-called *Minispiral* whose *Central Bar* overlaps in projected position with Sgr A*. If Sgr A* is located in the center of Sgr A West the emission measure of the extended component $0.5E_\mathrm{ext} \sim 10^6\,\mathrm{pc\,cm}^{-6}$ (Beckert et al. 1996) could account for the free-free absorption. The Bar, however, with an estimated emission measure of $E_\mathrm{Bar} \sim 9 \times 10^6 - 4 \times 10^7\,\mathrm{pc\,cm}^{-6}$ at the projected position of Sgr A* must then be located behind the source.

Currently available observations do not allow Beckert et al. (1996) to decide between synchrotron self-absorption and free-free absorption as the physical process that gives rise to the low-frequency turnover. Nonetheless the spectrum gives at least upper limits for the respective characteristic turnover frequencies, namely $\sim 1\,\mathrm{GHz}$. Array observations at and below $\sim 1\,\mathrm{GHz}$ may eventually resolve the question.

## 3. Relevance for the X-ray Emission of Sgr A*

Several groups claim to have detected X-ray emission from Sgr A*, most recently Predehl & Trümper (1994) with ROSAT in the energy range $1.2 - 2.5\,\mathrm{keV}$, and Koyama (1994) and Tanaka (private communication) with ASCA in the energy range $\sim 1 - 10\,\mathrm{keV}$ (for a review, see, e.g., Skinner 1993 and Predehl et al. 1994). The point source detected by ROSAT with an angular resolution of $25''$ coincides with the radio position of Sgr A* to within $10''$. Previous detections at higher energies ($3 - 30\,\mathrm{keV}$) with coded mask telescopes on Spacelab-2 and ART-P, but with considerably lower positional accuracy, were reported by Skinner et al. (1987) and Pavlinsky et al. (1994).

The ART-P source exhibits a non-thermal spectrum. If the ROSAT flux density is corrected for an X-ray absorption corresponding to the standard extinction $A_V \sim 30^{\mathrm{m}}$ ($\cong N_{\mathrm{H}} \sim 5.4 \times 10^{22}\,\mathrm{cm}^{-2}$) between Sun and Galactic Center, it lies $\sim 2$ to 3 orders of magnitude below the extrapolated ART-P spectrum. To bring the ROSAT flux densities into agreement with the ART-P spectrum requires an additional absorption correction corresponding to a column density of interstellar matter of $N_{\mathrm{H}} \sim 1 - 1.5 \times 10^{23}\,\mathrm{cm}^{-2}$ and to a visual extinction of $A_V \sim 56^{\mathrm{m}} - 83^{\mathrm{m}}$ (Predehl & Trümper 1994). This could be explained if the ROSAT source were located deep in the Sgr A East Core GMC against which Sgr A* and Sgr A West are seen in projection (see, e.g., figure 1a in Mezger 1994, and remember that at $\lambda\,870\,\mu\mathrm{m}$ $1\,\mathrm{Jy}\,(8''\mathrm{beam})^{-1}$ corresponds to $N_{\mathrm{H}} \sim 2 \times 10^{23}\,\mathrm{cm}^{-2}$). In this case, however, the ROSAT source would not be identical with Sgr A*, which our observations place at the center of the HII region Sgr A West and in front of the Bar. Sgr A West, on the other hand, is located in front of the extended synchrotron source Sgr A East (see Beckert et al. 1996, section 2) which, in turn, is located in front of the Sgr A East Core GMC (Mezger et al. 1989).

Alternatively, Predehl & Trümper suggest that the ROSAT source in fact coincides with Sgr A* but is subject to a very local or intrinsic absorption. NIR observations of the central 0.5 pc do not indicate localized dust absorption on scales of $\sim 1''$ (Krabbe et al. 1995). To comply with these observations the X-ray absorption would have to be provided by an ionized shell surrounding Sgr A* where most of the central dust has evaporated. The low-frequency turnover of the Sgr A* spectrum yields an upper limit to the emission measure of this shell of $E_{\mathrm{shell}} = \Lambda_{\mathrm{shell}} n_{\mathrm{e,shell}}^2 \lesssim E_{\mathrm{t}} \sim 10^6\,\mathrm{pc\,cm}^{-6}$. Here $\Lambda_{\mathrm{shell}}$ and $n_{\mathrm{e,shell}}$ are thickness and electron density, respectively, of the ionized shell. The corresponding column density of (ionized) hydrogen would be $N_{\mathrm{H,shell}} = \Lambda_{\mathrm{shell}} n_{\mathrm{e,shell}} = E_{\mathrm{shell}}\, n_{\mathrm{e,shell}}^{-1} \lesssim 3.1 \times 10^{24}\,\mathrm{cm}^{-5}\, n_{\mathrm{e,shell}}^{-1}$.

Since the electron density of the extended ionized component of Sgr A West, $n_{\rm e,ext} \sim 1.4 \times 10^3\,{\rm cm}^{-3}$ (see table 1 of Beckert et al. 1996), represents a lower limit to the electron density of the shell one obtains as an upper limit for its column density of $N_{\rm H,shell} \lesssim 2.2 \times 10^{21}\,{\rm cm}^{-2}$, which is by two orders of magnitude too low to account for the X-ray absorption quoted above.

## 4. Conclusions

In agreement with Koyama (1994) and Predehl and Tanaka (private communication) we conclude that the *hard* X-ray ART-P source and the ROSAT source are not identical. The *soft* X-ray source observed with ASCA (Koyama 1994) and the ROSAT source could be identical. This source then would coincide to within 10″ with the Sgr A* radio position and could be located in the HII region Sgr A West but not behind the Sgr A East Core GMC. In this case we estimate $L_{\rm X}[0.5-10\,{\rm keV}] \lesssim 2\times 10^2\,L_\odot$ as an upper limit of the X-ray luminosity. This source could be the X-ray counterpart of Sgr A* but may as well be associated with one of the HeI/HI stars found in the immediate vicinity of the Galactic Center (see, e.g., Krabbe et al. 1995, and references therein). The prototype of this stellar class, $\eta$ Car, is known to be a strong X-ray source (e.g., Koyama et al. 1990, and references therein). In the hard X-ray domain ($E \gtrsim 10\,{\rm keV}$), due to the lack of positional resolution, the currently available observations yield only a (very rough) upper limit.

## References

Beckert T., Duschl W. J., Mezger P. G., Zylka R. 1996, A&A in press
Duschl W. J., Lesch H. 1994, A&A 286, 431
Koyama K., 1994, in New Horizon of X-ray Astronomy - First Results from ASCA (Universal Academy Press, Tokyo), p181
Koyama K., Asaoka I., Ushimaru N., Yamauchi S., Corbet R. H. D. 1990, ApJ 362, 215
Krabbe A. et al. 1995, ApJL submitted
Krichbaum T. P. 1993, A&A 274, L37
Mezger P. G. 1994, in The Nuclei of Normal Galaxies – Lessons from the Galactic Center, eds R. Genzel, A. I. Harris, NATO ASI Series Vol. 445, (Kluwer Academic Publishers, Dordrecht), p415
Mezger P. G. et al. 1989, A&A 209, 337
Pavlinsky M. N., Grebenev S. A., Sunyaev R. A. 1994, ApJ 425, 110
Predehl P., Genzel R., Trümper J., Zinnecker H. 1994, in The Nuclei of Normal Galaxies – Lessons from the Galactic Center, eds. R. Genzel, and A. I. Harris, NATO ASI Series Vol. 445 (Kluwer Academic Publishers, Dordrecht), p21
Predehl P., Trümper J. 1994, A&A 290, L29
Skinner G. K. et al. 1987, Nature 330, 554
Skinner G. K. 1993, A&AS 97, 149

# Line Spectra from Advection-Dominated Disks

Etsuko OHNA and Jun FUKUE
*Astronomical Institute, Osaka-Kyoiku University,*
*Asahigaoka, Kashiwara, Osaka 582, Japan*

## Abstract

Emission line profiles expected from advection-dominated accretion disks are calculated. In a non-relativistic case, the line profiles are generally *double peaked.* During eclipse, however, the profiles of advection-dominated disks are remarkably different from those of a Keplerian disk. We briefly examine a relativistic case, in which the profiles are remarkably modified.

## 1. Introduction

Emission-lines originating from accretion-disk systems have been extensively investigated, from both the observational and theoretical view points. Observationally, various types of emission-lines were detected in accretion-disk systems such as cataclysmic variables, X-ray stars, and active galactic nuclei. Theoretically, emission-line profiles from accretion disks have been calculated by many researchers. In all of the current models, a *Keplerian rotating disk* is assumed.

In recent years, however, a new type of model of accretion disks – *advection-dominated disks* – has been proposed and examined by several groups. We thus examined emission-line profiles expected from advection-dominated accretion disks to distinguish them from traditional Keplerian ones.

## 2. Models and Calculation Procedures

We consider two types of non-Keplerian disks. The one is the self-similar solutions found by Narayan & Yi (1994). In their solutions the radial velocity $v_r$ and the azimuthal velocity $v_\varphi$ are respectively given as $v_r = -c_1 \alpha v_{\rm K}$, $v_\varphi = c_2 v_{\rm K}$, where $v_{\rm K}$ $(= \sqrt{GM/r})$ is the Keplerian velocity and $c_1$ and $c_2$ are some constants. We here set as follows; the ratio of specific heats $\gamma_0 =$

*S. Kato et al. (eds.), Physics of Accretion Disks, 95–98.*
© 1996 OPA (Overseas Publishers Association) Amsterdam B.V.

1.5, the standard viscosity parameter $\alpha = 1.0$, and the fraction of viscously dissipated energy which is advected $f = 0.5$; in this case $c_1\alpha = 0.42991$ and $c_2 = 0.43712$.

The other is the $\beta$-disk model developed by Fukue & Umemura (1995). Their solutions are described by a unique parameter, the effective normalized luminosity $\Gamma$, which expresses the luminosity of the central object normalized by the Eddington luminosity. Due to the effect of external radiation drag, the angular momentum of the gas is quickly lost inside the characteristic radius $r_0$, which is expressed as $r_0 = \frac{\Gamma^2}{1-\Gamma} r_g$, where $r_g$ is the Schwarzschild radius of the central object. Furthermore, in the vicinity of the central object, when the effective gravity is balanced with radiation drag, the infall velocity attains a terminal value $v_\infty$, which is expressed as $v_\infty = -\frac{1-\Gamma}{2\Gamma} c$. We set the parameter $\Gamma = 0.99$; in this case $r_0 = 98.01 r_g$ and $v_\infty = -0.00505c$.

As for the emissivity, we assume the radial dependence of emissivity $j$ as $j(r) \propto r^{-b}$ at $r_{in} \le r \le r_{out}$, where $b$ is constant, $r_{in}$ $(= 3r_g)$ the inner radius, and $r_{out}$ the outer radius of emitting region.

In order to calculate emission-line profiles numerically, we divided the disk plane into many segments, each having different line-of-sight velocities, which are Doppler-shifted with different magnitudes. Let $\boldsymbol{v}$ be the velocity vector of each segment and $\boldsymbol{s}$ the line-of-sight vector. The redshift $z$ of each segment then becomes $1 + z = L^{-1}\gamma(1 - \boldsymbol{v} \cdot \boldsymbol{s}/c)$, where $L$ is the lapse function representing the gravitational redshift and $\gamma$ is the Lorentz factor representing the transverse Doppler effect. Once the velocity fields $\boldsymbol{v}$ on the disk is given, the redshift of each segment is calculated. In addition, the intensity enhancement due to the Doppler-beaming effect and others is calculated by $I_\nu = (\nu/\nu_e)^3 I_{\nu_e} = I_{\nu_e}/(1+z)^3$, where $I_\nu$ is the observed intensity and $I_{\nu_e}$ the emitted intensity.

The global emission-line profile was finally obtained by summing the local line profiles emitted from each segment.

## 3. Results

In the non-relativistic case ($L = 1$, $\gamma = 1$, and $I_\nu = I_{\nu_e}$), blueshift-/redshift-areas of advection-dominated disks are remarkably different from those of a Keplerian disk (figure 1).

Typical line profiles are shown in figure 2 for several parameters. As easily seen from figure 2, they are generally *double peaked* features.

Profile variations during eclipse are shown in figure 3. In the Keplerian disk, a blue peak disappears at first and then a red peak disappears. In the advection-dominated disks, on the other hand, both peaks appear and disappear more or less simultaneously.

In the relativistic case, the profiles are remarkably modified. An example is shown in figure 4.

In order to specify several models for advection-dominated disks, we should adopt the observational strategy via line profiles and their variations during eclipse.

## References

Fukue J., Umemura M. 1995, PASJ 47, 429
Narayan R., Yi I. 1994, ApJL 428, L13

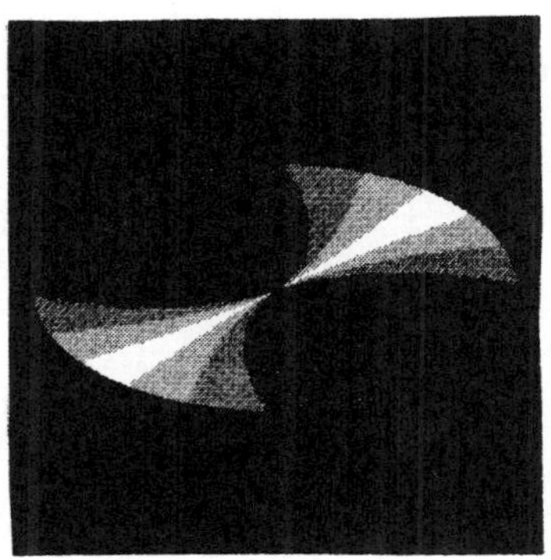

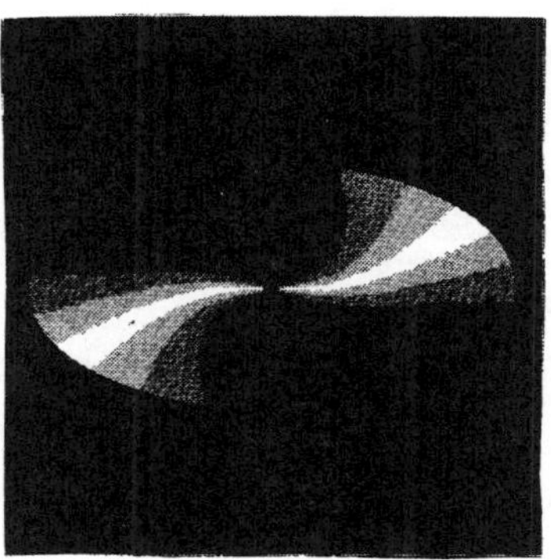

Fig. 1. The velocity fields expected. The left panel is an example for the self-similar disk and the right one for the $\beta$-disk. The parameters are $r_{\rm out} = 100r_{\rm g}$, $i = 60°$, and $b = 0$.

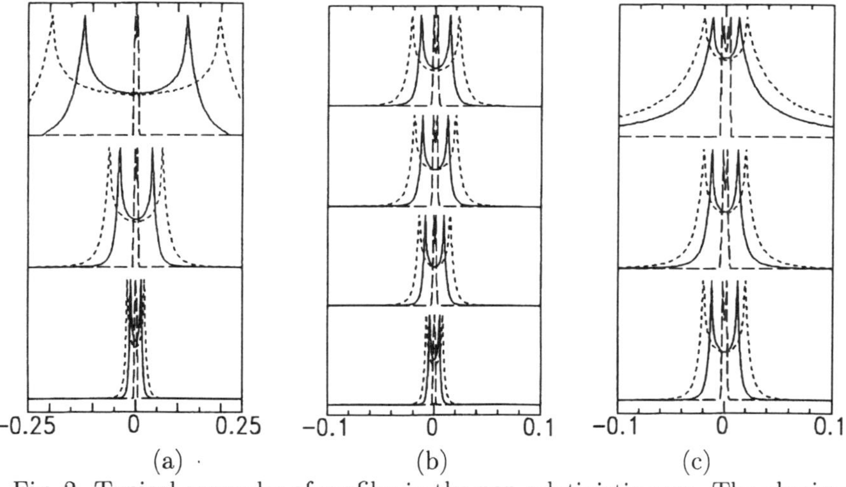


Fig. 2. Typical examples of profiles in the non-relativistic case. The abscissa is the redshift and the ordinate is the normalized flux. The solid curves are for the self-similar case and the dashed ones for the $\beta$-disk case (the dotted ones are for the Keplerian case). (a) From bottom to top $r_{\rm out} = 1000r_{\rm g}$, $100r_{\rm g}$, $10r_{\rm g}$. Other parameters are $i = 60°$ and $b = 0$. (b) From bottom to top $i = 20°$, $40°$, $60°$, $80°$. Other parameters are $r_{\rm out} = 1000r_{\rm g}$ and $b = 0$. (c) From bottom to top $b = 0$, 1, 2. Other parameters are $r_{\rm out} = 1000r_{\rm g}$ and $i = 60°$.

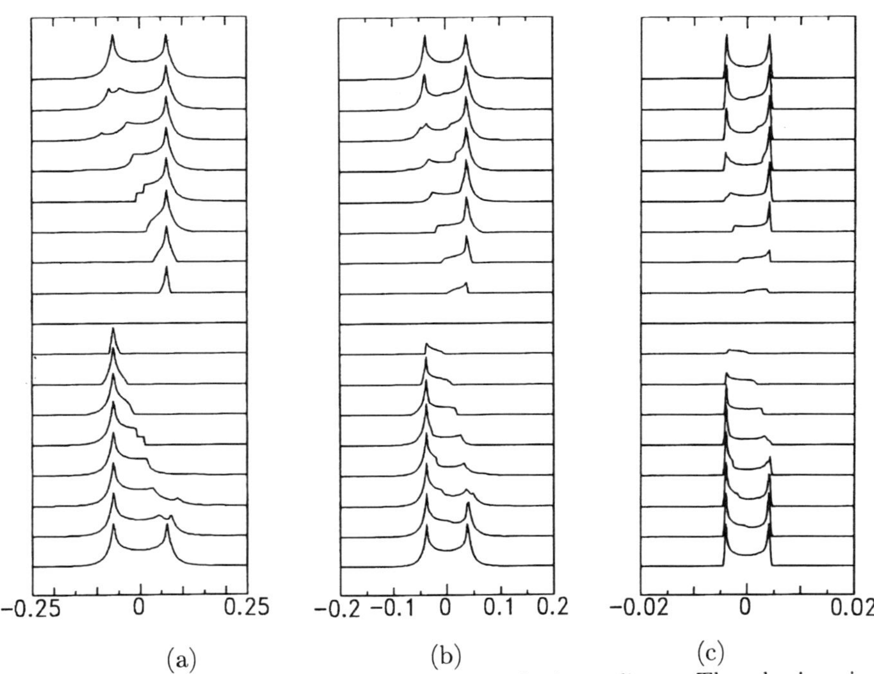


Fig. 3. Examples of line-profile variations during eclipse. The abscissa is the redshift and the ordinate is the normalized flux. (a) The Keplerian disk. (b) The self-similar disk. (c) The $\beta$-disk. The parameters are $r_{\rm out} = 100r_{\rm g}$, $i = 60°$, and $b = 0$.

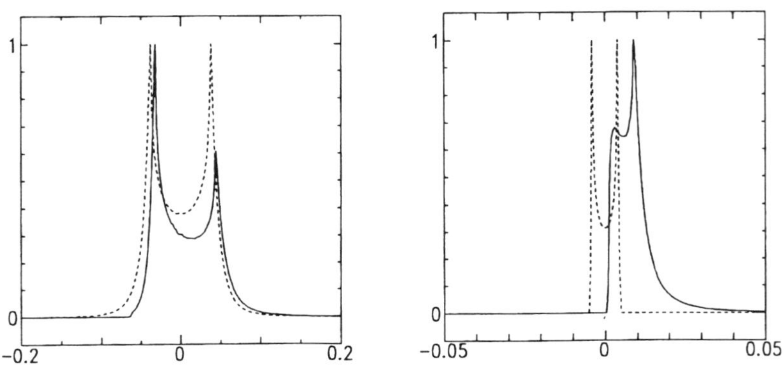


Fig. 4. Examples of profiles in the relativistic case (the dotted curves are for the non-relativistic case). The left panel is for the self-similar disk and the right one for the $\beta$-disk. The parameters are $r_{\rm out} = 100r_{\rm g}$, $i = 60°$, and $b = 0$.

# High-Speed Photometry of Dwarf Novae

Taichi KATO
*Department of Astronomy, Faculty of Science, Kyoto University, Sakyo-ku, Kyoto 606-01, Japan*

**Abstract**

Recent results of CCD time-resolved photometry of dwarf novae are reviewed especially in connection with superhumps in SU UMa stars and "super"-QPOs observed during superoutbursts. The report includes the discovery of ER UMa (a prototype of peculiar SU UMa stars), interpretation of peculiar light curves of WZ Sge stars as double superoutbursts based on observations of AL Com, and two examples of "super"-QPOs with implications on the source of QPOs.

## 1. Introduction

Dwarf novae (DNe) are a class of non-magnetic cataclysmic variables (CVs) which are semi-detached binaries containing white dwarfs and low-mass main-sequence stars. The Roche-overflown gas from a secondary star forms an accretion disk around a white dwarf. The chief source of the visual light of DNe is the accretion disks. Several sorts of disk instabilities dramatically affect the luminosity of the disk, and then become observable as a variation in the optical flux.

SU UMa stars are a sub-group of DNe and characterized by presence of short (normal) outbursts and long (super) outbursts. During superoutbursts, periodic light modulations (superhumps) with a period a few percent longer than the orbital period are observed (for a review, see Warner 1985). Superoutbursts and superhumps are currently considered to be defining characteristics of SU UMa stars, and their driving mechanisms seem to be understood by the combination of thermal and tidal instabilities in the accretion disk (Osaki 1989).

Among various sorts of intrinsic variations of DNe, superhumps, coupled with the outburst behavior, have now become one of the best observational

*S. Kato et al. (eds.), Physics of Accretion Disks, 99–104.*
© 1996 OPA (Overseas Publishers Association) Amsterdam B.V.

tools in probing the tidal effect on accretion disks and outburst mechanism itself. With the help of current development of CCD photometric techniques, a wide variety of applications of 'superhump photometry' has been exploited to accretion disk systems ranging from classical SU UMa stars to black hole transients.

## 2. ER UMa

ER UMa (=PG 0943+521) was independently discovered by several groups as a peculiar CV. This object shows unprecedented light curve among CVs, which is characterized by long-lasting bright states and frequently outbursting states which alternate every 43 days. These alternations were first assumed to represent nova-like state (thermally stable disk) and dwarf nova state (thermally unstable disk) caused by periodic modulation of the mass-transfer rate from the secondary (Honeycutt et al. 1994). ER UMa was thus first thought to be the best observational evidence of periodically changing mass-transfer rate to produce the outburst pattern. However, time-resolved CCD photometry has established the SU UMa-type nature of ER UMa by the discovery of superhumps during bright states (figure 1; Kato & Kunjaya 1995). This object actually represents one of extreme SU UMa-type DNe with very short supercycles. The stability of supercycles requires a constant mass-input from the secondary, which is perfectly contrary to what was first thought.

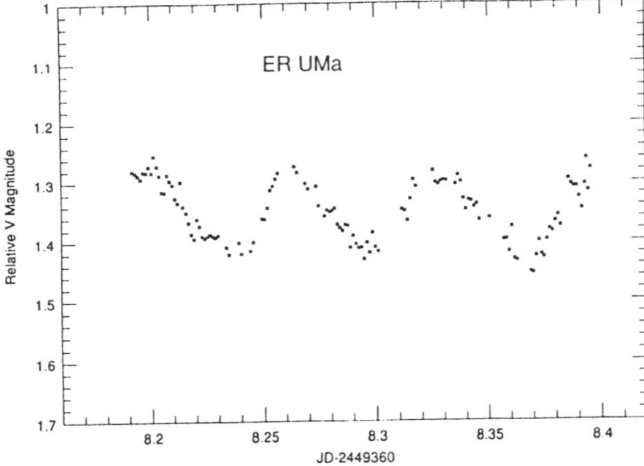


Fig. 1. *V*-band light curve of ER UMa on 1994 January 14. The light curve shows single-peaked humps with a steeper rise and gradual decline, which are very characteristic of superhumps of the usual SU UMa-type dwarf novae. The best estimate of the period is $0.06573 \pm 0.00005$ d.

## 3. AL Com as a WZ Sge Star

Although presently classified as an SU UMa-type dwarf nova, WZ Sge is well known as one of the most peculiar objects in that it shows only superoutbursts with exceptional duration and amplitude, and no normal outburst (Patterson et al., 1981, and references therein). Furthermore, in its decline from the 1978 outburst, WZ Sge showed a deep temporal dip. All of these characteristics have puzzled both theoreticians and observationalists with regards to its outburst mechanism. Among them the two competing theories have been most surviving. One is the mass-transfer burst theory, which is supported by the discovery of large-amplitude periodic humps in the early stage of the 1978 outburst of WZ Sge. These humps were considered to reflect the hot spot enhanced by the mass-transfer burst (Patterson et al., 1981). The other is an extension of thermal and tidal instability theory of SU UMa-stars towards the lowest mass-transfer rate (Osaki 1995a).

A dwarf nova AL Com was photometrically observed during the outburst in 1995 April, which occurred for the first time since 1975. The striking similarity of AL Com with WZ Sge, as demonstrated by the present observation (figure 2), provides a plentiful of materials in interpreting the enigmatic nature of WZ Sge-type dwarf novae.

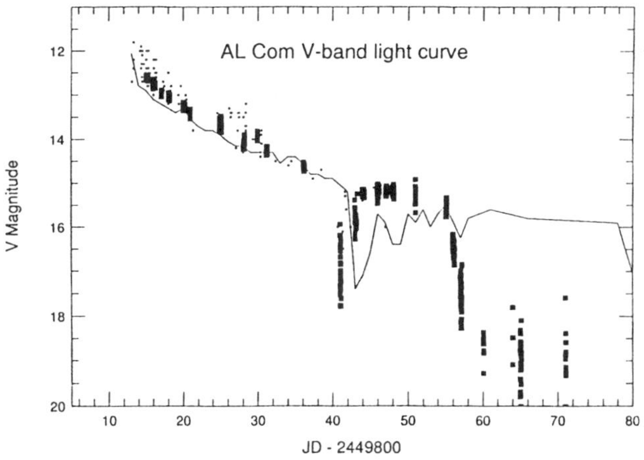


Fig. 2. The overall V-band light curve of the 1995 outburst of AL Com. Note the exceptionally long duration of the outburst interrupted by a deep dip, both of which are clearly reminiscent of the 1978 outburst of WZ Sge (solid line).

In the early stage of the outburst, two distinct types of periodic variation were observed: (1) 0.05666-day variation with decaying amplitudes (figure 3a), and (2) 0.0572-day superhumps which grew in ~ 11 days after the onset of the outburst. We attributed the former period as the orbital period of AL Com, which seems to be later confirmed by a period analysis of quiescent light variation. The double-humped profile of this variation seems to preclude the explanation by the enhanced hot spot by the mass-transfer burst from the secondary. Later immediately after the "dip", AL Com showed a short-living maximum followed by a rapid decline. After this "secondary dip" the star entered a new stage of the "plateau phase". The observation has shown that the superhumps again grew during this phase (figure 3b).

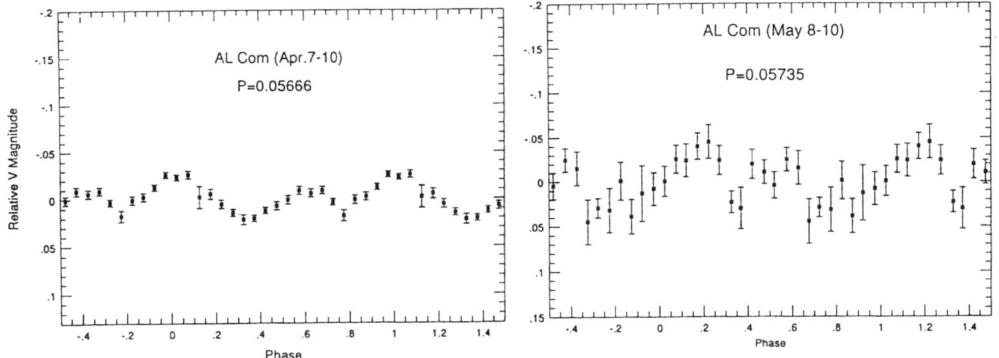


Fig. 3. a) "Orbital" superhumps observed in the earliest stage of the outburst. The double-humped profile of this variation seems to preclude the explanation by the enhanced hot spot. b) Folded light curve during the "plateau" phase, after the dip. The best period agrees with the superhump period within the error of the estimate. From the fact that the superhumps grew after the dip, we interpret the "plateau" phase as a new superoutburst.

We can therefore identify this stage as a start of a new superoutburst. A short-living maximum after the main dip may represent a normal outburst which triggered this superoutburst. The peculiar outburst light curve common to WZ Sge and AL Com now first seems to be naturally understood by occurrence of successive two superoutbursts, which are powered by usual combination of thermal and tidal disk instabilities.

## 4. Super-Quasi-Periodic Oscillations

Quasi-periodic oscillations (QPOs), oscillations of short coherence lengths, are a widely observed phenomenon in close binary systems, such as cataclysmic binaries (CVs; see Patterson 1981) and X-ray binaries (XBs; a review by van der Klis 1989). It is possible that QPOs in CVs and XBs might be of common origin. In fact, typical oscillation periods are 30 – 500 s for CVs and 10 – 200 ms for XBs, which are roughly ten – a few ten times the Keplerian time-scales at the inner edge of the disks for both cases. Several mechanisms have been proposed to account for the QPOs in CVs, including a reprocessing of the light by the oribiting blobs (Patterson 1979) and radial oscillation of the accretion disk (see e.g., Okuda et al. 1992). However, it is still premature to conclude what structure produces such a variation and periods.

We report on the discovery of QPOs with gigantic amplitudes ($\sim 20\%$) found in SU UMa stars during superoutbursts – called "super-QPOs".

The first case is the 1992 superoutburst of SW UMa (Kato et al. 1992). Super-imposed with the superhumps, one can see prominent QPOs with a mean period of 6.1 minutes and an unprecedented large amplitude of 0.2 mag (figure 4a). The QPOs comprise of two components: a dip component with a typical duration of one minute, and a nearly sinusoidal component (figure 4b). We suggest that the dip component was likely to be caused by the eclipse of the central light source by a rotating blob or by a vertical inflation of the accretion disk with a horizontal dimension of $6 \times 10^9$ cm. Since the periods of the sinusoidal component and the dip component are the same, it is natural to think that this eclipsing body is also responsible for the sinusoidal variation. Since the sinusoidal component has common characteristics with usually observed QPOs (cf. Patterson 1981 for SW UMa), the sinusoidal component can therefore be interpreted as being an enhancement of normal QPOs in scale, and the source of QPOs chanced to eclipse the central object to produce the dip component.

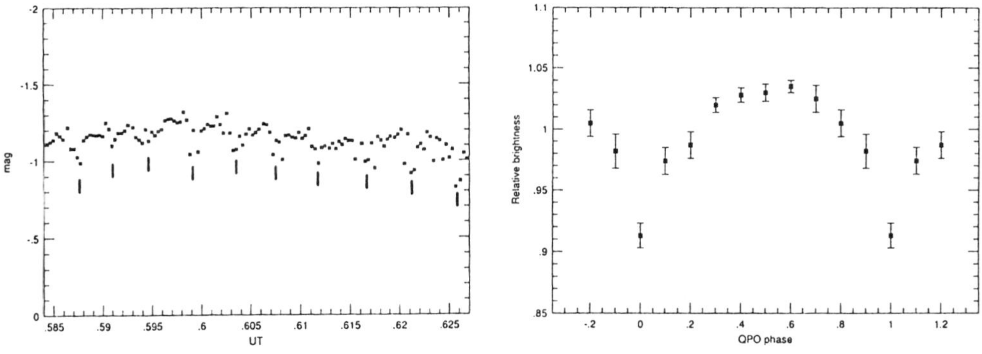


Fig. 4. a) *V*-band light curve of SW UMa during the 1992 superoutburst. Super-imposed with the superhumps, one can see prominent QPOs with a mean period of 6.1 minutes and an unprecedented large amplitude of 0.2 mag. b) Coadded QPO profile. The QPOs comprise of two components: a dip component with a typical duration of one minute, and a nearly sinusoidal component.

The second case is the 1991 superoutburst of EF Peg. Again super-imposed with superhumps, large-amplitude QPOs (again super-QPOs) were observed with rapidly decaying periods from 18 to 7 min within two hours (figure 5). Such a rapid change in periods can not be explained any pre-existing oscillation models of the accretion disk, but only reasonably explained by a spirally falling blob. In these two cases, appearance of superhumps or the growth of tidal instability may be somehow responsible for this class of specially enhanced QPOs. It is possible that a violent motion of the gas resulting from the tidal disturbance could be responsible for making a (rapidly falling) blob or the inflated structure of the disk.

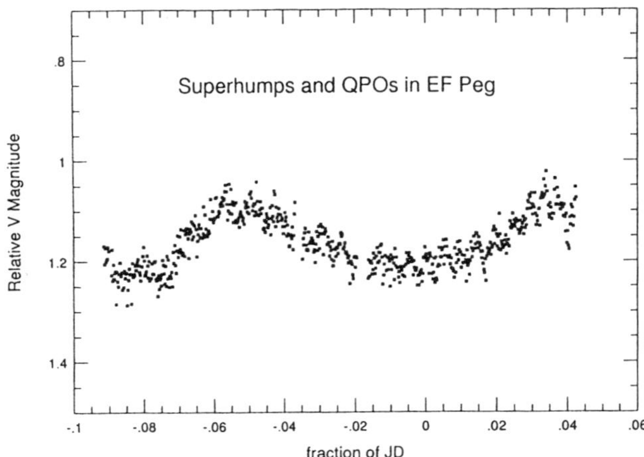


Fig. 5. $V$-band light curve of EF Peg during the 1991 superoutburst. Superimposed with superhumps, large-amplitude QPOs were observed with rapidly decaying periods from 18 to 7 min within two hours.

In conclusion, CCD time-resolved photometry, coupled with the development of theories, has proven to be one of the most powerful tools in probing the basic structures and instabilities of accretion disks. And dwarf novae will continue to play a first-class important role in observational research in basic physics of accretion disks.

## References

Kato T., Kunjaya C. 1995, PASJ 46, 163
Kato T., Hirata R., Mineshige S. 1992, PASJ 44, L215
Okuda T., Ono K., Tabata M., Mineshige S. 1992, MNRAS 254, 427
Osaki Y. 1989, PASJ 41, 1005
Osaki Y. 1995a, PASJ 47, 47
Osaki Y. 1995b, PASJ 47, L11
Patterson J. 1979, ApJ 234, 978
Patterson J., McGraw J. T., Coleman L., Africano J. L. 1981, ApJ 248, 1067
van der Klis M. 1989, AnnRevA&A 27, 517
Warner B. 1985, Interacting Binaries, ed. P. P. Eggelton, J. E. Pringle (D. Reidel Publishing Company, Dordrecht) p367

# A Unified Scheme for Dwarf Nova-Type Variables

Yoji OSAKI
*Department of Astronomy, School of Science,*
*University of Tokyo, Bunkyo-ku, Tokyo 113, Japan*

**Abstract**

A unified scheme for dwarf nova-type variables is proposed within the basic framework of the disk instability model in which outburst behaviors of these stars are classified into four regions in orbital-period versus mass-transfer rate diagram. In particular, rich varieties of outburst behaviors of cataclysmic variables below the period gap are understood by the thermal-tidal instability model in which the coupling of the two intrinsic instabilities in the accretion disk plays a unique role.

## 1. A Unified Scheme of Dwarf Nova Outbursts

There is a rich variety of outburst behaviors in non-magnetic cataclysmic variable stars (abbreviated as CVs), starting from non-erupting nova-like stars to various sub-classes of dwarf novae (DNe). The purpose of this paper is to demonstrate that almost all of outburst behaviors in non-magnetic CVs can be understood in a unified way within the basic framework of the disk instability model. In particular, rich varieties of outburst behaviors in CVs below the period gap are understood by the thermal-tidal instability model.

It is fairly widely accepted that the outbursts of dwarf novae can basically be explained by the disk instability model, at least for those above the well-known CV period gap. In this model, the mass transfer rate from the secondary star is assumed to be constant but accretion from the disk to the central white dwarf is not steady but intermittent because of intrinsic instability in accretion disks. The instability itself is of a kind of thermal instability associated with ionization/recombination of the most abundant element of hydrogen. This instability gives rise to a thermal relaxation oscillation in the disk and this model is thus called the thermal limit-cycle instability model (see, e.g., a review by Cannizzo 1993).

*S. Kato et al. (eds.), Physics of Accretion Disks, 105–110.*
**© 1996 OPA (Overseas Publishers Association) Amsterdam B.V.**

Another intrinsic instability that operates in accretion disks, was later discovered by Whitehurst (1988), which is called the "tidal instability" or "tidally-driven eccentric instability". In this instability the accretion disk is deformed to an eccentric form and its eccentric pattern slowly precesses in the prograde direction in the inertial frame of reference. The so-called "superhump" phenomenon observed during "superoutbursts" in SU UMa stars is now believed to be explained by this tidal instability. The tidal instability is found to occur when the disk's outer edge reaches the 3:1 resonance radius (Whitehurst 1988; Hirose & Osaki 1990; Lubow 1991), a condition that is possible only in binary systems with extremely low mass-ratio, $q$, with $q < 0.25$, where $q = M_2/M_1$ and $M_1$ and $M_2$ are masses of the mass-accreting primary star and the secondary star, respectively. The latter condition is in turn satisfied exclusively in CVs below the period gap.

The present author (Osaki 1989a,b) has proposed the so-called "thermal-tidal instability model" to explain the superoutburst phenomenon of SU UMa stars, in which the two intrinsic instabilities, the thermal instability and tidal instability, are properly coupled. This model is basically within the general frame-work of the disk instability model of dwarf novae, because mass transfer rate from the secondary star is assumed to be constant in this model and all of outbursting activity is thought to be caused by intrinsic instabilities within accretion disks.

An intriguing possibility has now opened in that almost all variety of outburst light curves of dwarf novae may be explained by a single principle: the disk instability model. In this unification model, different outbursting behaviors among non-magnetic cataclysmic variables are basically classified by two-parameters characterizing accretion disks in these systems; that is, orbital period of the system and mass transfer rate from the secondary star. For a given orbital period, mass transfer rate from the secondary determines the thermal stability nature of accretion disks (Smak 1983) in a sense that CV systems with high mass transfer rate yield hot "stable" disks corresponding to nova-like systems while those with mass transfer rate below the critical one give rise to thermally unstable disks, producing dwarf nova outbursts and the border-line case between the two yields Z Cam stars showing occasional standstill.

In the CV binary systems the orbital period is closely related to the mass of the secondary star in a sense that a longer (shorter) orbital-period system has higher (lower) mass secondary as far as the secondary star is near the low-mass main sequence. In fact, the mass of the secondary star is estimated to be $M_2 \simeq 0.2\ M_\odot$ at the period gap. Since the mass of the primary white dwarf is of the order of one solar mass, the period gap gives approximately a dividing line between the tidally stable- (above the gap) and tidally unstable systems (below the gap).

Figure 1 schematically illustrates different classes of non-magnetic CVs in this unification model. They are basically divided into four regions depending on different combinations of the two intrinsic instabilities.

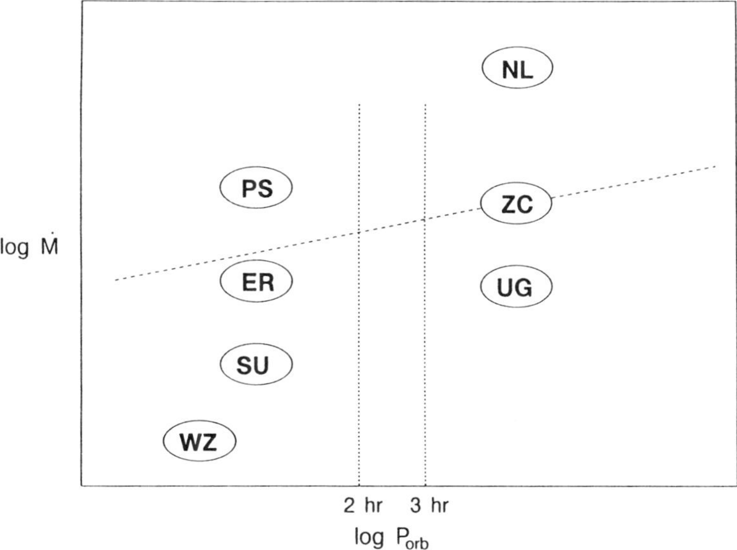


Fig. 1. $P_{\rm orb}$-$\dot{M}$ diagram showing different outburst behaviors of non-magnetic CVs. The region surrounded by dotted vertical lines shows that of the CV's period gap with 2-3 hours. Dashed line shows the borderline between the thermally stable and unstable disks. Symbols in the figures are; NL: nova like stars, ZC: Z Cam stars, UG: U Gem stars, PS: "permanent superhumpers", ER: ER UMa stars, SU: SU UMa stars, and WZ: WZ Sge stars. See the text for more details.

The CVs above the period gap can be classified into nova-like, Z Cam-type DNe, and U Gem-type DNe as already mentioned above. It has turned out that a similar classification of non-magnetic CVs below the period gap is possible.

Stars designated as "PS" in figure 1 are permanent superhump systems (hereafter called "permanent superhumpers") which are nova-like stars and thus thought to have hot thermally stable disks but still exhibit permanent superhumps (e.g., Skillman & Patterson 1993). These stars are located in the upper left region and thus they are thermally stable but tidally unstable.

The CV systems that are located in the lower left corner are in general classified to SU UMa stars which exhibit superoutbursts and superhumps. These stars can be both thermally and tidally unstable. Stars designated by "ER" in figure 1 are ER UMa stars (or RZ LMi stars) discovered very recently (Kato & Kunjaya 1995; Robertson et al. 1995), which exhibit extremely short supercycles of 19 days to 45 days. As will be demonstrated below, these stars are SU UMa stars with high mass transfer rates, a borderline case between permanent superhumpers and ordinary SU UMa stars and thus they are the Z

Cam counterpart below the period gap. Another extreme SU UMa stars are WZ Sge stars designated by "WZ" which exhibit extremely large-amplitude outbursts on rare occasions. These stars are also understood as extreme SU UMa stars with extremely low mass transfer rate.

## 2. Thermal-Tidal Instability Model

The present author (Osaki 1989a) has proposed the so-called thermal-tidal instability model to explain the supercycle of SU UMa stars. In this model, both the normal outburst and superoutburst are caused by the thermal instability in the accretion disk. During the early phase of the supercycle, the disk is compact and the thermal instability produces quasi-periodic episodes of accretion, which are observed as normal outbursts but the accreted mass in each normal outburst is less than that transferred during quiescence because of inefficient tidal removal of angular momentum from the disk. Both the mass and angular momentum of the disk are gradually built up. The disk radius expands further with each successive outburst until it eventually exceeds the critical radius for the tidal instability; this final normal outburst triggers the tidal instability, producing a precessing eccentric disk (observed as "superhumps"). The resulting outburst greatly clears the disk mass (producing "superoutburst") because of greatly enhanced tidal torques due to the eccentric disk. After the end of the superoutburst, the disk returns to the starting compact state. This is the basic idea of the thermal-tidal instability model for SU UMa stars. The supercycle of an SU UMa star is understood in this model as a relaxation cycle of the disk radius.

Observations show a wide variety in activity within SU UMa stars. In fact, Vogt (1993) has classified SU UMa stars into three groups by their activity : group A, "active" stars showing frequent outbursts (e.g., VW Hyi), group B, "intermediate" in activity between two extremes (e.g., OY Car), and group C, very inactive stars or "WZ Sge" stars that exhibit large outbursts in a very rare occasion. We may add to this activity sequence two more groups: permanent superhumpers and ER UMa stars, and these two new groups should be put above Vogt's group A. Warner (1995) summaries the same observational property of SU UMa stars in a diagram exhibiting relationship between the recurrence time of normal outburst $T_{\rm N}$ and the superoutburst recurrence time $T_{\rm S}$.

We first discuss the case of low mass transfer rates with $\dot{M} < 10^{16}$ g s$^{-1}$ which applies to the classical SU UMa stars and WZ Sge stars. The present author has already presented in Garching conference (Osaki 1994) and in Padova conference (Osaki 1995d) that Vogt's (1993) activity sequence may be explained as a sequence of decreasing mass transfer rate in the thermal-tidal instability model. In this model, the supercycle length $T_{\rm S}$ is given for appropriate binary parameters by

$$T_{\rm S} \simeq 120 \text{ day}/\dot{M}_{16},$$

while the recurrence time, $T_{\rm N}$, of normal outbursts is given by

$$T_{\rm N} \simeq 14 \text{ day}/\dot{M}_{16}^2.$$

Here $\dot{M}_{16}$ is mass transfer rate in units of $10^{16}$g s$^{-1}$. Thus if the mass transfer rate is decreased along this sequence, the supercycle length is increased in proportion to the inverse of mass transfer rate while the number of normal outbursts in a supercycle decreases in proportion to the inverse of the supercycle length.

However, in order to explain an extremely long recurrence time of WZ Sge itself of 30 years, it is not enough to decrease the mass transfer rate but also we need to decrease the viscosity parameter in the cold state, $\alpha_{\rm cold}$. The reason for this is that the so-called inside-out type outburst always occurs within the viscous diffusion time irrespective to mass transfer rate and the viscous diffusion time is determined by the viscosity in the cold state or the viscosity parameter $\alpha_{\rm cold}$. In order to avoid an earlier occurrence of normal outburst in WZ Sge, we must choose an extremely low $\alpha_{\rm cold} \leq 0.001$, a conclusion reached by Smak (1993), by Osaki (1994, 1995a) and by Howell et al. (1995).

Let us now discuss the case for high mass transfer rate with $\dot{M} > 10^{16}$ g s$^{-1}$. As already noted, there is a critical mass transfer rate above which an accretion disk is always in hot and thermally stable state, and it is approximately given by

$$\dot{M}_{\rm crit} \simeq 8.3 \times 10^{16} \ (P_{\rm orb}/100 \text{ min})^{1.7} \text{ g s}^{-1}.$$

If the mass transfer rate is higher than the critical one, the system remains in hot state. These CVs below the period gap are identified as permanent superhumpers, nova-like stars exhibiting permanent superhumps because they are thermally stable but tidally unstable.

Osaki (1995b,c) has studied outburst light-curves of the ER UMa stars. He has examined the supercycle length as a function of the mass transfer rate under fixed model parameters. It is found that the supercycle length is inversely proportional to the mass transfer rate if $\dot{M}_{16} < 1$ as expected. However, mass transfer rate is further increased above $\dot{M}_{16} \sim 1$, the supercycle length takes a broad minimum and it finally goes to infinity when the mass transfer rate becomes above the critical value for the thermal instability. The star ER UMa itself is found to be simulated very well as a system having a mass transfer rate near the broad minimum mentioned above (see, figure 2 of Osaki 1995b).

For model parameters used to simulate ER UMa, the minimum supercycle length was about 45 days and thus the supercycle length as short as 20 days observed for RZ LMi could not be reproduced by these model parameters. It was found (Osaki 1995c) that the minimum supercycle length is sensitive to one model parameter that describes the strength of tidal torques during the supermaximum when the disk becomes eccentric. It has turned out that the supercycle length as short as 19 days of RZ LMi can be reproduced if the tidal torques during the superoutburst in RZ LMi are significantly weaker than those of ER UMa and ordinary SU UMa stars. The light curve of RZ LMi is also found to be simulated very well in the case of weaker tidal torques (see, figure 2 of Osaki 1995c).

The wide variety in activity of CVs below the period gap indicates that the mass transfer rates from the secondary stars should have much wider range than that expected from the standard scenario for CV evolution based on the gravitational-wave radiation. A cyclic variation in mass transfer rate in time scale of $10^3 \sim 10^4$ years is thus inferred (e.g. the hibernation scenario for CV evolution).

A more detailed account of this model will be presented elsewhere (Osaki 1996).

**References**

Cannizzo J. K. 1993, in Accretion Disks in Compact Stellar Systems, ed J. C. Wheeler (World Scientific Publishing, Singapore) p6

Hirose M., Osaki Y. 1990, PASJ 42, 135

Howell S. B., Szkody P., Cannizzo J. K. 1995, ApJ 439, 337

Kato T., Kunjaya C. 1995, PASJ 47, 163

Lubow S. H. 1991, ApJ 381, 259

Osaki Y. 1989a, PASJ 41, 1005

Osaki Y. 1989b, in Theory of Accretion Disks, ed F. Meyer et al. (Kluwer Academic Publishers, Dordrecht) p183

Osaki Y. 1994, in Theory of Accretion Disks-2, ed W. Duschl, et al. (Kluwer Academic Publishers, Dordrecht) p93

Osaki Y. 1995a, PASJ 47, 47

Osaki Y. 1995b, PASJ 47, L11

Osaki Y. 1995c, PASJ 47, L25

Osaki Y. 1995d, in Proc. Padova-Abano Conference on Cataclysmic Variables: Inter Class Relations, (Kluwer Academic Publishers, Dordrecht) in press

Osaki Y. 1996, PASP in press

Robertson J. W., Honeycutt R. K., Turner G. W. 1995, PASP 107, 443

Skillman D. R., Patterson J. 1993, ApJ 417, 298

Smak J. 1983, ApJ 272, 234

Smak J. 1993, Acta Astronomica 43, 101

Vogt N. 1993, in Cataclysmic Variables and Related Physics, ed O. Regev, G. Shaviv (The Israel Physical Society, Jerusalem) p63

Warner B. 1995, preprint

Whitehurst R. 1988, MNRAS 232, 35

# The Observations of Eclipses in Dwarf Novae during Outbursts

Daisaku NOGAMI
*Department of Astronomy, Faculty of Science, Kyoto University, Sakyo-ku, Kyoto 606, Japan*

**Abstract**

We observed a deeply eclipsing dwarf nova (DN), IP Peg, between Aug. 26 and Sep. 7, 1994 during a long outburst. The maximum luminosity was reached on August 28. An analysis of the eclipse light curve revealed an expansion of the accretion disk during the premaximum phase. This finding clearly agrees with what is expected by the disk instability model of DN outburst. We also find the asymmetry of the eclipse profile developing with the time.

## 1. Introduction

Dwarf Novae (DNe) are a kind of the accretion disk (AD) systems. The observed activities mostly originate in the AD. DNe are far superior to other AD systems in that they have ADs directly observable in the optical range and that the variability time scales are appropriate to observe. These two properties make DNe the best laboratory to study ADs.

The outbursts of DNe are caused by global structural changes in ADs. In the course of an outburst, thus, the eclipse light curve varies depending on the physical situation of the AD. There are hence good reasons to believe that the eclipse profiles during outbursts are indeed a treasure house containing rich information regarding the dynamical behavior of ADs.

DNe known to show deep eclipses include V2051 Oph, OY Car, HT Cas, Z Cha, DV UMa, IP Peg, U Gem and HS 1804+6753 in order of shortness of the orbital period (see Ritter & Kolb 1995). We observed IP Peg, DV UMa and HS 1804+6753 during outbursts between 1994 January and 1995 October. Among them, the observations of IP Peg between Aug. 26 and Sep. 7, 1994 is the most important, since the eclipses were caught in the premaximum phase of outburst, which is a very rare case. In the following, we only report the results of IP Peg observations.

*S. Kato et al. (eds.), Physics of Accretion Disks, 111–114.*
© 1996 OPA (Overseas Publishers Association) Amsterdam B.V.

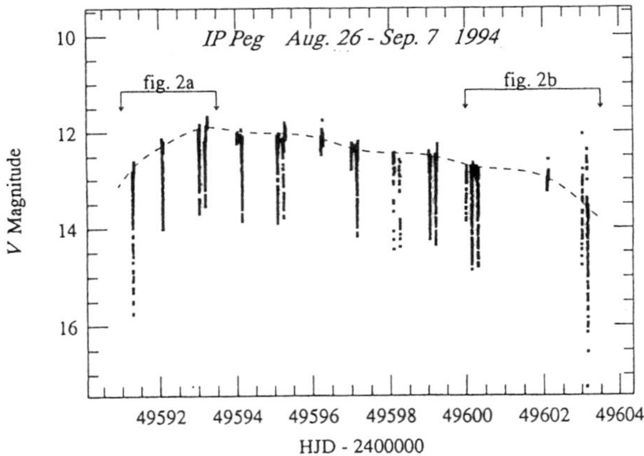


Fig. 1. The light curve of IP Peg during the long outburst.

## 2. Observation

We observed IP Peg between August 26 and September 7, 1994 during a long outburst at Ouda Station, Kyoto University, using 60 cm reflector with CCD camera (Thomson TH 7882, 576 × 384 pixels) in $V$ band. The exposure time was 20 or 30 s, depending on the transparency of the sky. The read-out and saving dead time is 13 s. The frames were, after a correction for standard debiasing and flat fielding, processed by a personal-computer-based automatic PSF photometry packages developed by Taichi Kato.

The light curve is shown in figure 1. The abscissa is the $V$ magnitude measured based on the brightness of a nearby star, BD+17 4907 ($V$=$10.^{m}927$, Skiff 1994). The ordinate is HJD − 2400000. The dashed line is the estimated magnitude out of eclipse. Data points are vertically spreading at each eclipse.

We totally caught 17 eclipses from the early rise to the end of the outburst. Figure 2 displays the light curves of the first four eclipses during the rising phase and at the maximum (left panel), and those of the last three eclipses (right panel). The abscissa represents the flux normalized by that out of eclipse. Eclipse profiles are shifted by 0.5 in flux in figure 2a and by 0.7 in figure 2b. The ordinate is the phase calculated by the following improved ephemeris deduced from our data,

$$ephemeris(HJD) = 2449591.29361 + 0.15820616 \times E, \qquad (1)$$

where $E$ is the number of the eclipse.

## 3. Analysis

In our analysis, we assumed that 1) the radiation is emitted only from the

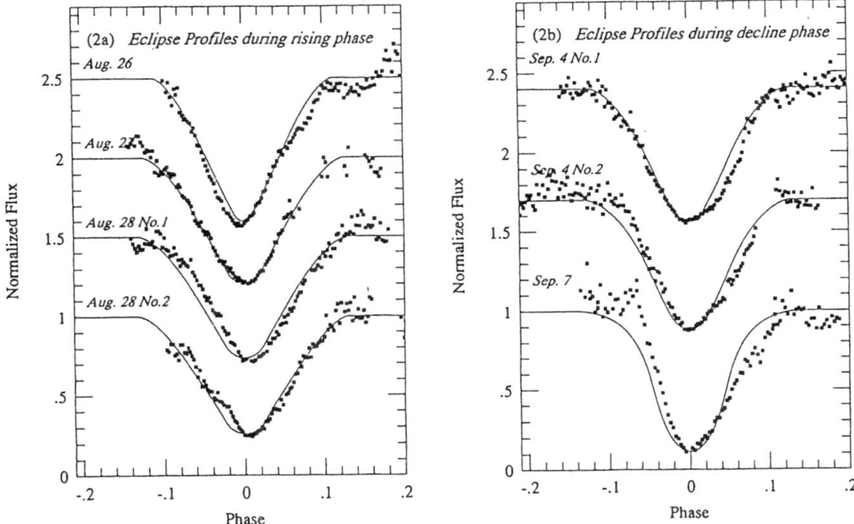


Fig. 2. The eclipse profiles obtained during the rising phase (a) and the late decay phase (b). Solid lines represent the best-fit light curve based on the parameters obtained by our analysis.

AD; 2) the temperature at a distance ($r$) from WD, $T(r)$, is given by a power law:

$$T(r) = T_{\text{out}} \left( \frac{r}{r_{\text{disk}}} \right)^{-\alpha}, \tag{2}$$

where $T_{\text{out}}$, $r_{\text{disk}}$ and $\alpha$ are the temperature at the outer edge of the AD, the radius of the AD, and a power index, respectively; 3) the AD locally emits blackbody radiation, and 4) the shape of the secondary is spherical.

In addition, we adopted the inclination of $i = 79.3°$ and the secondary radius of $r_2/A = 0.334$, where $A$ is the binary separation, following the quiescent light curve analysis by Marsh (1988). Under these assumptions, the various eclipse light curve can be synthesized with $T_{\text{out}}, r_{\text{disk}}, \alpha$ being free parameters. Then, we made the least-square fit to the observed eclipse profile. The results are listed in table 1.

The errors in $r_{\text{disk}}/A$ is 0.01 for the rising phase. However, this must be much larger for the last three eclipses because of the growing asymmetry in the eclipse light curve which cannot be treated by the present analysis. The other parameters also include relatively large errors, since these parameters are not so sensitive to the variation of the eclipse profile.

## 4. Discussion

The analysis revealed that the AD had already expanded at least two days before the maximum of the outburst, since the radius of the AD was $r_{\text{disk}}/A \simeq$

Table 1. The results of the analysis.

| Date | | | $T_{\rm out}$ (K) | $\alpha$ | $r_{\rm disk}/A$ |
|---|---|---|---|---|---|
| Aug. | 26 | | 8000 | 0.0 | 0.34 |
| | 27 | | 10000 | 0.0 | 0.40 |
| | 28 | No.1 | 10000 | 0.0 | 0.42 |
| | 28 | No.2 | 11000 | −0.1 | 0.42 |
| Sep. | 4 | No.1 | 8000 | 0.1 | 0.39 |
| | 4 | No.2 | 10000 | 0.2 | 0.41 |
| | 7 | | 7000 | 0.6 | 0.45 |

0.25 - 0.3 in IP Peg at quiescence (Wood et al. 1988). Ichikawa & Osaki (1992) calculated the behavior of AD for the entire outburst cycle based on the disk instability (DI) model (e.g., Meyer & Meyer-Hofmeister 1981) and on the mass-transfer burst (MTB) model (e.g., Bath 1973). Our results support the disk instability model, since according to the MTB model the AD should shrink once at the beginning of the outburst.

If the MTB model were correct, the eclipse profile in the early rise should be asymmetric owing to an enhanced mass transfer. On the contrary, the profile was more symmetric (figure 2a) than in quiescence (e.g., Wood et al. 1988). The symmetric profile implies the axisymmetric temperature distribution without any excess at the hot spot. However, the growing asymmetry, clear in the profile of Sep. 7 (a steep ingress and the ingress phase of $\sim -0.6$) indicates that the AD structure was already close to the quiescent structure at that time. Taking into account that the magnitude out of eclipse was $\sim$ 13.5 on Sep. 7, only by 1.5 mag brighter than that at quiescence, the very weak feature due to the eclipse of the hot spot excludes the possibility of an enhanced mass-transfer occurring during the late stage of the outburst.

## References

Bath G. T. 1973, Nature Phys. Sci. 246, 84

Ichikawa S., Osaki Y. 1992, PASJ 44, 15

Marsh T. R. 1988, MNRAS 231, 1117

Meyer F., Meyer-Hofmeister E. 1981, A&A 104, L10

Ritter H., Kolb U. 1995, in X-ray Binaries, eds W. H. G. Lewin, J. van Paradijs, E. P. J. van den Heuvel (Cambridge University Press, Cambridge) p578

Osaki Y. 1974, PASJ 26, 429

Rutten R. G., Kuulkers E., Vogt N., van Paradijs J. 1992, A&A 265, 159

Skiff B. 1994, private communication

Vogt N. 1983, A&A 128, 29

Wood J. H., Marsh T. R., Robinson E. L., Steining R. F., Horne K., Stover R. J., Schoembs R., Allen S. L., Bond H. E., Jones D. H. P., Grauer A., Ciardullo R. 1988, MNRAS 239, 805

# Disk Instability Model for AM CVn Stars

Motohiko TSUGAWA and Yoji OSAKI
*Department of Astronomy, School of Science,*
*University of Tokyo, Bunkyo-ku, Tokyo 113, Japan*

## Abstract

We examine AM CVn stars, ultra-short-period variables thought to have helium disks, from the standpoint of the disk instability model. By calculating the vertical structure of the helium accretion disks, we obtain the S-shaped thermal equilibrium curves similar to those found in hydrogen-rich disks. We apply these results to AM CVn stars. We have found that the observed large amplitude photometric variations in AM CVn stars can be explained by the thermal-tidal instability model of helium accretion disks.

## 1. Introduction

The AM CVn stars are enigmatic blue variables with very peculiar characteristics (see a review by Warner 1995). Six stars have so far been classified to this class. Their main characteristics are: (1) hydrogen deficient spectra, and (2) photometric variations with very short periods of 1000-3000 s.

Although the exact nature of these stars is not yet known, the most promising model for these stars is an Interacting Binary White Dwarf model (abbreviated as "IBWD"). The IBWD systems are twin white-dwarf binary systems in which a very low-mass He white dwarf secondary transfers H-deficient gas onto another white dwarf (the primary) and a He accretion disk is formed around the primary.

In addition to these characteristics, three of six AM CVn stars, CR Boo, V803 Cen, and CP Eri, show large amplitude photometric variations. Their amplitudes reach 4 mag and the time scale is of the order of days. The cause of the large-amplitude photometric variations has not yet been clarified but they are thought to be a similar phenomenon to the dwarf nova outbursts. Two models have been discussed in the past for this phenomenon

*S. Kato et al. (eds.), Physics of Accretion Disks, 115–118.*
© 1996 OPA (Overseas Publishers Association) Amsterdam B.V.

just in parallel to the hydrogen-rich accretion disk system of the dwarf novae. The thermal instability was suggested by Smak (1983a) and examined by Cannizzo (1984) while Warner (1995) proposed the irradiation-induced mass-transfer instability.

In this paper, we examine the disk instability model for the helium accretion disk system. To do so, we have calculated the vertical structure of the helium accretion disks. By using these results we discuss thermal equilibrium of the helium disks and their stability. Then we apply these results to AM CVn stars.

## 2. Vertical Structure of Helium Accretion Disk

We calculate the vertical structure of the helium disk by using the model presented by Mineshige & Osaki (1983) in which the standard alpha model is used for viscosity and the convective energy transport is treated by the mixing length formulation.

Figure 1 shows one of examples of our calculations in which the thermal equilibrium curves for the He disk and the H-rich disk are exhibited on the surface density $\Sigma$–the effective temperature $T_{\rm eff}$ plane. Our results basically confirm Smak's (1983a), and the equilibrium curve for the He disk shows the familiar S-shaped structure just like the hydrogen disk and thus the helium accretion disks can be thermally unstable as well.

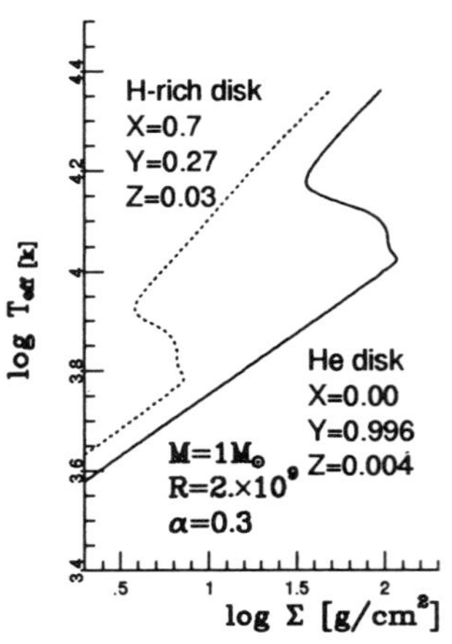


Fig. 1., Thermal equilibrium curves for H-rich disk (dotted curve) and for He disk (solid curve).

## 3. Application to AM CVn stars

In this paper, we adopt the stand-point that AM CVn stars are the IBWD systems having He accretion disks. We also adopt the binary evolutionary scenario presented by Warner (1995) in which mass transfer is driven by gravitational radiation. As discussed by Warner (1995), masses of secondary stars in AM CVn stars are considered to be very low (i.e., less than $0.1 M_\odot$ and their mass ratios are estimated very small ($q < 0.1$). In such a low mass-ratio system, the accretion disk may suffer the tidal instability (Whitehurst 1988), by which an accretion disk is deformed to a precessing eccentric disk. The tidal instability is thought to be responsible for the superhump phenomenon in the SU UMa stars with H-rich disks. In the case of AM CVn stars, there is observational evidence that supports the existence of the eccentric disks (Patterson et al 1993).

We now apply the disk instability model to AM CVn stars. The disk instability model adopted here is the thermal-tidal disk instability model (Osaki 1989) to explain the normal- and super-outbursts of SU UMa stars in which the ordinary thermal instability is coupled with the tidal instability. In our model, the short period photometric variations of AM CVn stars with periods of 1000-3000 s are interpreted as either superhump periods or orbital periods (Warner 1995) while the large amplitude photometric variations are interpreted as dwarf-nova type outbursts.

Let us first estimate the critical accretion rate for the thermal stability. The critical mass accretion rate below which no hot steady state exists is related to the critical temperature if we set the outer disk radius (Smak 1983b). In tidally unstable systems, the accretion disk is truncated at the 3:1 resonance radius which is given by 0.48×(binary separation). If we adopt this radius as the outer radius of accretion disks, the critical accretion rate is

$$\dot{M}_{\rm crit} = 1.3 \times 10^{-9} \left(\frac{T_{\rm e,crit}}{10^{4.2}\ {\rm K}}\right)^4 \left(\frac{P_{\rm orb}}{1000\ {\rm s}}\right)^2 , \qquad (1)$$

where the critical effective temperature for thermal stability is roughly given by $T_{\rm e,crit} \simeq 10^{4.2}$K for the helium disk.

We next estimate the mass transfer rate of IBWD systems. In parallel to CV evolution, we can determine the mass transfer rate in IBWD systems if we know the orbital angular momentum loss rate from the binary and the mass-radius relation of the secondary star. Here we assume that the angular momentum loss occurs via the gravitational wave radiation.

As for the mass-radius relation is concerned, the simple mass-radius relation of the helium white dwarf in thermal equilibrium is not applicable in this case because the mass transfer rate is very high. In such a case, we need to consider non-thermal equilibrium mass-radius relation because the secondary star is not in thermal equilibrium. Here we use the following non-thermal equilibrium mass radius relation due to Savonije et al (1986): $R_2 = 2.0 \times 10^9 (M_2/M_\odot)^{-0.19}$.

We then obtain the mass transfer rate as a function of orbital period:

$$\dot{M}[M_\odot\ {\rm yr}^{-1}] = \frac{2.6 \times 10^{-10}}{1.65 - 2(M_2/M_1)} \left(\frac{M_1 + M_2}{M_\odot}\right)^{2/3} \left(\frac{P_{\rm orb}}{1000\ {\rm s}}\right)^{-5.21} . \qquad (2)$$

Let us now compare these results with observations. Figure 2 illustrates the critical mass accretion rate for the thermal stability given by equation (1) and the mass transfer rate estimated from the IBWD evolutionary scenario given be equation (2) as functions of orbital period.

The orbital periods of AM CVn stars are estimated from the periods of short photometric variation because difference between the superhump period and the binary orbital period is small. We use the periods of short photometric variations of the six AM CVn stars compiled by Warner (1995). We find that AM CVn itself and EC 15330–1403 are in the thermally stable region while others are in unstable region. Thus the large amplitude variations of CR Boo, V803 Cen and CP Eri could well be explained by the thermal-tidal instability of the disk.

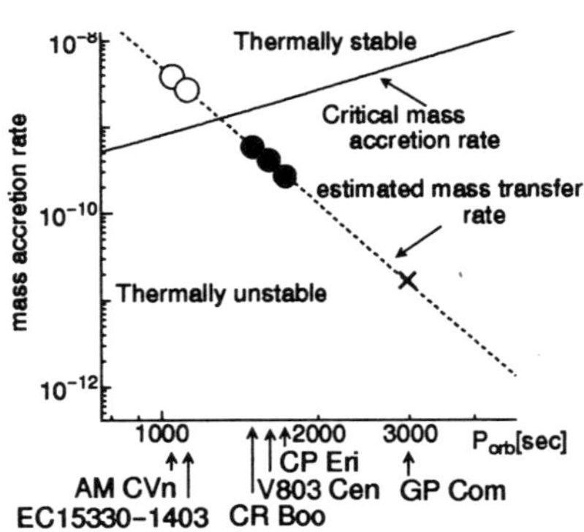


Fig. 2., The critical mass transfer rate for thermal stability (solid line) and mass transfer rate estimated from binary evolution ( dotted line).

Warner (1995) suggested the irradiation-driven mass-transfer instability for the large-amplitude variations in these three stars. We cannot exclude this possibility at present. More detailed theoretical and observational works will be required to clarify light variations in these stars.

GP Com is located in the thermally unstable region but it has so far shown no outburst phenomenon. However, some observational evidence such as its emission-line spectrum indicates that GP Com is in low state similar to quiescence in dwarf novae. GP Com may be mostly in low state and the frequency of outbursts might be very low (Smak 1983a; Warner 1995).

## References

Cannizzo J. 1984, Nature 311, 443
Mineshige S., Osaki Y. 1983, PASJ 35, 377
Osaki Y. 1989, PASJ 41, 1005
Osaki Y. 1996, PASP in press
Patterson, J., Halpern, J., Shambrook, A. 1993, ApJ 419, 83
Smak J. 1983a, Acta Astron 33, 333
Smak J. 1983b, ApJ 272, 234
Warner B. 1995, Ap&SS 225, 249
Whitehurst R. 1988, MNRAS 232, 35

# Non-Local Effects of Turbulence on Excitation of Trapped Oscillations in Dwarf-Nova Accretion Disks

Tatsuya YAMASAKI and Shoji KATO
*Department of Astronomy, Faculty of Science, Kyoto University, Sakyo-ku, Kyoto 606-01, Japan*

**Abstract**

In a previous paper (Yamasaki et al. 1995) we showed, in order to suggest one possible mechanism of quasi-periodic oscillations (QPOs) in dwarf novae, that some axially symmetric radial oscillations can be trapped and excited in dwarf nova accretion disks. This study was done by using a diffusion-type turbulent stress tensor. The adoption of a diffusion-type stress tensor is, however, not suitable, since the relaxation time of turbulence in Keplerian disks is comparable with the timescale of the oscillations, and a delay of time response of the turbulence to the oscillations has non-neglegible effects on excitation of the oscillations. In this study, we reexamine the excitation of trapped oscillations on dwarf novae disks by taking the effects of the above-mentioned time delay of stress tensor into account.

## 1. Introduction

Dwarf Novae (DNe) are one subgroup of cataclysmic variables (CVs). They are distinguished from other CVs in terms of their outburst activities which are generally explained by a thermal limit-cycle instability of accretion disks.

DNe are also famous, since they exhibit QPOs during outburst. There is no model of QPOs which is widely accepted. Here, we pay attention to axisymmetric radial oscillations of accretion disks trapped in a finite region as a possible origin of QPOs (cf. figure 1).

The purposes of the present study are i) to examine the behaviors of the trapped oscillations by linear theory; ii) to examined whether viscous and

*S. Kato et al. (eds.), Physics of Accretion Disks, 119–122.*
© 1996 OPA (Overseas Publishers Association) Amsterdam B.V.

thermal effects overcome any damping effects, so that the trapped oscillations can indeed be excited in dwarf nova disks.

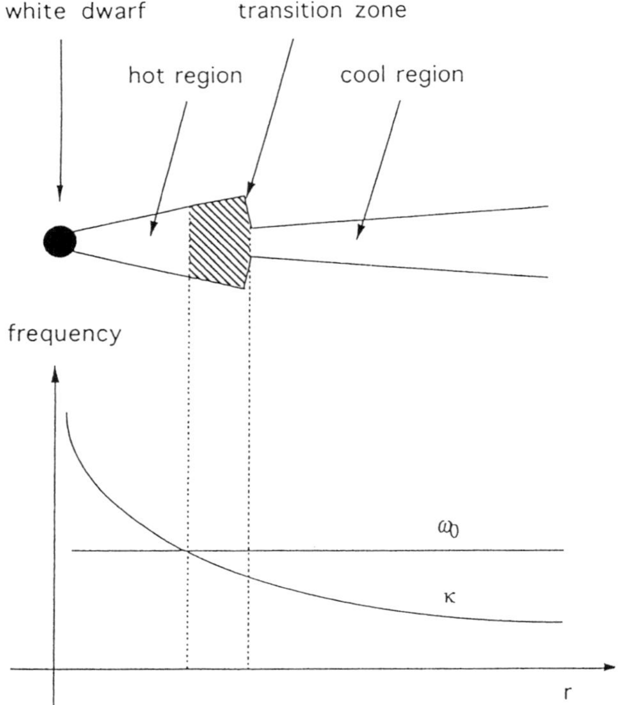


Fig. 1. Radial dependence of disk thickness of the dwarf-nova accretion disk during outburst, and the schematic relation between the functional form of the epicyclic frequency $\kappa(r)$ and the wave frequency $\omega_0$. The geometrically thick region near to the central star is hot. At the radius where the wave frequency $\omega_0$ becomes equal to the epicyclic frequency $\kappa$, a wave propagating inwards is reflected back. The outer thin region is cool. A wave propagating outwards cannot penetrate to this region beyond the transition layer. The waves are thus trapped in the shaded region.

## 2. Stress Tensor Modeling

Viscous effects can excite or damp the oscillations. The viscosity is assumed to be attributed to turbulence. To describe the time change of viscous-stress tensor, we adopt the transport equations of turbulence-stress tensor:

$$\frac{\partial t_{ij}}{\partial t} + U_k \frac{\partial t_{ij}}{\partial x_k} = -t_{ik}\frac{\partial U_j}{\partial x_k} - t_{jk}\frac{\partial U_i}{\partial x_k} + \Pi_{ij} - \frac{2}{3}\epsilon\delta_{ij}. \tag{1}$$

Here, $t_{ij}$ is the stress tensor per unit mass, $U_i$ the $i$-component of the flow velocity, $\Pi_{ij}$ the pressure-strain correlation, and $\epsilon$ the energy dissipation rate per unit mass. We apply here a modified second-order closure model introduced by Kato (1994a,b). In this model the pressure-strain correlation is approximated as $\Pi_{ij} = -C_1 \kappa b_{ij} + C_2 K S_{ij}$, where $C_1$ and $C_2$ are dimensionless universal constants of the order of unity, and $b_{ij}$ and $S_{ij}$ are the anisotropy tensor and the mean rate-of-strain tensor, respectively. Here, $\kappa$ is the epicyclic frequency, and $K$ is the turbulent energy given by $K = t_{ii}/2$. In the case of the usual second-order closure modeling (in this modeling, $\Pi_{ij} = -C_1(\epsilon/K)b_{ij} + C_2 K S_{ij}$), a comparison with experiments shows that adopting $C_1 \sim 3.6$ and $C_2 \sim 0.8$ is resonable (Speziale 1991).

## 3. Results

Our results show that the periods of the trapped oscialltions are in the range 70-600 s, which is equivalent to the range of the observed QPO periods. The frequencies of the trapped modes increase as the transition layer propagates inwards in the decline phase of an outburst. Such a tendency is, however, not always clear observationally (figure 2).

The results further show that the growth rate of trapped oscillations decrease compared with the case of a diffusion-type stress tensor, if non-local behaviors of turbulent stress tensor are taken into account. The oscillations are, however, still excited except for cases when the viscosity is small (figure 3). The more details are published in Yamasaki & Kato (1996).

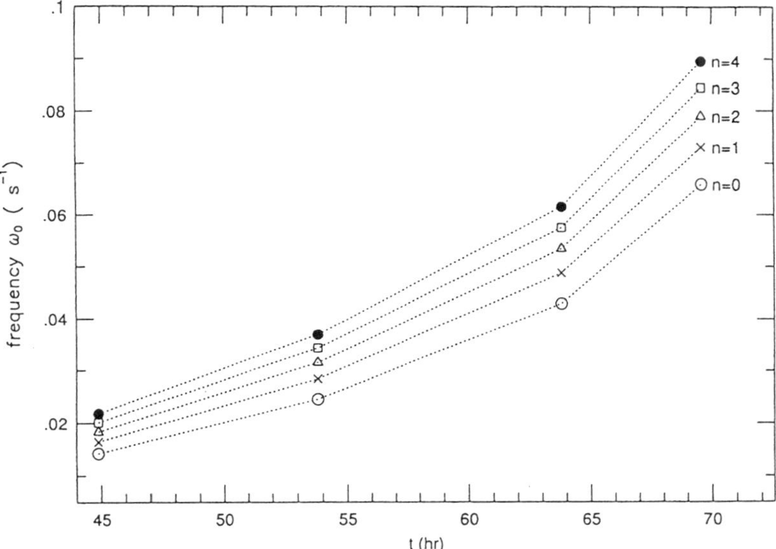


Fig. 2. The frequencies $\omega_0$ for five adiabatic oscillations. Since QPOs are observed mainly in the decaying phase of an outburst, we pick up four time segments $t = t_1$, $t_2$, $t_3$, $t_4$ during the decaying phase. They are epochs of the elapsed times after the maximum light being 44.88hr, 53.81hr, 63.79hr, and 69.57hr. The quantity $n$ denotes the number of nodes of oscillation in the radial direction.

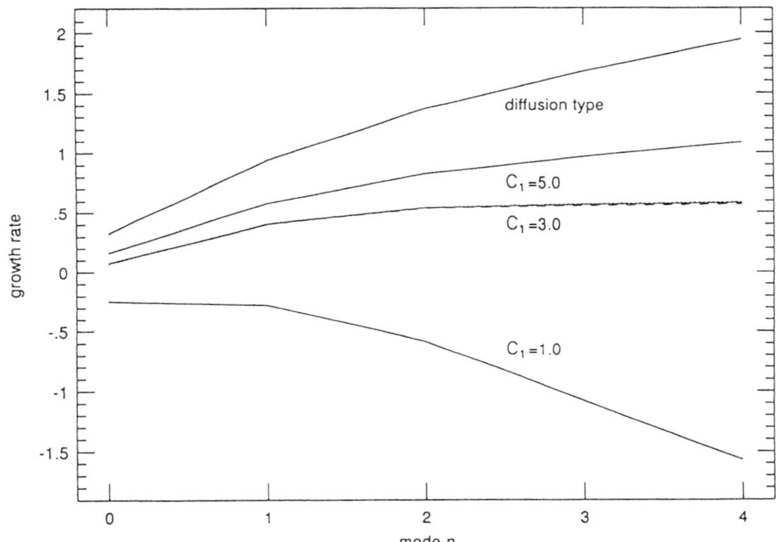


Fig. 3. Dimensionless growth rate (in units of the angular velocity of rotation at the outer boundary) at the stage $t = t_1$. The solid curves show the case of diffusion-type viscosity, and the cases of non-local stress tensor of $C_1 = 1.0$, 3.0, 5.0 with $C_2 = 0.5$. The dashed and dotted curves show the case of $C_2 = 0.2$ and $C_2 = 0.8$, respectively, with $C_1 = 3.0$. The results in the other cases ($t = t_2 - t_4$) are similar to those shown here.

## Acknowledgements

The authors are grateful to Dr. S. Mineshige for useful conversations.

## References

Kato S. 1994a, PASJ 46, 415
Kato S. 1994b, PASJ 46, 589
Speziale C. G. 1991, Ann. Rev. Fluid Mech. 23, 107
Yamasaki T., Kato S., Mineshige S. 1995, PASJ 47, 59
Yamasaki T., Kato S. 1996, PASJ in press

# Time Dependent Calculations of Accretion Disks Boundary Layers

Patrick GODON
*NASA/Jet Propulsion Laboratory, California Institute of Technology, 4800 Oak Grove Dr., MS 238-332, Pasadena, CA 91109, USA*

## Abstract

One-dimensional time dependent calculations of geometrically thin accretion disks boundary layers are presented for various systems: pre-main sequence stars (FU Orionis, T Tauris), main sequence stars (symbiotic AX Per & CI Cyg), and white dwarfs (in cataclysmic variables). At very low mass accretion rates (T Tauri systems) the boundary layer becomes optically thin ($\tau \approx 1$). At very large accretion rates (FU Orioni and symbiotic systems accreting close to the Eddington limit), advection becomes important in the inner part of the disk.

## 1. Boundary Layers

One-dimensional time-dependent simulations of Accretion Disks Boundary Layers (BL) are calculated around white dwarfs (Godon 1995; Godon et al. 1995), young stellar objects (YSO) (T Tauri and FU Orioni stars, Godon 1996b, c), and main sequence stars (Godon 1996a). The vertically averaged (time-dependent) equations are solved for a 1D thin accretion disk in the radial direction, including radiation (in both optically thick and thin cases), advection of energy, partial ionization of the Hydrogen gas, and an improved opacity law (Godon 1996c, based on Bell & Lin 1994). The spatial dependence of the equations is treated with the use of the Chebyshev method of collocation while the temporal scheme of the equations is treated with an explicit 4th order Runge-Kutta method. For more details see Godon et al. (1995) and Godon (1995, 1996c).

All the models relax toward steady state through oscillations of $\dot{M}$, due to acoustic waves trapped between the two boundaries which decay on a viscous time scale.

*S. Kato et al. (eds.), Physics of Accretion Disks, 123–126.*
© 1996 OPA (Overseas Publishers Association) Amsterdam B.V.

Table. 1. Boundary Layer models for various systems.

| star | $M_*$ ($M_\odot$) | $R_*$ ($R_\odot$) | $\Omega_*$ ($\Omega_{*K}$) | $\dot{M}$ ($M_\odot/y$) | $\alpha$ | $H$ ($R_*$) | $\delta_{BL}^{Th}$ ($R_*$) | $\delta_{BL}^{dyn}$ ($R_*$) | $T_{eff}^{Max}$ ($10^3$K) |
|---|---|---|---|---|---|---|---|---|---|
| WD | 1. | .013 | .33 | 6.7(−10) | ≈ .01 | .03 | .08 | .003 | 125 |
| | 1. | .013 | .33 | 1.6(−9) | ≈ .01 | .03 | .08 | .003 | 135 |
| YSO | 1. | 4.3 | .1 | 5.0(−7) | .30 | .10 | .30 | .10 | 8.3 |
| | 1. | 4.3 | .1 | 2.5(−6) | .30 | .11 | .30 | .10 | 10.5 |
| | 1. | 4.3 | .1 | 8.5(−6) | .30 | .12 | .50 | .10 | 11.7 |
| | 1. | 4.3 | .1 | 1.0(−4) | .30 | .40 | ... | .50 | 14.0 |
| | 1. | 4.3 | .1 | 2.5(−6) | .005 | .25 | ... | .50 | 7.9 |
| | 1. | 4.3 | .1 | 2.5(−6) | .30 | .11 | .30 | .10 | 10.5 |
| | 1. | 4.3 | .1 | 5.0(−6) | .02 | .26 | ... | .50 | 8.2 |
| | 1. | 4.3 | .1 | 5.0(−6) | .06 | .12 | .50 | .10 | 10.1 |
| | 1. | 4.3 | .1 | 5.0(−7) | .10 | .12-.05 | .40 | .10 | 7.3 |
| | 1. | 4.3 | .1 | 2.5(−6) | .02 | .17 | .80 | .20 | 9.1 |
| | 1. | 4.3 | .1 | 1.5(−5) | .015 | .30 | ... | .40 | 8.0 |
| Sym | .5 | .4 | .1 | 2.0(−5) | ≈ .1 | .17 | .30 | .10 | 60 |
| | .5 | .3 | .1 | 4.0(−5) | ≈ .1 | .24-.16 | .35 | .15 | 97 |
| | .5 | .2 | .1 | 2.5(−5) | ≈ .1 | .25-.16 | .50 | .10 | 120 |
| | .5 | .2 | .1 | 4.0(−5) | ≈ .1 | .30-.20 | .70 | .20 | 130 |
| | .5 | .2 | .1 | 1.0(−5) | ≈ .1 | .15-.12 | .40 | .05 | 100 |
| | .5 | .2 | .1 | 2.5(−5) | ≈ .1 | .25-.16 | .50 | .10 | 120 |
| | .5 | .2 | .1 | 4.0(−5) | ≈ .1 | .30-.20 | .70 | .20 | 130 |
| | .5 | .2 | .1 | 7.0(−5) | ≈ .1 | .35-.20 | 1.0 | .20 | 136 |
| | .5 | .2 | .1 | 9.0(−5) | ≈ .1 | .26-.20 | .70 | .20 | 126 |
| | .5 | .2 | .1 | 1.8(−4) | ≈ .1 | .37-.23 | .70 | .40 | 135 |
| | .5 | .2 | .1 | 3.0(−4) | ≈ .1 | .40-.25 | .70 | .50 | 135 |
| | .5 | .3 | .1 | 4.0(−5) | ≈ .1 | .24-.16 | .35 | .15 | 97 |
| | .4 | .3 | .1 | 7.0(−5) | ≈ .1 | .26-.20 | .50 | .15 | 88 |
| | .3 | .3 | .1 | 2.0(−5) | ≈ .1 | .18 | .40 | .10 | 69 |
| | .3 | .3 | .1 | 2.0(−4) | ≈ .1 | .40-.25 | 1.0 | .40 | 94 |

The oscillations are not associated with the viscous instability, since only the very inner part of the disk is considered, where the viscous instability is not expected to appear for the low value of $\alpha$ used here (Papaloizou & Stanley 1986). However, the *acoustic nature* of the oscillations agree well with the characteristics of observed QPOs (Godon 1995).

Another common feature of all the models is the presence of a rather 'cool' thermal BL much larger than the dynamical BL, since in the BL region the radial diffusion of energy is an important process. All the time dependent models we have computed, in table 1, exhibit the same general features of optically thick and geometrically thin steady state BLs (Narayan & Popham

1993; Popham et al. 1993; Regev & Bertout 1995; Popham & Narayan 1995).

Table. 2. Accretion disks boundary layers for classical T Tauris.

| $M_*$ $(M_\odot)$ | $R_*$ $(R_\odot)$ | $\Omega_*$ $(\Omega_{*K})$ | $\dot{M}$ $(M_\odot/y)$ | $\alpha$ | $H$ $(r)$ | $\delta_{BL}^{Th}$ $(r)$ | $\delta_{BL}^{dyn}$ $(r)$ | $T_{eff}^{Max}$ $(10^3\mathrm{K})$ | $\tau_{BL}$ |
|---|---|---|---|---|---|---|---|---|---|
| .8 | 2.15 | .1 | 5(−9) | .05 | .06-.02 | .20 | .04 | 6.4 | 30 |
| .8 | 2.15 | .1 | 1(−8) | .01 | .06-.01 | .20 | .04 | 6.1 | $\gg 1$ |
| .8 | 2.15 | .1 | 3(−8) | .03 | .06-.02 | .20 | .05 | 5.7 | 35 |
| .8 | 2.15 | .1 | 3(−8) | .10 | .06-.07 | .10 | .03 | 6.1 | 5-20 |
| .8 | 2.15 | .1 | 6(−8) | .20 | .06-.07 | .10 | .03 | 6.0 | 2-10 |
| .8 | 2.15 | .1 | 5(−7) | .16 | .05-.07 | .15 | .04 | 6.0 | 30 |
| 1. | 1.6 | .1 | 1(−8) | .05 | .04-.02 | .15 | .04 | 5.7 | 4-10 |
| 1. | 4.3 | .1 | 5(−70 | .05 | .07-.03 | .15 | .05 | 4.2 | 20 |

At low accretion rate, typical of T Tauris, the optical depth in the BL approaches $\approx 10$ (table 2). The thermal BL is completely ionized, while the cool disk is completely neutral. An ionization transition separates the two distinct regions. For large value of the $\alpha$ viscosity parameter, the optical depth in the BL decreases to $\approx 1$, its effective temperature slightly increases and the disk becomes partially ionized. The BL effective temperature agrees well with the model of Regev & Bertout (1995).

Table. 3. Advection in boundary layers for various systems.

| System | $\alpha$ | $\dot{M}$ $(M_\odot/y)$ | $H$ $(R_*)$ | $\delta_{BL}^{Th}$ $(R_*)$ | $\delta_{BL}^{dyn}$ $(R_*)$ | $\zeta = \frac{L_{adv}}{L_{acc}}$ |
|---|---|---|---|---|---|---|
| Sym | 0.05 | 2.0(-5) | 0.17 | 0.30 | 0.10 | $\ll 1$ |
| | 0.11 | 4.0(-5) | 0.24-0.16 | 0.35 | 0.15 | $\ll 1$ |
| | 0.14 | 1.0(-5) | 0.15-0.12 | 0.40 | 0.05 | $\ll 1$ |
| | 0.07 | 4.0(-5) | 0.30-0.20 | 0.70 | 0.20 | $\ll 1$ |
| | 0.1 | 7.0(-5) | 0.35-0.20 | 1.00 | 0.20 | $\ll 1$ |
| | 0.22 | 9.0(-5) | 0.26-0.20 | 0.70 | 0.20 | 0.07 |
| | 0.18 | 1.8(-4) | 0.37-0.23 | 0.70 | 0.40 | 0.11 |
| | 0.16 | 3.0(-4) | 0.40-0.25 | 0.70 | 0.50 | 0.12 |
| YSO | 0.3 | 5.0(-7) | 0.10 | 0.30 | 0.10 | $\ll 1$ |
| | 0.1 | 5.0(-7) | 0.05-0.12 | 0.40 | 0.10 | $\ll 1$ |
| | 0.3 | 2.5(-6) | 0.11 | 0.30 | 0.10 | $\ll 1$ |
| | 0.06 | 5.0(-6) | 0.12 | 0.50 | 0.10 | $\ll 1$ |
| | 0.3 | 8.5(-6) | 0.12 | 0.50 | 0.10 | $\ll 1$ |
| | 0.005 | 2.5(-6) | 0.25 | ... | 0.50 | 0.14 |
| | 0.02 | 5.0(-6) | 0.26 | ... | 0.50 | 0.23 |
| | 0.015 | 1.5(-5) | 0.30 | ... | 0.40 | 0.21 |
| | 0.3 | 1.0(-4) | 0.40 | ... | 0.50 | 0.27 |

The optical thickness is also in good agreement with previous works (Basri & Bertout 1989; Hartigan et al. 1991).

At accretion rate close to the Eddington limit (FU Orionis; symbiotic AX Per & CI Cyg), the optical thickness in the BL region increases and the BL cannot cool efficiently. Consequently, a large fraction of the energy $\zeta = L_{adv}/L_{acc} \approx 0.1 - 0.2$ is advected into the inner boundary (table 3). The advection $\zeta$ in YSOs depends mainly on $\rho$, while advection in symbiotics is only a function of $\dot{M}$. This is due to the fact that the opacity is a function of $\rho$ in YSOs, but is constant (electron scattering) in symbiotics (the temperature does not vary very much).

**Acknowledgements**

The Cray Supercomputer (JPL/CalTech Cray Y-MP2E/232) used in this investigation was provided by funding from the NASA Offices of Mission to Planet Earth, Aeronautics, and Space Science. This work was performed while the author held a National Research Council - (NASA Jet Propulsion Laboratory) Research Associateship. This research was carried out at the Jet Propulsion Laboratory, California Institute of Technology, under contract with the National Aeronautics and Space Administration.

## References

Basri G., Bertout C. 1989, ApJ 341, 340
Bell K. R., Lin D. N. C. 1994, ApJ 427, 987
Godon P. 1995 MNRAS, 274, 61
Godon P. 1996a ApJ in press
Godon P. 1996b MNRAS in press
Godon P. 1996c ApJ submitted
Godon P., Regev O., Shaviv G. 1995, MNRAS 275, 1093
Hartigan P., Kenyon S. J., Hartmann L. W., Storm S. E., Edwards S., Welty A. D., Stauffer J. 1991, ApJ 382, 617
Narayan R., Popham R. 1993, Nature 362, 820
Papaloizou J. C. B., Stanley G. Q. G. 1986, MNRAS 220, 253
Popham R., Narayan R., Hartmann L., Kenyon S. J. 1993, ApJL 415, L127
Popham R., Narayan R. 1995, ApJ 442, 337
Regev O., Bertout C. 1995, MNRAS 272, 71

# The Physics of Black Hole X-ray Transients

J. Craig WHEELER[1], Soon-Wook KIM[1], Michael MOSCOSO[1], Masaaki KUSUNOSE[1], and Shin MINESHIGE[2]
*1. Department of Astronomy, University of Texas, Austin TX, 78712, USA*
*2. Department of Astronomy, Faculty of Science, Kyoto University,* Kyoto 606-01, Japan

**Abstract**

The most plausible mechanism for triggering the outburst of black hole candidate X-ray transients is the ionization thermal instability. The disk instability models can give the observed mass flow in quiescence, but not the X-ray spectrum. Self-irradiation of the disk in outburst may not lead to X-ray reprocessing as the dominant source of optical light, but may play a role in the "reflare." The hard power-law spectrum and radio bursts may be non-thermal processes driven by the flow of pair–rich plasma from the disk at early times and due to the formation of a pair–rich plasma corona at late times. The repeated outbursts suggest some sort of clock, but it is unlikely that it has anything to do with a simple X-ray heating of the companion star.

## 1. Introduction

Many of the X-ray novae discovered in recent years are black hole candidates by direct measure of their mass function or by shared properties (McClintock & Remillard 1986; Tanaka & Lewin 1995 and references therein). Among the commonly observed features are a primary outburst maximum, a "secondary reflare", 50–80 days after maximum, and a "third broad bump" in the decay, a few hundred days later. The first two features are especially marked in the soft X-ray and are probably associated with the geometrically thin, optically thick accretion disk. The latter is associated with the hard power-law source that may be a signature of black hole accretion. Nova Per and Nova Vela revealed repeated "mini-outbursts" after the third, broad bump (Callanan *et al.* 1995; Della Valle et al. 1996). Two "super-luminal sources", Nova Oph

*S. Kato et al. (eds.), Physics of Accretion Disks, 127–132.*
© 1996 OPA (Overseas Publishers Association) Amsterdam B.V.

1993 (GRS 1915+105) and Nova Sco 1994 (GRO J1655−40) display jet-like outflow in the radio (Mirabel & Rodriguez 1994; Harmon et al. 1994) and radio activity is commonly associated with all outbursts (Han & Hjellming 1992).

The power-law flux rises in Nova Muscae before the soft X-ray, but declines rapidly so that the soft flux dominates the total power at its maximum (Miyamoto et al. 1993). There is a strong dip in the hard flux just before the observations of a transient line at 480 Kev (Sunyaev et al. 1992; Goldwurm et al. 1992). This feature has been associated with a red–shifted positron annihilation line. Although it occurs at the rest wavelength of a Li de-excitation line (Martín et al. 1994), the lack of other de-excitation lines and the association of the feature with the modulation of the hard power-law flux suggests that annihilation is still a reasonable interpretation. The secondary reflare is essentially a feature of the soft flux and hence of the accretion disk. It occurs just as the hard power-law component reaches a minimum, for reasons that are not understood. At about 150 days in Nova Muscae, the disk component of the soft flux plummets and the power, including that in soft X-rays, becomes dominated by the power-law source at the "third broad bump."

## 2. Disk Irradiation

The disk instability (Mineshige & Wheeler 1989) naturally gives a rapid rise and slower decline in the soft X-ray and optical. It can give an exponential decay as observed in some sources, but the origin of this is under debate (Cannizzo et al. 1996). The mass transfer from the companion in A 0620−00 greatly exceeds that attributed to the soft X-rays from the inner disk (Marsh et al. 1994; McClintock et al. 1995). This shows that the disks are not in steady state in quiescence, a principle prediction of the disk instability models.

In principle, strong irradiation can keep the disk ionized and prevent the disk instability. We find that for models that approximately match the light curve of A 0620−00 and similar sources the irradiation can not be that severe (Kim et al. 1996a,b, and the associated contributed paper to this conference). The irradiation even at modest levels can subtly affect the disk evolution by prolonging the disk in the intermediate, metastable partially ionized, "stagnation" state. One especially interesting manifestation of this is that some regions of the disk can be driven from the metastable state back to the hot, ionized state. This causes them to be brighter intrinsically and to intercept more irradiation. In our current models the associated optical flare can match the time and amplitude of the optical component of the reflare in A 0620−00. We do not get appreciable modulation of the mass flow through the inner edge of the disk and so do not get an obvious increase in the soft X-ray flux, one of the defining characteristics of the reflare.

The models that match the light curve of A 0620−00 have the ignition of the thermal instability in the outer portions of the disk. This implies that the optical light curve should always rise before any harder flux associated with the increase of the mass flow in the inner portions of the disk. The question of whether the hard or soft X-ray flux should increase first depends on a better understanding of the origin of the hard flux.

These models also have the interesting property that the cooling wave slows as the density declines, but never reaches the inner edge of the disk. The inner disk thus always remains in the hot ionized state. The mass flow rates are $\sim 10^{12}$ g s$^{-1}$, close to those constrained by observations in quiescence (Marsh et al. 1994; McClintock et al. 1995). The temperatures we derive, are, however, less than 300,000 K, whereas the quiescent X-ray spectrum implies a temperature of about $2\times10^6$ K.

Van Paradijs & McClintock (1994) have argued that, as for the LMXB, the optical flux of the black hole candidates is due to reprocessing of X-rays. This is not clear since the black hole systems are transient and the flux ratio will vary in time. In addition, the disk instability models for X-ray transients give an adequate amount of optical light even with no irradiation. For the same orbital period, black hole disks are larger than neutron star disks. This mass-dependent factor alone could make a black hole disk a factor of five to ten larger in optical emitting area than for a neutron star. Van Paradijs & McClintock (1994) argue that $L_{opt}$ is proportional to $L_x^{1/2}$. We have constructed steady state models with no irradiation and found that the optical luminosity depends more steeply on $L_x$ than that with increasing disk radius. Furthermore, we have computed time-dependent, non-irradiated models and found that on the decline the locus of optical and X-ray flux is even slightly steeper than the steady state models. This subject requires study in greater depth.

## 3. The Reflare and Other Bumps

The "secondary reflare" is a rather common (though not universal) phenomenon of the black hole candidates and has never been observed in a neutron star system. It is thus worthy of understanding even though it is a secondary effect compared to the primary outburst.

Chen et al. (1993) discuss the possibility that the mass transfer rate from the companion can be modulated by irradiation from the inner disk. It is not clear that a burst of mass transfer would give either the observed optical or X-ray features. Even a sharp burst of added mass will be spread by the finite viscous response of the disk (especially when the outer parts of the disk are in the cold state) so that any later effect in the X-rays will be delayed with respect to the optical and very spread out in time. There are also questions of whether the disk blocking invoked by Chen et al. (1993) to account for the delay of the secondary reflare is consistent with their estimates of mass transfer and energetics that depend on irradiating the companion. Similar issues arise in their model for the third broad bump.

Augusteijn et al. (1993) suggest an oscillation of the light curve in the decay in which each successive burst is a "reflection" of the previous burst that heats the companion and drives more mass transfer after some time delay. This model seems to be remarkably reminiscent of the "mini-outbursts" in Novae Per and Nova Vela. Augusteijn et al. even predicted bursts in Nova Per in August 1993 and December 1993, (but also 21 April) as observed. Augusteijn et al. deserve great credit for drawing attention to the fact that there may be some "clock" underlying the bursts in Nova Per and perhaps other objects, but there are still open questions concerning their particular model. Augusteijn et al. did not clearly differentiate the "second reflare" from the "third broad bump" as we are defining them here. They adjust parameters of their model to fit the second reflare of GS 2000+25 in one illustration of their model, but then calibrate the model of Nova Per on the third broad bump in order to "predict" the later outbursts in that system. It is not at all clear that the second reflare and the third broad bump involve similar physics. The models of Augusteijn et al. also do not consider the state of the disk, especially in its cool, quiescent, low-viscosity state, in a self-consistent way.

We do not yet have a complete understanding of the physical mechanism of the secondary reflare (or subsequent flares). No model yet proposed can naturally account for why the secondary reflare seems to coincide with the drop in the hard X-ray flux. Nevertheless, the irradiated models we have investigated show that effects in the disk alone can give optical outbursts that may be related to the optical flares seen. They also give us a new perspective from which to consider questions of the irradiation of the companion.

Unlike the pictures proposed by Chen et al. (1993) and Augusteijn et al. (1993), our current models show that the direct X-ray irradiation of either the outer disk or the $L_1$ point is blocked by the inner disk throughout the decay phase prior to the secondary reflare. The hypothesis of the X-ray-irradiated mass transfer burst models, that the $L_1$ point be irradiated, therefore, seems to be contradicted by the shadowing given in the current models (Kim et al. 1996a,b).

## 4. Advection

Narayan et al. (1996) obtain a fit to both the optical and X-ray spectra of A 0620−00 in quiescence by invoking a hot two-temperature advective disk solution in the inner disk matched to a steady state disk in the outer portions that provides the optical luminosity. The advective solution, however, is of low efficiency and requires a mass flow rate of $4 \times 10^{14}$g s$^{-1}$, much higher than the estimates based on steady state, geometrically thin, optically thick disks by Marsh et al. (1994) and McClintock et al. (1995) and much higher than the quiescent flow rates we obtain in these models. The steady-state disks appended to the advection solutions are not consistent with the quiescent state of the disk being modeled. The advection solutions require some means of severely depleting the surface density of the inner portions of the disk as the disk approaches quiescence. It is difficult to see how such a solution matches physically in terms of the surface density and angular momentum with the outer geometrically thin, quiescent, Keplerian disk.

## 5. The Hard Power Law, Radio Outbursts, and Positrons

The hard power law component is commonly assumed to be a Comptonized thermal spectrum. Such a simple model can fit some objects at some epochs, but that does not make it unique or correct. Such models ignore the obvious evidence for non-thermal particles and magnetic fields implied by the common radio outbursts that are frequently associated with the X-ray bursts (Han & Hjellming 1992). The recent super-luminal sources are only the most extreme example. It is most likely that the non-thermal particles and magnetic fields arise in the disk and hence must be incorporated into models of the hard power-law emission.

The soft X-ray component that probably arises in the accretion disk peaked more slowly than the hard power law flux in Nova Muscae. This may mean that the inner disk was incomplete in quiescence or the early phase of the outburst. The radius of the geometrically thin, optically thick disk may have shrunk in response to increased mass flow attendant with the disk instability in the outer disk, thus giving rise to the delayed rise of the soft flux.

The first flare of the hard flux in systems like Nova Muscae can be associated with a non-thermal, magnetic, pair-rich outflow (Moscoso & Wheeler 1993). This phase shows QPO's, correlated radio synchrotron bursts, and at least in Nova Muscae, the line feature that is plausibly associated with annihilation. If this is the annihilation line, then it is much too narrow to represent annihilation in the region where positrons are created and hence implies flow of some kind.

The "third bump," as we have defined it here, may more closely resemble a quasi-static corona of the sort frequently modeled in the literature (Mineshige et al. 1995). The fact that the disk component of the soft X-ray flux declines as this late hard component comes in strongly suggests that the corona is displacing the geometrically thin disk. With a larger effective inner radius, the disk simply becomes too cool to emit soft X-rays.

Moscoso is constructing a model to better understand the source of the outflow in the primary outburst. This model consists of an inner hot, pair-rich corona represented by a single zone. Above this corona, photon annihilation will generate electron/positron pairs and associated annihilation. The parallel component of the average momentum of the photons that produce pairs is assumed to represent the bulk outflow momentum of pairs. The remaining momentum is randomized to provide the thermal component of the pair energy. Account will be taken of both the isotropic and anisotropic Comptonization. This simple model will give an estimate of the typical flow time scales, speeds, and the optical depth so that annihilation line profiles can be estimated.

## Acknowledgements

JCW, SWK, and MK thank the organizers of the meeting for hospitality and a very stimulating meeting. This research is supported in part by NASA Grants.

## References

Augusteijn T., Kuulkers E., Shaham J. 1993, A&A 279, L13

Callanan P. J. et al. 1995, ApJ submitted

Cannizzo J. K., Chen W., Livio M. 1996, ApJ in press

Chen W., Livio M., Gehrels N. 1993, ApJL 408, L5

Della Valle M., Benetti S., Wheeler J. C. 1996, in preparation

Goldwurm A. et al. 1993, ApJL 389, L79

Han X., Hjellming R. M. 1992, ApJ 400, 304

Harmon B. A. et al. 1995, Nature 374, 703

Kim S.-W., Wheeler J. C., Mineshige S. 1995a, in preparation

Kim S.-W., Mineshige S., Wheeler J. C. 1996b, in preparation

Marsh T. R., Robinson E. L., Wood J. H. 1994, MNRAS 266, 137

Martín E. L., Rebolo R., Casares J., Charles P. A. 1994, ApJ 791, 1994

McClintock J. E., Horne K., Remillard R. A. 1995, ApJ 442, 358

McClintock J. E., Remillard R. A. 1986, ApJ 308, 110

Mirabel I. F., Rodriguez L. F. 1994, Nature 371, 46

Mineshige S., Kusunose M., Matsumoto R. 1995, ApJL 445, L43

Mineshige S., Wheeler J. C. 1989, ApJ 343, 241

Miyamoto S. et al. 1993, ApJL 403, L39

Moscoso M. D., Wheeler J. C. 1993, in Interacting Binary Stars, ed A. W. Shafter (San Francisco: ASP), p100

Sunyaev R. et al. 1992, ApJL 389, L75

Tanaka Y., Lewin W. H. G. 1995 in X-Ray Binaries, eds W. H. G. Lewin, J. Van Paradijs, & E. P. J. Van Den Heuvel (Cambridge University Press, Cambridge), p126

Van Paradijs J., McClintock J. E. 1994, A&A 290, 133

# Time Variabilities and Disk Instabilities

Ronald E. TAAM
*Northwestern University, Dearborn Observatory, 2131 Sheridan Road, Evanston, IL 60208, USA*

## Abstract

The global evolution of accretion disks is investigated as a model for quasi-periodic oscillations in binary systems containing either a neutron star or black hole. Attention is focused on the nonlinear evolution of disks which are linearly unstable to thermal viscous modes. The numerical results for an accretion disk-corona system illustrate that low amplitude variations associated with thermal viscous instabilities can be produced provided that the disk remains near a marginally stable state.

## 1. Introduction

The discovery of time variability in the form of quasi-periodic oscillations (QPOs) in low mass X-ray binary systems (LMXBs) and in black hole candidate (BHC) systems in the last decade has opened a new window for the study of these objects. Numerous observational and theoretical studies have focused on the phenomenon since the identification of its origin offers the potential for significantly enhancing our fundamental understanding of compact objects and of accretion processes in these systems. The combination of observational and theoretical investigations will be essential to fully realize the implications of this discovery.

QPOs have been observed from a number of compact sources. Of interest has been the discovery of QPOs at similar frequencies from different classes of objects. Specifically, the analysis of data from the BATSE experiment on the Compton Gamma Ray Observatory has revealed evidence for peak noise in the power density spectrum of Cyg X-1 centered around 70 mHz in the energy range from 45 to 140 keV; in other observations of Cyg X-1, the peak noise is centered around 30-45 mHz (see Kouveliotou 1994). These low frequency QPOs have also been detected in Cyg X-1 by the GRANAT/SIGMA satellite

*S. Kato et al. (eds.), Physics of Accretion Disks, 133–138.*
© 1996 OPA (Overseas Publishers Association) Amsterdam B.V.

as reported by Vikhlinin et al. (1994) who observed peak noise at $\sim$ 40 mHz in the energy range of 40-70 keV. QPOs from other BHCs have also been detected. For example, QPOs at 35 mHz and 200 mHz have been seen from GRO J0422+32 and QPOs at 26 mHz have been discovered in GX 339-4 (see Kouveliotou 1994). These low frequency oscillations are not restricted to BHC systems, however, since they have also been observed in the Rapid Burster (RB) as well (Lubin et al. 1992, 1993). This source which is known to be a neutron star (see Lewin et al. 1993, 1995), reveals the presence of QPOs at a frequency of $\sim$ 40 mHz and at $\sim$ 4 Hz.

Since the oscillations are observed from both neutron star and black hole candidate systems, considerable theoretical interest has focused on accretion disk models. Consequently, it is natural to consider the possibility that disk instabilities are relevant to the phenomenon. Among the various models proposed for the general QPO phenomenon is a model suggested by Abramowicz et al. (1989) who suggested that thermal and viscous instabilities in accretion disks could be important for QPOs with frequencies $\sim$ 1 Hz for LMXBs.

As is well known, the study of accretion disks has been central to our understanding of X-ray binaries. In spite of its phenomenological success, the fundamental physics of accretion in disks is poorly understood since their structure and evolution is determined by the magnitude and functional form of some anomalous (turbulent) viscosity. Although there have been some promising developments involving the role of magnetic instabilities (see Chandrasekhar 1961; Balbus & Hawley 1991), significant research remains before these concepts can be applied to real accretion disks. Thus, the model description for the time variability associated with QPOs in terms of accretion disk instabilities must, by necessity, be phenomenological.

## 2. Disk Instabilities

The local stability of an accretion disk has been the subject of study by a number of investigators since the pioneering work of Pringle et al. (1973). The instabilities can occur as a result of thermal, viscous, or dynamical perturbations (see Lightman & Eardley 1974; Shakura & Sunyaev 1976; Kato 1978; Bath & Pringle 1982, Blumenthal et al. 1984) on a steady state $\alpha$ disk structure (Shakura & Sunyaev 1973). The global investigation of the thermal and viscous instabilities have revealed a wide diversity of disk behaviors. For example, the nonlinear and time dependent calculations of Taam & Lin (1984), Lasota & Pelat (1991), and by Honma et al. (1991) reveal that very large amplitude outbursts (greater by more than a factor of ten above the quiescent level of emission) are produced for a viscous stress proportional to the total pressure in the disk. The observations of small amplitudes in the QPO data from the RB and BHCs suggest that such models are incomplete or require revision. A recent development in this direction has been the recognition that accretion disk coronae could play an important role in affecting time variability. Coronae had been invoked early in the development of accretion disk theory (see Liang & Price 1977) to account for the presence of optically thin hot matter in X-ray binary systems. More recently, they have been invoked in the interpretation of spectra from Seyfert galaxies and galactic black holes (Haardt & Maraschi 1991; Haardt et al. 1994). In the context of time variability and disk instabilities, Chen (1995) suggested that a disk corona system is naturally weakly unstable due to the stabilizing effects associated with coronal energy dissipation (see also Ionson & Kuperus 1984; Svensson & Zdziarski 1994). In this picture, a hot optically thin region is not only present above an optically thick disk, but may also be present in the innermost radial region of the disk.

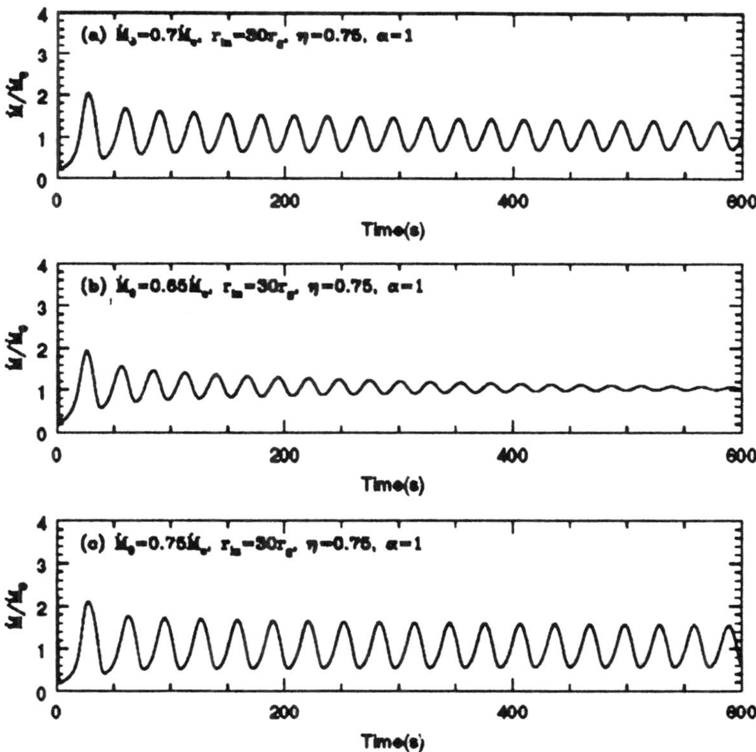


Fig. 1. Time variations of the mass flow rate relative to the input rate at the inner edge of the disk. In all three sequences the fraction of energy dissipated in the corona is 0.75, the inner edge of the optically thick region is located at 30 Schwarzschild radii and $\alpha$ is unity. The top, middle and lower panels correspond to mass accretion rates of 0.7, 0.65, and 0.7 times the critical value (see Abramowicz et al. 1995).

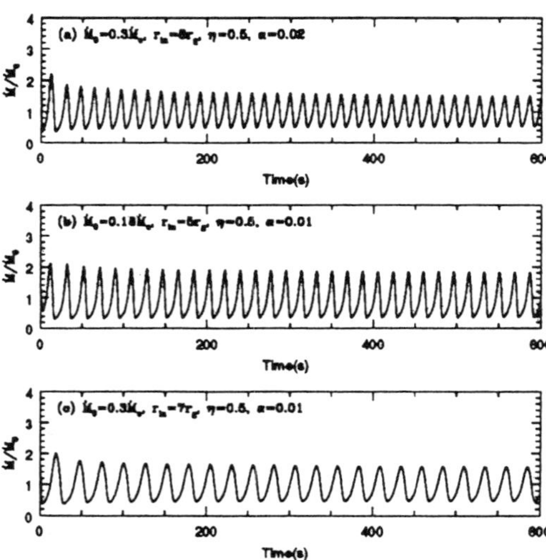


Fig. 2. Time variations of the mass flow rate relative to the input rate at the inner edge of the disk. In all three sequences the fraction of energy dissipated in the corona is 0.5. The top, middle and lower panels correspond to mass accretion rates ($\alpha$ values) of 0.3 (0.02), 0.18 (0.01), and 0.3 (0.01) times the critical value (see Abramowicz et al. 1995).

Recent time dependent work by Abramowicz et al. (1995) has shown that small amplitude oscillations are, indeed, possible in this model provided that the disk remains close to a marginally stable state. In essence, the coronal energy dissipation effectively reduces the amplitude of the mass flow modulations generated from the thermal and viscous instabilities (in comparison to models without such a corona). An example of the temporal variability of the mass flow modulations is shown in figures 1 and 2 for the case of a $10M_\odot$ black hole and a $1.4M_\odot$ neutron star respectively. Within the context of this model, low frequency oscillations of $\sim$ 40 mHz would be present in BHC systems provided that significant energy is dissipated in the corona ($> 70\%$), the rate of mass accretion is greater than about 0.6 times the Eddington rate, and the optically thin inner disk region extends to less than about 30 Schwarschild radii. Similar frequencies can be found for accretion disks surrounding neutron stars provided that the fraction of energy dissipated in the corona is greater than about 50%, the mass accretion rates are greater than about 0.1 times the Eddington rate, the inner edge of the optically thick disk lies close to the neutron star surface, and $\alpha$ is reduced from unity to 0.01. For a greater fraction of energy dissipated in the corona, the disk is stabilized whereas for significantly lower fractions, large amplitude oscillations can develop. The tendency for the accretion disk to lie near a marginally stable state (to produce small amplitude oscillations) may reflect the fact that the fraction of energy dissipated in the corona may be a function of the strength of the instability. For example, the coronal energy dissipation may be increased for greater instability. This would have the effect of increasing the fraction of energy dissipated in the corona, thereby reducing the tendency toward instability.

## 3. Conclusion

It has been shown that accretion disks surrounding compact objects can exhibit nonsteady behavior. The amplitude of the variability as well as the timescales of variation both depend on the model. The disk evolves to a limiting cyclic state and mass flow modulations must be introduced into the thermally unstable region to produce QPO behavior. The frequencies and amplitudes of oscillations are both functions of the mass accretion rate and the viscosity prescription with the global oscillations occurring in a restricted range of mass accretion rates.

Because the nature of viscosity is not well understood, further progress requires close contact between both observers and theorists. Observationally, sources can be observed over a broad energy range to determine the relationship between the QPO parameters and the level of emission (i.e., the rate of mass accretion). It is essential to observe sources over a broad energy range and levels of intensity so that correlations between the properties of the QPOs and the bolometric luminosity can be established. Observations of transient sources are especially useful in this regard since the behavior of the sources can be observed over a wide range of luminosities. Theoretically, there is a need to determine the generality of the nonsteady behavior. Fundamental studies remain to assess the role of magnetic instabilities on angular momentum transport. The rapidly advancing computer technology should enable one to make significant progress via multi-dimensional magnetohydrodynamical simulations in the future.

## Acknowledgements

This research was supported in part by NASA under grant NAGW-2526. The author would like to express his thanks to the local organizing committee for financial support.

## References

Abramowicz M. A., Chen X., Taam R. E. 1995, ApJ 452, 379

Abramowicz M. A., Szuszkiewicz E., Wallinder F. 1989, in Theory of Accretion Disks, ed F. Meyer, W. J. Duschl, J. Frank, E. Meyer-Hofmeister, (Kluwer, Dordrecht), p141

Balbus S. A., Hawley J. F. 1991, ApJ 376, 214

Bath G. T., Pringle J. E. 1982, MNRAS 199, 267
Blumenthal G. R., Yang L. T., Lin D. N. C. 1984, ApJ 287, 774
Chandrasekhar S. 1961, Hydrodynamic and Hydromagnetic Instability (Clarendon, Oxford)
Chen X. 1995, ApJ 448, 803
Haardt F., Maraschi L. 1991, ApJL 380, L51
Haardt F., Maraschi L, Ghisellini G. 1994, ApJL 432, L95
Honma F., Matsumoto R., Kato S. 1991, PASJ 43, 147
Ionson J. A., Kuperus M. 1984, ApJ 284, 389
Kato S. 1978, MNRAS 185, 629
Kouveliotou C. 1994, in Proceedings of the Second Compton Symposium, eds C. E. Fichtel, N. Gehrels, J. P. Norris (AIP Press, New York), p202
Lasota J. P., Pelat D. 1991, A&A 249, 574
Lewin W. H. G., van Paradijs J., Taam R. E. 1993, SpSciRev 62, 223
Lewin W. H. G., van Paradijs J., Taam R. E. 1995, in X-Ray Binaries, ed W. H. G. Lewin, J. van Paradijs, & E. P. J. van den Heuvel (Cambridge University Press, Cambridge), p175
Liang E. P. T., Price R. H. 1977, ApJ 218, 247
Lightman A. P., Eardley D. M. 1974, ApJL 187, L1
Lubin L. M., Lewin W. H. G., Rutledge R. E., van Paradijs J., van der Klis M., Stella L. 1992, MNRAS 258, 759
Lubin L. M., Lewin W. H. G., van Paradijs J., van der Klis M. 1993, MNRAS 261, 149
Pringle J. E., Rees M. J., Pacholczyk A.G. 1973, A&A 29, 179
Shakura N. I., Sunyaev R. A. 1973, A&A 24, 337
Shakura N. I., Sunyaev R. A. 1976, MNRAS 175, 613
Svensson R., Zdziarski A. 1994, ApJ 436, 599
Taam R. E., Lin D. N. C. 1984, ApJ 287, 761
Vikhlinin A., et al. 1994, ApJ 424, 395

# Time Variabilities of Black Hole Candidate X-Ray Stars

Sigenori MIYAMOTO [‡] [§]
*Department of Earth and Space Science, Faculty of Science, Osaka University, Machikaneyama 1-1, Toyonaka, Osaka 560, Japan*

## Abstract

Firstly, we described present status of observational investigation on short term variabilities of X-ray energy-spectral components of black-hole candidate X-ray binaries (BHC-XBs) together with weak magnetic-field neutron star X-ray binaries (NS-XBs). The hard and soft power-law components and the disk-blackbody component of the BHC-XBs have their own inherent normalized power-spectrum densities and phase lags. The variabilities of corresponding X-ray energy-spectral components of the NS-XBs are similar to those of BHC-XBs, although X-ray energy spectra of these two kinds of X-ray stars are different in their high intensity state. There is no X-ray component from the neutron surface in the low intensity state of NS-XBs. Secondly, long term variability of BHC and NS XBs is described. There are two states in the BHC-XBs; the power-law hard and power-law soft states. Some of BHC-XBs show large hysteretic behavior of the states. There is no observation of large hysteretic behavior of NS-XBs. Finally, a model is assumed that in the X-ray emitting region of the accretion disk, the power-law hard state corresponds to the advection-dominated accretion disk and the power-law soft state corresponds to the cooling-dominated accretion disk, and this model is compared with observations and discussed. There are problems to explain the observations by the model. Magnetic field seems to play an important role to explain the observations.

[‡]**Emeritus Professor of Osaka University**
[§]**Present address: Kiyoshikojin 1-2-30-613, Takarazuka 665, Japan**
***S. Kato et al. (eds.), Physics of Accretion Disks, 139–146.***
**© 1996 OPA (Overseas Publishers Association) Amsterdam B.V.**

## 1. Introduction

The black-hole candidate X-ray binaries (BHC-XBs) have been known to have two states; the high state and the low state (for reviews see, for example, Oda 1977; Liang & Nolan 1984; Tanaka 1989). Figure 1 shows schematic summary of the X-ray energy spectra in these two states of the BHC-XBs and the non-pulsating neutron star X-ray binaries (NS-XBs) together with their short term variabilities (Miyamoto 1993, 1994). In the high state X-rays consist of two energy-spectral components; the disk-blackbody component and the soft power-law component. The disk-blackbody component is emitted from the accretion disk. In the low state X-ray consists of one component; the hard power-law component, which has a power-law shape in the X-ray energy region. The photon index of the power-law component is about 2.2–2.7 in the high state, and about 1.6–1.7 in the low state.

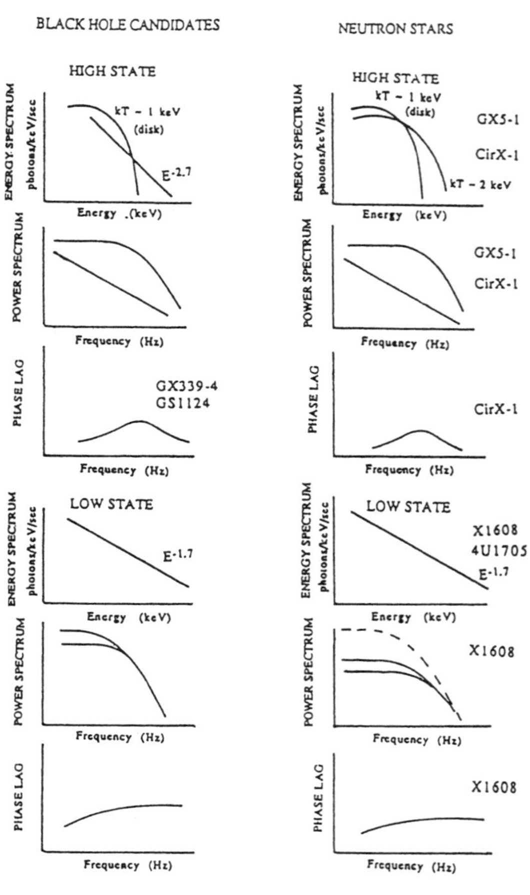


Fig. 1. A schematic summary of the energy spectra and the short term variability of X-rays from BHC-XBs and NS-XBs (Miyamoto 1993, 1994).

## 2. Short Term Variability of X-Ray Components

To see short term variability of X-rays, the Normalized Power Spectrum Density (NPSD), the power-spectrum density normalized to the X-ray intensity, has been used (Miyamoto et al. 1991). The NPSDs of BHC-XBs in the high state consist of two components: the flat-top noise and the power-law noise. The flat-top noise is due to the soft power-law component and the power-law noise is due to the disk-blackbody component. The NPSDs of these components have their own inherent shapes and values. Typical NPSDs of these components are shown in figure 2.

In the high state, on 1991 January 22, a few days after the maximum of the X-ray flux of an X-ray nova of GS1124-683, we observed a large peak in the phase lag as shown in figure 1 (Miyamoto et al. 1993). This is the same peaked phase lag observed in GX 339−4, just after the maximum of its X-ray outburst in 1988. This large peak of the phase lag corresponds to the time lag of about 0.1 s and was explained due to large hot electron clouds (Miyamoto et al. 1991). The phase lag in Nova Muscae 1991 increased gradually from 1991 January 11 to 22 (Miyamoto et al. 1994b). Almost simultaneously, on 1991 January 17, a radio burst was observed in GS 1124−683 by Ball et al. (1995), and on 1991 January 20–21, a broad Gaussian line of 474 keV (the positron annihilation line) was observed with GRANAT by Sunyaev et al. (1992).

Non-pulsating neutron star X-ray binaries (NS-XB) are believed to have a weak magnetic-field neutron star. X-ray energy spectra and short term variabilities of NS-XBs are shown in figure 1. X-ray spectra in the high intensity state have two components; the disk-blackbody component and a Comptonized-blackbody component. In the low intensity state the energy spectrum has only one component; i.e., the hard power component and the photon index is about 1.7 (Mitsuda et al. 1989). These are consistent with the observation that in the weakly magnetized NS-XBs, the brightest soft X-ray sources have no hard X-rays larger than 30 keV, while hard power-law tails extending up to about 100–200 keV (or more) are likely to arise in X-ray bursters, if they reach sufficiently low intensity state (Barret & Vedrenne 1994). It is curious that there is no X-ray component from the neutron star surface in the low intensity state.

In the high intensity state of NS-XBs, although the energy spectrum is different from that of BHC-XBs, the NPSDs are similar to those of BHC-XBs as shown in figure 1 (Miyamoto et al. 1993, 1994b). Cir X-1, a NS-XB, showed the same large peaked phase lag just after its flaring up (Miyamoto et al. 1994b). At that time, the energy spectrum of Cir X-1 was similar to those of NS-XBs in the high intensity state and the NPSDs was the flat-top type. Thus short term variabilities of X-rays from Cir X-1 are quite similar to those of BHC-XBs in the high state.

In the low intensity state of NS-XBs, the NPSDs of X 1608−522 are similar to those of the NPSDs of BHC-XBs (Yoshida et al. 1992; Miyamoto et al. 1994b). The phase lag of X 1608−522 in the low intensity state is also similar to those of BHC-XBs in the low state. The values are about a factor of two smaller than those of BHC-XBs (Miyamoto 1993, 1994).

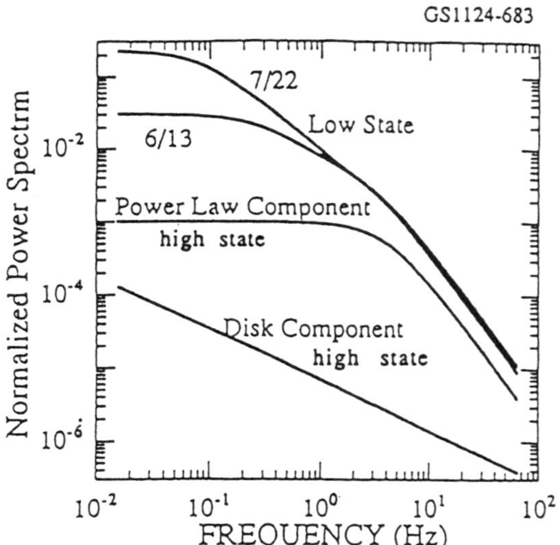


Fig. 2. Typical values of the normalized power-spectrum densities of the X-ray energy-spectral components of a BHC-XB (GS 1124−683) (Miyamoto et al. 1994a).

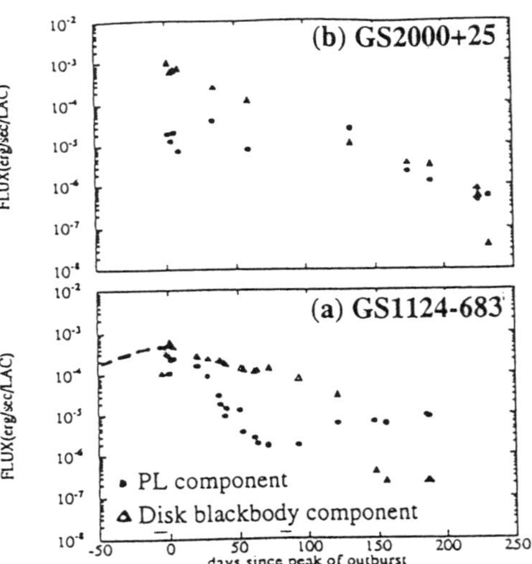


Fig. 3. Long term variability of the X-ray components. (a) Nova Muscae 1991 or GS 1124−683. The hard PL component (2–100 keV) at the initial part of the transient is added by a dashed line. (b) GS 2000+25 observed with GINGA (Miyamoto 1993; Terada et al. 1996).

## 3. Long Term Variability of X-Ray Components

Figure 3 shows long term variability of the X-ray component of Nova Muscae 1991 (GS 1124−683) and GS 2000+25 observed with GINGA (Miyamoto et al. 1993; Terada et al. 1996). Until recently, terminology of the high state and the low state has been used to represent the states of BHC-XBs. However this terminology is not suitable if we take into account X-rays up to 100 keV. GX 339−4 and Cyg X-1 increased their X-ray intensity with the X-ray energy spectrum of the low state, and near their X-ray (2–100 keV) flux maximum, these BHC-XBs changed their energy spectrum to those in the high state, and then decreased their X-ray flux. Thus these X-ray stars showed large hysteretic behavior, and we call the PL hard state and the PL soft state instead of the low state and the high state, respectively.

This large hysteretic behavior of the states was noticed for the first time by Miyamoto et al. (1995) on recent observations of GX 339−4 with BATSE in GRO and ASM in GINGA. Combining the observations of GX 339−4 and GS 1124−683, one can imagine a schematic transition of the states of BHC-X-ray transients as shown in figure 3a, where the hard PL component (2–100 keV) at the initial part of the transient is shown by a dashed line.

As for the BHC X-ray transients, two types of the rise time (the fast-rise-type and the slow-rise-type) (Harmon et al. 1994b) and three ways of transition between the two states have been observed with BATSE and GINGA. In the slow-rise-type X-ray transits increases their X-ray flux (2–100 keV) with the time constant of about 30–60 days. GX 339−4, Cyg X-1, and GRS 1915+105 (Harmon et al. 1994b) belong to this type. In the fast-rise-type X-ray transients increases their X-ray flux (2–100 keV) with the time constant of about 2–6 days. GRO J1719−24 and GRS 1009−45 are this type. Three ways of the transition between the states are as follows. The first is those which increase its X-ray flux in the PL hard state, change to the PL soft state near its X-ray flux maximum and decrease the X-ray flux and then change to the PL hard state. GX 339−4, Cyg X-1, and GRO J1719−24 (figure 4a) are this kind (Harmon et al. 1994b). The second is those which are always in the PL hard state. GS 2023+338 (Terada et al. 1991; 1994; Miyamoto 1993, 1994) and GRO J0422+32 (Harmon et al. 1994b) are this kind. The third is those which are always in the PL soft state. GRS 1009−45 (figure 4b) is this kind (Harmon et al. 1994b). It is of interest to note that there seems to be no BHC-XBs which increase their X-ray flux in the PL soft state and change the state to the PL hard state near their X-ray flux maximum.

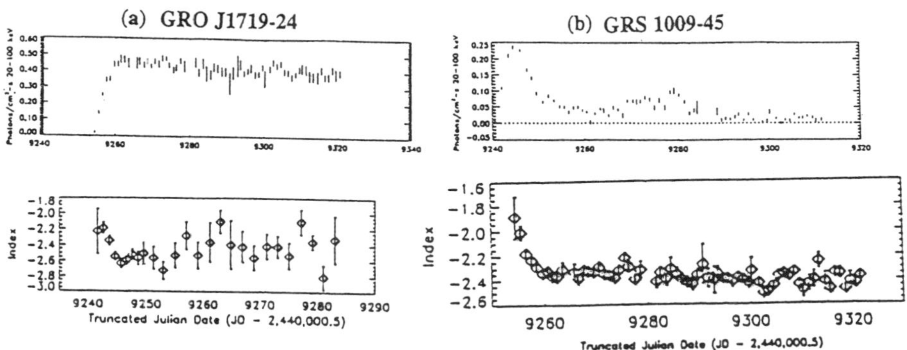


Fig. 4. Examples of fast rise type X-ray transients. (a) Fast rise transients which increase the X-ray flux in the PL hard state, change to the PL soft state near the X-ray flux maximum and decrease the X-ray flux and then change to the PL hard state (GRO J1719−24). (b) Fast transients which are always in the PL soft state (GRS 1009−45).

## 4. Comparison with the Advection-Dominated Accretion Disk Model

Let us compare long term variabilities of BHC-XBs with AD and CD disks model that in the X-ray emitting region of the accretion disk, the cooling-dominated accretion disk (CD disk) (see Frank et al. 1992 for a review) corresponds to the PL soft state and the advection-dominated accretion disk (AD disk) (e.g., Abramowicz et al. 1988, 1995; Narayan & Yi 1994, 1995a,

b) corresponds to the PL hard state. The $\log(\dot{M}/\dot{M}_{\rm Edd})$ vs $\log\Sigma$ relation of the accretion disk is shown schematically in figure 5, where $\Sigma$ is the matter density of the accretion disk and $\dot{M}/\dot{M}_{\rm Edd}$ is the ratio of the matter accretion rate to that at the Eddington limit.

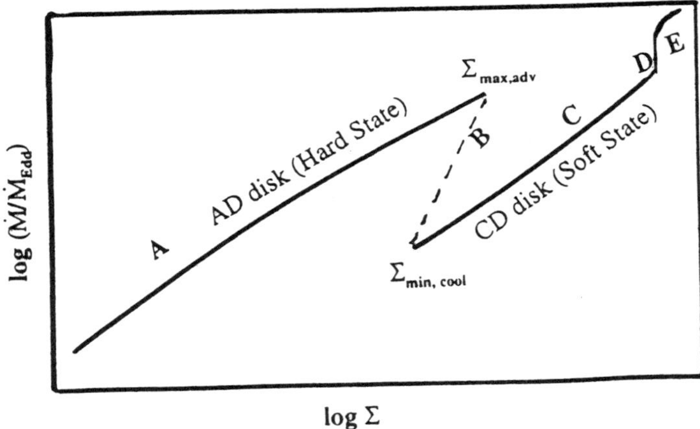


Fig. 5. Schematic relation between $\log(\dot{M}/\dot{M}_{\rm Edd})$ and $\log\Sigma$ in the X-ray emitting region of the accretion disk of the black hole.

Two rise types of the BHC X-ray transients may be explained as follows. Most of the fast-rise-type transients may be induced by the accretion disk instability. On the other hand, the slow rise transients may be induced by the instability of the companion star surface. Some of these may have a fast rise time when accreting matter falls directly to the central part of the accretion disk. One example of this may be GS 2023+338 (Miyamoto 1993, 1994).

The hysteretic behavior of BHC-XBs by the AD and CD disk model can be explained as follows. The X-ray intensity of the BHC X-ray transients increases in the AD disk, if the initial matter density of the disk is below the lower surface density limit $\Sigma_{\rm min,cool}$ (see figure 5). With increase of the accreting rate, when the surface density becomes higher than the upper density limit $\Sigma_{\rm max,adv}$, the disk changes to the CD disk. The low energy X-rays emitted from the inner edge of the accretion disk are Compton up-scattered in the hot electron clouds existed at the central part of the disk, and become the soft PL component (Miyamoto et al. 1991, 1992). When the matter density of the accreting disk decreases less than $\Sigma_{\rm min,cool}$, the CD disk changes to the AD disk, which is the start of the PL hard state. If the density of the accretion disk does not exceed $\Sigma_{\rm max,adv}$, the X-ray transients are always in the PL hard state. X-ray transients are always in the PL soft state, if these have transited to the CD disk at their high X-ray intensity, and since then the density of the accretion disk has not decreased less than $\Sigma_{\rm min,cool}$.

We define the parameters $f_{\rm AD}$ and $f_{\rm CD}$, the fractions of the gravitational energy which are advected with the flow onto the black hole and not emitted as luminosity in the AD disk and the CD disk respectively. The value of $f_{AD}$ is expected to be almost one, and $f_{CD}$ is small. We also define the parameters $A_{AC}$ and $A_{CA}$, the multiplication factors of the accretion rate after the transitions from the AD disk to the CD disk and from the CD disk to the AD disk, respectively. If the CD disk changes to the AD disk, the luminosity will change by a multiplication factor of $A_{CA}(1 - f_{AD})/(1 - f_{CD})$. In the reverse, the luminosity will change by the factor of $A_{AC}(1 - f_{CD})/(1 - f_{AD})$. From observed results of the long term variability of X-ray binary stars, luminosities corresponding to the values of $\Sigma_{max,adv}$, $\Sigma_{min,cool}$, and the values of $f_{CD}$, $f_{AD}$, $A_{CA}$, $A_{AC}$ was estimated.

In the case of GS 1124−683 (see figure 3a), the transition from the PL soft state to the PL hard state occurred at the luminosity of about a factor of 100 less than the Eddington limit in the CD disk. The value of $A_{CA}(1-f_{AD})/(1-f_{CD})$ is estimated to be about 0.2. In the case of GS 2000+25 (figure 3b), the transition occurred at the luminosity of about a factor of 10000 less than the Eddington limit in the CD disk, and the value of $A_{CA}(1 - f_{AD})/(1 - f_{CD})$ is estimated to be about 0.4.

The transition from the PL hard state to the PL soft state occurred at the luminosity of $7 \times 10^{-8}$ erg s$^{-1}$ cm$^{-2}$ in the case of Cyg X-1 (Ling et al. 1987) and $1.1 \times 10^{-8}$ erg s$^{-1}$ cm$^{-2}$ in the case of GX 339−4 (Maejima et al. 1984; Miyamoto et al. 1995). Thus the luminosities which correspond to the density $\Sigma_{max,adv}$ in the AD disk are about a factor of 24 less than the Eddington limit luminosity ($L_{Edd}$) in both Cyg X-1 and GX 339−4. The X-ray luminosities did not change largely before and after these transitions. Thus the value of $A_{AC}(1 - f_{CD})/(1 - f_{AD})$ is of the order of one at the transition from the AD disk to the CD disk at $\Sigma_{max,adv}$.

In the NS-XBs case, the transition from the high intensity state (CD disk) to the low intensity state (AD disk) and its reverse were observed in X 1608−522 (Mitsuda et al. 1989) and in 4U 1705−44 (Langmeier et al. 1987), respectively. The X-ray energy spectra of these NS-XBs are the two components thermal type in the high intensity state. Transitions in both directions occurred when the X-ray luminosity was about $10^{37}$ erg s$^{-1}$, i.e., about 1/10 of $L_{Edd}$, and there seems to be no large hysteretic behavior such as observed in BHC-XBs. In the low intensity state, the energy spectrum has only one component: the hard power component and there is no X-ray component from the neutron star surface. Thus $f_{AD}$ seems to be quite small, and $A_{AC}$ in BHC-XBs seems to be of the order of one.

Can we explain the quite small value of $f_{AD}$ with the existing AD disk model? Why is the neutron star surface component not observed in the AD disk of the NS-XBs, if $f_{AD}$ is about one? Can we explain short term variability by the AD and CD disk model? Why does the disk-blackbody component increase its X-ray flux after the X-ray energy spectrum has changed from the PL hard one to the soft one? The large peaked phase lags observed in

BHC-XBs and NS-XBs can be explained by large hot electron clouds, which are produced by rapid rotation of the magnetic field trapped in the CD disk (Miyamoto & Kitamoto 1991). Why does the magnetic field in the AD disk not play a important role to solve above questions?

## References

Abramowicz M. A. et al. 1988, ApJ 332, 646
Abramowicz M. A. et al. 1995, ApJL 438, L37
Ball et al. 1995, MNRAS 273, 722
Barret D., Vedrenne G. 1994, ApJS 92, 505
Frank J., King A. R., Raine D. J. 1985, Accretion Power in Astrophysics (Cambridge University Press, Cambridge)
Harmon B. A. et al. 1994a, ApJL 425, L17
Harmon B. A. et al. 1994b, AIP Conf 304, 210
Langmeier et al. 1987, ApJ 323, 288
Liang E. P., Nolan P. L. 1984, SpaceSciRev 38, 353
Ling J. C. et al. 1987, ApJL 321, L117
Maejima Y. et al. 1984, ApJ 285, 712
Mitsuda K. et al. 1989, PASJ 41, 97
Miyamoto S. 1995, Proc. of IIAS Workshop on Mathematical Approach to Fluctuations Vol. II (held in Kyoto, Japan, 13-23 September, 1993 by International Institute for Advanced Study, Kyoto, Japan), ed T. Hida (World Scientific Publish Co., Singapore) p254
Miyamoto S. 1994, ISAS RN 548
Miyamoto S. et al. 1988, Nature 336, 450
Miyamoto S., Kitamoto S. 1989, Nature 342, 773
Miyamoto S., Kitamoto, S. 1991, ApJ 374, 741
Miyamoto S. et al. 1991, ApJ 383, 784
Miyamoto S. et al. 1992, ApJL 391, L21
Miyamoto S. et al. 1993, ApJL 403, L39
Miyamoto S. et al. 1994a, ApJ 435, 398
Miyamoto S. et al. 1994b, New Horizon of X-ray Astronomy, ed F. Makino, T. Ohashi (Universal Academy Press Inc., Tokyot), p47
Miyamoto S. et al. 1995, ApJL 442, L13
Narayan R., Yi I. 1994, ApJL 428, L13
Narayan R., Yi I. 1995a, ApJ 444, 231
Negoro H. et al. 1994, ApJL 423, L127
Oda M. 1977, SpaceSciRev 20, 757
Sunyaev R. A. et al. 1992, ApJL 389, L75
Tanaka Y. 1989, Proc. 23$^{rd}$ ESLAB Symp. on Two Topics in X-ray Astronomy, ed J. Hunt
Terada K. et al. 1991, Proc. Frontiers of X-ray Astronomy, ed Y. Tanaka and K. Koyama (Universal Academy Press, Tokyo) p323
Terada K. et al. 1994, PASJ 46, 677
Terada K. et al. 1996, PASJ in press
Yoshida K. et al. 1993, PASJ 45, 605

# Advection-Dominated Disks in Soft X-Ray Transients

Shin MINESHIGE
*Department of Astronomy, Faculty of Science, Kyoto University, Sakyo-ku, Kyoto 606-01, Japan*

**Abstract**

Black-hole X-ray transients exhibit hysteretic spectral transitions during outbursts; hard X-ray peaks generally precede soft X-ray peaks. To account for such behavior we propose a new model, in which the disk behavior is distinct across a critical radius. The outer portions suffer a thermal limit-cycle instability thereby modulating mass flow into the inner portions. In order to produce weak X-rays during the quiescence, however, the instability should be suppressed in the inner portions, probably due to small optical depths. Instead, the inner portions undergo forced bimodal transitions between the optically-thick, cooling-dominated branch, and the optically-thin, advection-dominated branch in response to variable mass input from the outer portions.

## 1. Introduction

Recent successive discoveries of black-hole X-ray transients have invoked a fundamental question as to the origin of the outbursts and of the observed complex spectral behavior during the outbursts. In typical black-hole transients, such as Nova Muscae (GS 1124–683), the outbursts began with a hard precursor, stayed in the soft state around the maximum light, and then exhibited a clear soft-hard transition in the decay, when the luminosity was 1% of the maximum (Lund 1993). Certainly, there exists a hysteretic relation between disk luminosity and the spectral states (Miyamoto et al. 1995). Interestingly, low-mass X-ray binaries containing neutron stars also exhibit soft-hard transitions when their X-ray luminosity decreases to be $L_{\rm x} \sim 10^{36}$ erg $\rm s^{-1}$ (Mitsuda et al. 1989).

*S. Kato et al. (eds.), Physics of Accretion Disks, 147–152.*
© 1996 OPA (Overseas Publishers Association) Amsterdam B.V.

According to the disk instability model (hereafter DI model), the disk becomes thermally unstable when the disk becomes too cool for hydrogen to be ionized (Cannizzo et al. 1985). The disk then spontaneously exhibits a bimodal state transition. Although the original DI model successfully reproduced the basic features of the light curves of X-ray transients (Huang & Wheeler 1989; Mineshige & Wheeler 1989), there remain two unresolved issues. First, the complex spectral behavior as mentioned above was unaccounted for. Second, the presence of weak X-ray emission from Nova Muscae during the quiescence (McClintock et al. 1995) is also difficult to explain, since the DI model predicts practically no X-ray emission during the quiescence. The DI model needs some modifications.

## 2. Distinct Local Behavior

It has been established that besides the usual optically thick (standard-type), and thin branches, there exists a distinct branch, in which advective energy transport is important (Abramowicz et al. 1995; Narayan & Yi 1995; see also articles by Narayan and by Abramowicz in this volume). Narayan et al. (1995, hereafter NMY) proposed a (radially) two-zone model for accretion disks in quiescent black-hole transients. According to their model, the inner portions at $R \lesssim (3000 - 5000) R_S$ are advection-dominated, surrounded by a standard-type disk at large radii during the quiescence. Their calculated spectra nicely reproduced the observed optical, UV, and X-ray spectra during the quiecence (cf. Wagner et al. 1994). NMY assumed that a transition from the outer to the inner parts may be induced by evaporation of disk material (cf. Meyer & Meyer-Hofmeister 1994). Honma (1996), on the other hand, calculated the global (radial) disk structure and found that such a bimodal transition is possible, if the radial thermal conduction is included. We construct a new disk-instability model based on these results.

We first illustrate in figure 1 the proposed behavior of the accretion disks in soft X-ray transients. There exists a critical radius, $R_{cr}$, separating inner and outer portions, and the disk behavior plotted in the $(\Sigma, \dot{M})$ diagrams is distinct across the critical radius. Here, $\Sigma$ is surface density and $\dot{M}$ is mass flow rate.

A key fact is that the **S** curve should disappear at small radii, $R < R_{cr}$, due probably to a small optical depth (cf. Mineshige & Wheeler 1989). The outer portions (see the right panel) spontaneously exhibit bimodal transition between Branches I and I* due to the thermal ionization instability. In the inner portions (see the left panel), on the other hand, the disk instability is suppressed due to the disappearance of the **S** shape, but instead, the **N**-shaped equilibrium curve appears (Matsumoto et al. 1985; Narayan & Yi 1995). It is interesting to note that when the mass-input rate is modulated externally, the disk will undergo forced transitions between Branch I and III,

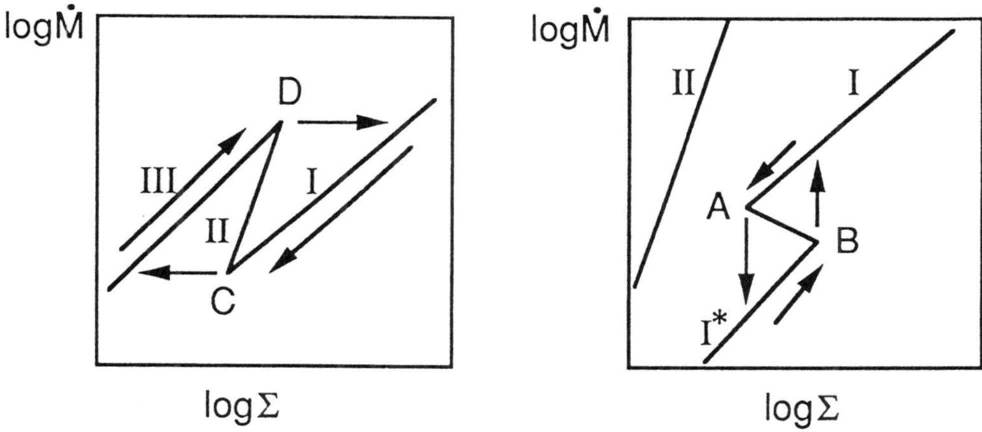


Fig. 1. Distinct disk evolution in the surface density ($\Sigma$) vs. mass-transfer rate ($\dot{M}$) diagrams in the inner (left) and outer (right) portions of the disk, respectively. There are several thermal equilibrium solutions: the optically thick, standard-type branch (Branch I), the hydrogen neutral branch (Branch I*), the optically thin, cooling-dominated branch (Branch II), and the optically thin, advection-dominated branch (Branch III). The outer portions exhibit bimodal transition between Branches I and I* for typical model parameters relevant in X-ray transients, while the inner portions undergo forced, bimodal transitions between Branches I and III in response to variable mass inflow from the outer portions.

thereby exhibiting bimodal spectral behavior (Mineshige et al. 1995). The disk staying on Branch I will produce blackbody radiation, while optically thin bremsstrahlung and the inverse Compton are main emission mechanisms on Branch III. Therefore, the disk will exhibit the hard-state spectrum during the quiescence and in the rise phase, while it will produce the soft-state spectrum in the early decay phase.

## 3. Implication for Global Evolution

Figure 2 schematically depicts the global disk evolution of black-hole X-ray transients in the rise (upper) and decay (lower) phases, respectively. The quiescence disk stays on Branch I* at $R > R_{\rm cr}$ and on Branch III at $R < R_{\rm cr}$, respectively, thereby predominantly emitting optical fluxes and hard power-law X-rays. The mass flow rate is $\min(\dot{M}_{\rm B}, \dot{M}_{\rm in})$, where $\dot{M}_{\rm B}$ and $\dot{M}_{\rm in}$ are mass-flow rate correspoinding to point B and mass-input rate, respectively. This situation is very similar to that invoked by NMY, although NMY assumed steady flow at the outer portions. Note that the calculated spectra by NMY well reproduced the observed X-ray spectra during the quiescence, which can be fit either by a blackbody or an optically thin bremsstrahlung spectrum both with a few tens of keV (Wagner et al. 1994; Verbunt et al. 1994).

As transferred mass from a companion star piles up in the disk, the outer parts of the disk move right- and upward along Branch I* in figure 1. Surface density ($\Sigma$) thus increases with the time, and when the critical value ($\Sigma$ at point B) is reached, a thermal instability is triggered. A transition to Branch I is excited locally and is propagated spatially, transforming outer parts into the hot (Branch I) state. Although $\dot{M}_{\rm in} < \dot{M}_{\rm B}$ at $R = R_{\rm out}$ (cf. figure 2), the outer edge seems also to experience an upward transition owing to the outward propagation of a heating wave, since otherwise a tidal instability, the most probable cause of superhump light variation, cannot be induced (Ichikawa et al. 1994). The difference between $\dot{M}_{\rm in}$ and $\dot{M}$ at $R_{\rm cr}$ during the quiescence should not be large, only by a factor of a few. Otherwise, the quiescent optical–UV–X-ray spectra will not be simultaneously reproduced (cf. NMY).

The upward transition induces an enhanced mass inflow into the inner parts. In the early rise phase, the inner portions are advection-dominated (Branch III) at radii for which $\dot{M} < \dot{M}_{\rm D}$, while they are cooling-dominated (Branch I) otherwise (Honma 1996). The interface moves inward as $\dot{M}$ grows (see the upper panel of figure 2). Eventually, the innermost region moves upward along Branch III in figure 1 and reach point D, where a transition to the Branch I is completed, since there are no advection-dominated branches at higher mass-input rates. Most of power then started to be emitted in soft X-rays. Theoretically, this hard-to-soft transition is expected to occur when the disk luminosity is barely below the Eddington, $L_{\rm d} \approx 0.3\alpha^2 L_{\rm Edd}$ (with

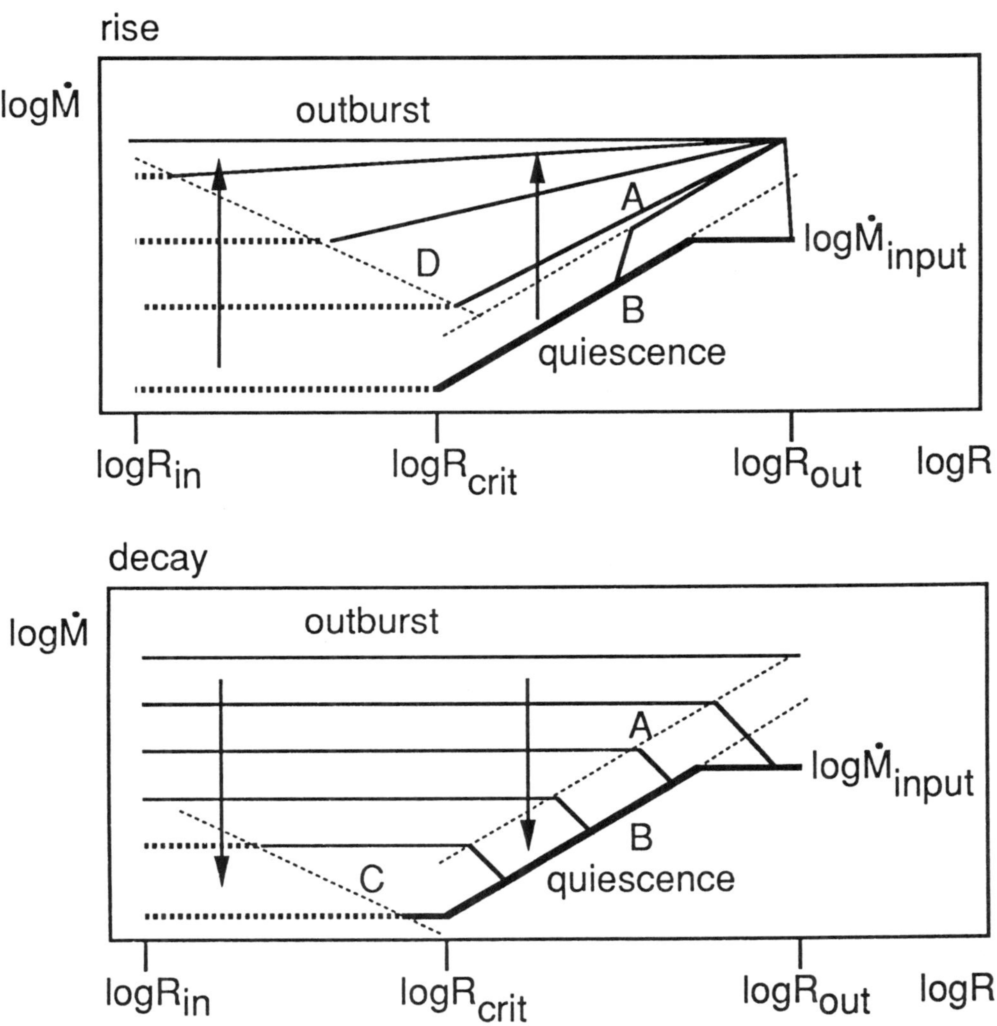


Fig. 2. The entire disk evolution in soft X-ray transients in rise (upper) and decay (lower) phases, respectively. In both panels, the thin solid, thick solid, and thick dashed lines, respectively, represent the disk being on Branch I, I*, and III. The thin dashed lines indicate the critical lines A, B, C, and D (see figure 1). The mass input rate into the disk is indicated by $\dot{M}_{\rm in}$. Note that the outer unstable portions of the disk are non-steady (in the sense that mass is accumulating) during the quiescence. The mass-inflow rate into the inner portions can be modulated by several orders of magnitude.

$\alpha$ being the viscosity parameter of order unity). This is in good agreement with the observations. Around the peak, hence, the entire disk stays on the standard branch.

After reaching the maximum, the mass-flow rate begins to decrease, as the disk material is depleted and a transition from Branch I to Branch I* is triggered successively inward in the outer portions (see the lower panel of figure 2). The inner portions then come to point C along Branch I (figure 1), where a transition to Branch III is induced, since there is no optically thick branch at lower luminosities. This soft-to-hard transition occurs at $L_d/L_{Edd} \sim 0.1 - 0.01$ observationally. This number is not easy to explain theoretically, so quantitative modeling is left as future work. More details of the present model will be discussed elsewhere (e.g., Mineshige 1996).

## References

Abramowicz M. A. 1996, in this volume

Cannizzo J. K., Wheeler J. C., Ghosh P. 1985, in Proc. Cambridge Workshop on Cataclysmic Variables and Low-Mass X-Ray Binaries, ed D. Q. Lamb, J. Patterson (Reidel, Dordrecht) p307

Honma F. 1996, PASJ in press; also in this volume

Huang M., Wheeler J. C. 1989, ApJ 343, 229

Lund N. 1993, A&AS 97, 289

Matsumoto R., Kato S., Fukue J. 1985, in Theoretical Aspects on Structure, Activity and Evolution of Galaxies III, ed S. Aoki, Y. Yoshii, (Univ. of Tokyo, Tokyo) p102

McClintock J. E., Horne K., Remillard R. A. 1995, ApJ in press

Meyer F., Meyer-Hofmeister E. 1994, A&A 288, 175

Mineshige S. 1996, PASJ in press

Mineshige S., Kusunose M., Matsumoto R. 1995, ApJL 445, L43

Mineshige S., Wheeler J. C. 1989, ApJ 343, 241

Mitsuda K., Inoue H., Nakamura N., Tanaka Y. 1989, PASJ 41, 97

Miyamoto S., Iga S., Kitamoto S., Kamado Y. 1993, ApJL 403, L39

Miyamoto S., Kitamoto S., Hayashida K., Egoshi W. 1995, ApJL 442, L13

Narayan R., McClintock J. E., Yi I. 1995, ApJ in press (NMY)

Narayan R., Yi I. 1995, ApJ 452, 710

Narayan R. 1996, this volume

Wagner R. M., Starrfield S. G., Hjellming R. M., Howell S. B., Kreidl T. J. 1994, ApJL 429, L25

# Shot Profiles of Cygnus X-1

Hitoshi NEGORO
*Institute of Space and Astronautical Science,*
*3-1-1 Yoshinodai, Sagamihara, Kanagawa 229, Japan*

## Abstract

The X-ray variability of black hole candidates in the hard (low) state is characterized by flare-like events, X-ray shots. Using *Ginga* data of Cyg X-1, average properties of the shots have been obtained by superposing a number of shots. The shots have nearly symmetrical profiles with long wings before and after the sharp peaks. The energy spectrum is softer than the average spectrum of this source, and dramatically hardens at the peak intensity. The rise profiles of the shots independent of the peak intensity and the spectral change at the peak intensity prefer a model that the shots arise from density fluctuations of accretion matter drifting to the inner region of the disk around a black hole. The advection-dominated disk theory supports this hypothesis.

## 1. Introduction

Observational studies of stellar black hole candidates (BHCs) show that the BHCs have two distinct states in the energy spectrum and time variability (e.g., Tanaka 1989). In the hard (or low) state, the energy spectrum can be roughly represented by a power law with a photon index 1.5–1.7, and the intensity largely and aperiodically varies on time scales from several tens of seconds to milliseconds (e.g., Miyamoto et al. 1992). The properties of the aperiodic time variability are hardly understood though extensive observational and theoretical studies have been carried out.

Shot-noise models have been utilized to connect the time variability to physical models quantitatively. In the shot-noise models, the time variability is assumed to consist of identical shots, which occur at random according to a Poisson distribution (Lochner et al. 1991 and references therein). It is also assumed that the shots have an exponential rise and/or decay. Their time

*S. Kato et al. (eds.), Physics of Accretion Disks, 153–158.*
© 1996 OPA (Overseas Publishers Association) Amsterdam B.V.

constants are determined by fitting, for instance, an observed power spectral density (PSD) with a model PSD derived from the shot profile assumed.

We should note, however, that the actual time variability may not consist of such shot events even if an observed PSD is successfully reproduced from some PSDs due to the shots assumed. This is mainly because the PSD has lost information on the phase components of the power spectrum (half of information on time-sequence data). As a result, properties obtained in the shot-noise models strongly depend on an assumption of the shot profile.

Recently, using *Ginga* data of Cyg X-1, *average* shot profiles have been directly obtained through superposing a number of shots by aligning their peaks. Here, the most important properties of the shots obtained with this new method (Negoro et al. 1994, 1996a; Negoro 1995) are summerized, and the origin of the shots is discussed.

## 2. Observations and Analyses

Three pointing observations of Cyg X-1 with the Large Area Proportional Counter (LAC) on board *Ginga* were carried out in 1987, 1990, and 1991. Data with the exposure time of 11.0 ksec in 1987 and 16.8 ksec in 1990 were used in this analyses. The LAC covered X-rays in 1.2–37.1 keV in 1987 and 1.2–58.4 keV in 1990. Average counting rates were $\sim 4000$ c/s/$S_{LAC}$ in 1987 and 4500 c/s/$S_{LAC}$ in 1990, where $S_{LAC}$ was the effective area of the LAC, $\sim 4000$ cm$^2$. In both the observations, Cyg X-1 was in the low state.

Figure 1 shows an example of light curves in the 1990 observation, plotted in different time scales. Highly aperiodic variations can be clearly seen. One can also notice that there are big flare-like events, called X-ray shots, with very sharp peaks in both the time scales. Note that, in general, the shots each do not have so similar profiles and not occur so frequently (Negoro et al. 1995) as shown in this figure.

To investigate average properties of such shots, the data were binned into 31.25 ms to suppress the effects of the counting statistics, and the local mean number of counts $\langle p \rangle$ was calculated at intervals of 32 sec. Next, shots were selected using the criteria that their peak number of counts $p$ should be large than 2.0–2.6 times the local mean number of counts $\langle p \rangle$ (c.f., *lower panel* in figure 1), and should have the maximum intensity within 8 sec on either side. Finally, these shots were superposed by aligning their peaks, and average profiles of the shots were obtained.

## 3. Results

The superposed shots in 1987 and 1990, obtained from shots with the peak intensities larger than $2.0\langle p \rangle$, show nearly symmetrical profiles with long wings before and after the sharp peaks (*upper panel* in figure 2). Here, increased shot components are defined as excess counts above the baseline counts and indicated by the dashed line in the upper panel of figure 2.

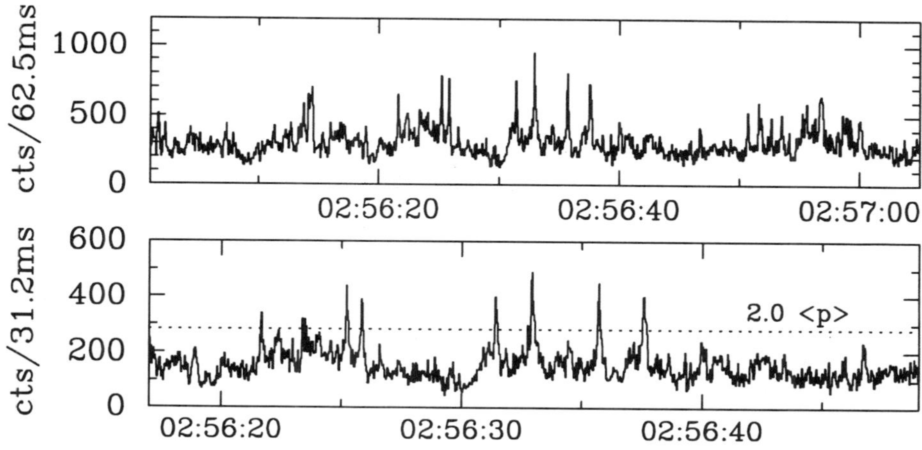


Fig. 1. Light curves of Cyg X-1 observed in 1990 with the LAC.

Each of the rise and decay profiles of the increased shot component is roughly represented by two empirical functions. However, there is no reason why the other models are ruled out. A sum of two exponential function gives much better fits to the profiles than a single exponential function. The time constants are $\sim 0.1$ s and $\sim 1$ s. Detailed analyses show that these time constants hardly change on time scales of hours or days but change by a factor of $\sim 1.5$ on time scale of years, except for the decay profiles in 1987 which also underwent changes on time scale of days.

The profiles are also described by functions of $A/(\tau - t)^{\alpha}$ in the rise phase and $A/(\tau + t)^{\alpha}$ in the decay phase, where $t$ is relative time set to 0 at the peak intensity and the other parameters are variable constants. $\tau$ and $\alpha$ obtained from the fits are about $\sim 0.05$ and $\sim 1$ in the rise phase, and $\sim 0.00$ and $\sim 0.7$ in the decay phase, respectively.

Note that the superposed shots also have a few sub-structures unable to be represented by these models, for instance, a small bump at the beginning of the shots.

Hardness ratios of the excess shot components in 1987 and 1990 are softer than the average ratio of this source, especially before the peak intensity (*lower panel* in figure 2). The ratios before the peak intensity do not change significantly. The ratios, however, dramatically harden at the peak intensity, and continue to harden for 0.2 s. After that, the ratios undergo complex, and slight changes. It should be noted that the ratios after the peak intensity are almost the same as the average ratios (dotted line).

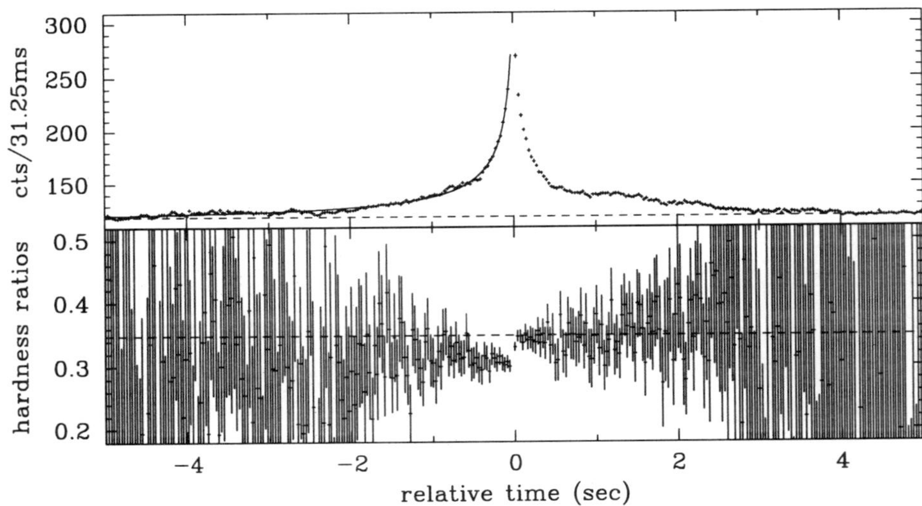


Fig. 2. (*upper*) The superposed shot in the energy band 1.2–36.8 keV in 1987. The best fitting model with the function of $A/(\tau - t)^{\alpha}$ and the average counting rate are indicated by the solid and the dashed line. (*lower*) The hardness ratios (6.9–13.8 keV/1.2–6.9 keV) of the excess shot component. The dotted line indicates the average ratio of this source.

These are no pronounced differences in the ratios in 1987 and in 1990 except for the behavior of the complex changes after the peak intensity.

Note that these properties of the shots are independent of the peak intensity except for the following two tendencies. The decay profiles depend on the peak intensity so that the asymmetry of the profiles is pronounced for bigger shots. Smaller shots tend to have hard spectra, but they are never harder than the average spectra.

## 4. Discussion

The average shot properties were obtained with no assumption of the profiles. All the results described here are free of the counting statistics (c.f., Negoro et al. 1994). Though the superposed-shot profiles are roughly represented by the sum of two exponentials, this does not mean that there are two kinds of shots with different time constants (Negoro et al. 1994). Furthermore, the obtained properties are well consistent with structures of observed PSDs and phase logs at frequencies below $\sim 1$ Hz (Negoro et al. 1996b), indicating that the time variations represented by the power spectrum at those frequencies are due to the shots just like the superposed shots.

However, one should remember that all rapid time variations with much smaller peak intensities are ignored in these analyses. Even if each shot has many sharp spikes in the rise and/or decay phase, those profiles will be smeared in superposing shots.

As for the origin of the shots, theoretical models based on the magnetic reconnections and disk instabilities have been proposed. They, however, do not explain any shot properties. Growing time scales of instabilities (e.g., Piran 1978) and reconnection time scales (Galeev et al. 1979; Pudritz et al. 1984) are much shorter than the observed ones, $\sim 1$ s. Furthermore, those time scales depend on the radius where an instability or a magnetic flare occurs. This leads to some relationships between time constants of the shots and their magnitude (peak intensity), which are not recognized in the superposed shots. The bright rotating spot model, predicting profiles similar to the superposed shots (Bao 1992), also has difficulties to explain the spectral properties and the long time scales observed (or energy generation rates of hot spots at large radii ($\sim 100 r_s$) corresponding to the long time scales).

The longest time scales observed suggest that the shots arise from dense accretion matter drifting to the inner region of an accretion disk (Rothschild et al. 1974; Miyamoto et al. 1992). In fact, such a model explains some observational properties of the shots on some assumptions that 1) the accretion rate fluctuates at some radius of the disk (c.f., Mineshige et al. 1995) and 2) the resultant dense matter drifts inward as $dr/dt = -Cr^{-\lambda}$ where $C$ and $\lambda$ are constant, and that 3) the gravitational energy is released as matter falls into the deep potential well of a black hole.

Setting the time to 0 when the matter crosses the innermost radius of the accretion disk $r_{\rm in}$ and using the second assumption, we obtain

$$r(t) = (r_{\rm in}^{\lambda+1} - (\lambda+1)Ct)^{\frac{1}{\lambda+1}}. \tag{1}$$

Using this with the third assumption (the virial theory here), the radiative power $L$ from the matter can be described as,

$$L = \frac{GMm}{2r^2}\left(-\frac{dr}{dt}\right) = \frac{2GMm}{2}\frac{1}{(r_{\rm in}^{\lambda+1} - (\lambda+1)Ct)^{\frac{\lambda+2}{\lambda+1}}}, \tag{2}$$

where $G$, $M$, and $m$ are the gravitational constant, the mass of the black hole, and the mass of the accretion matter, respectively. Taking $r_{\rm in}^{\lambda+1}/(\lambda+1)C \equiv \tau$, $(\lambda+2)/(\lambda+1) \equiv \alpha$, equation (2) is the same as the function describing the rise profile of the superposed shot. Furthermore, one can also obtain constants, $\tau$ and $\alpha$, similar to the observed ones using a standard disk parameters and assuming that, for instance, $m = m_o r^{\beta}$ $(\beta \sim 1)$[1].

Independence of $\tau$ and $\alpha$ from $m$ in this model explains why the rise profiles are independent of the peak intensity of the shots. The disappearance of the matter into the black hole at the time 0 also explains the rapid change in spectrum at the peak intensity, but not the emission after the peak intensity at all. The emission may be due to energy transferred from small to large

[1]It is essential that all the fundamental equations used are described by powers of $r$.

radii with the angular momentum (Shakura et al. 1973). The observed hard spectrum similar to the average spectrum suggests that the shots provide energy for the region where most of observed X-rays are emitted.

In the advection-dominated disk (Narayan et al. 1994; Abramowicz et al. 1995), some of the above assumptions may be justified. In that disk, if a perturbation occurs, the resultant fluctuation survives with falling into the black hole (Manmoto et al. 1995). Energy transferred from small to large radii is also expected (Narayan et al. 1994). In a similar avalanche (SOC state) model, suppression of the shot appearance after each shot event is expected (Mineshige et al. 1994), and observationally confirmed (Negoro et al. 1995). The SOC state is likely to occur in the advection-dominated disk (Takeuchi et al. 1995). The fact that the advection-dominated disk appears in low accretion rates (Narayan et al. 1995) is also consistent with observations.

The relation between the above models will be clarified by further theoretical studies and numerical simulations. Observational studies of the relationship between the luminosity and line profiles will also give us more information about the region of the emitting matter and its velocity.

**References**

Abramowicz M. A. et al. 1995, ApJL 438, L37
Bao G. 1992, A&A 257, 594
Galeev A. A., Rosner R., Vaiana G. S. 1979, ApJ 229, 318
Lochner J. C., Swank J. H. Szymkowiak A. E. 1991, ApJ 375, 295
Manmoto T. et al. 1996, this volume.
Mineshige S., Takeuchi M., Nishimori H. 1994, ApJL 435, L125
Mineshige S., Kusunose M., Matsumoto R. 1995, ApJ 445, 43
Miyamoto S. et al. 1992, ApJL 391, L21
Narayan T., Yi I. 1995, ApJL 428, L13
Narayan T., Yi I. 1995, ApJ in press
Negoro H., Miyamoto S., Kitamoto S. 1994, ApJL 423, L127
Negoro H. et al. 1995, ApJL 452, L49
Negoro H. 1995, Ph.D. thesis, Osaka Univ.
Negoro H. et al. 1996a, in preparation
Negoro H. et al. 1996b, in preparation
Piran T. 1978, ApJ 221, 652
Pudritz R. E., Fahlman G. G. 1982, MNRAS 198, 689
Rothschild R. E. et al. 1974, ApJL 189, L13
Shakura, N., Sunyaev, R. 1973, A&A 24, 337
Takeuchi M., Mineshige S., Negoro H. 1995, PASJ 47, 617
Tanaka Y. 1989, in 23rd ESLAB Symposium (ESA SP-296), ed. J. Hunt, B. Battrick (Noordwijk: ESA), Vol. 1, p3

# X-Ray Fluctuations from Black Hole Objects: Disk in a Self-Organized Criticality

Mitsuru TAKEUCHI and Shin MINESHIGE
*Department of Astronomy, Faculty of Science, Kyoto University, Sakyo-ku, Kyoto 606-01, Japan*

**Abstract**

It is well known that X-ray emissions from black hole candidates have aperiodic fluctuations with $1/f$-like power-spectral densities (PSDs), where $f$ is a frequency. To account for such intriguing features, we propose a self-organized criticality (SOC) model of accretion disk. With only one free model parameter, our model can reproduce various observational data of Cyg X-1.

## 1. Procedure of SOC Model

To explain the observed rapid X-ray variability, it is essential that physical quantities in the accretion disk vary rapidly. We consider this picture of a black hole accretion disk in the frame work of a simple cellular automata model based on the concept of Bak et al. (1988). Procedures are as follows (Mineshige et al. 1994a, b):

(1) Divide the accretion disk surface into 64 × 64 cells in the radial and azimuthal directions.

(2) Input a mass $m$ into one randomly selected cell in the outermost ring at $r = r_1$.

(3) Search for an unstable cell, at which $M_{i,j} > M_{\rm crit}$, where $M_{i,j}$ is the mass of the cell at $(r_i, \phi_j)$ and $M_{\rm crit}$ is an *a priori* given critical mass. If there is an unstable cell, let $3m$ fall from this cell into the inner three cells as

$$M_{i,j} \to M_{i,j} - 3m \tag{1}$$

$$M_{i+1,j-1} \to M_{i+1,j-1} + m \tag{2}$$

$$M_{i+1,j} \to M_{i+1,j} + m \tag{3}$$

*S. Kato et al. (eds.), Physics of Accretion Disks, 159–162.*
© 1996 OPA (Overseas Publishers Association) Amsterdam B.V.

$$M_{i+1,j+1} \to M_{i+1,j+1} + m. \tag{4}$$

For simplicity, we now assume that $M_{\rm crit} \propto r$. We call this process an "avalanche." In addition to an avalanche, we perform another process of mass accretion. Select one cell in each ring randomly and let a mass $m'$ fall into the inner cell as

$$M_{i,j} \to M_{i,j} - m', \quad \text{and} \quad M_{i+1,j} \to M_{i+1,j} - m'. \tag{5}$$

We calculate models with $m' = m/100, m/10$, and $m/5$. This process is called a "gradual mass flow," and should be accomplished regardless of $M_{i,j}$. Here, we do not specify the mechanism for an avalanche; it could be a Parker instability, magnetic recconection, and so on. And a gradual mass flow corresponds to usual mass accretion by viscousity.

(4) Assuming that a fraction of the dissipated gravitational energy turns to radiation, calucalate the energy dissipated by mass accretion. Repeat (1)-(4) in more than $10^5$ time steps, make a light curve, and calculate a PSD.

## 2. Results

The results obtained by the SOC model are in good agreement with the following observations of Cyg X-1 (cf. Takeuchi et al. 1995).

We first display the light curves (left) and PSDs (right) in figure 1, respectively, for a model with $m' = m/10$. As gradual mass flows increase, the PSD becomes flatter in the high frequency region. In case of $m' = m/10$, especially, we find PSD $\propto 1/f^{1.5}$, which agrees most with the observation of Cyg X-1.

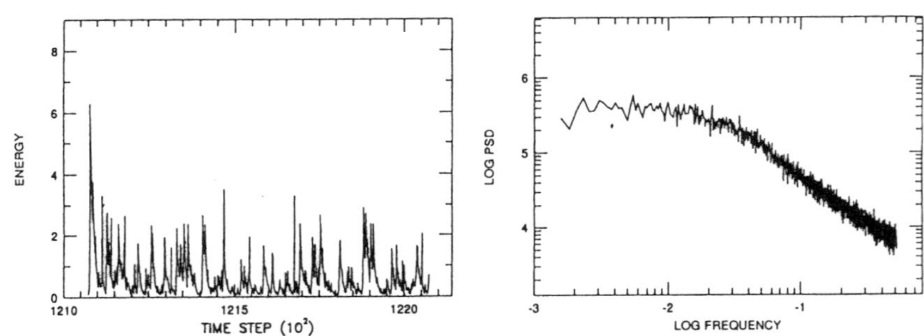


Fig. 1. Light curve (left) and PSD (right) of the SOC model with $m' = m/10$.

Second, we calculate the histograms of shot intervals for shots with similar values of the peak intensities, $p$. In figure 2, we plot the distributions of shots with $p \geq 0.50\langle p\rangle$ and $1.25\langle p\rangle$ in the left and right panels, respectively, where $\langle p\rangle$ is the averaged peak intensity. The open circles represent $F/F_{\rm poi}$, where $F$ is the shot frequency obtained by our simulations and $F_{\rm poi}$ is the frequency of the Poisson distribution, which is the expected distribution if shots would occur randomly. The error bars correspond to $1\sigma$. Figure 2 clearly shows that small shots tend to occur randomly and large shots occur after some waiting time. It is considerable that, in case of larger shots, a large amount of gas is dissipated by a large event and a longer time is needed to refill an accretion disk with gas untill the next large event takes place. A similar tendency is also found in the observational data (Negoro et al. 1995). So we explain that there should be "reservoirs" of mass on an accretion disk.

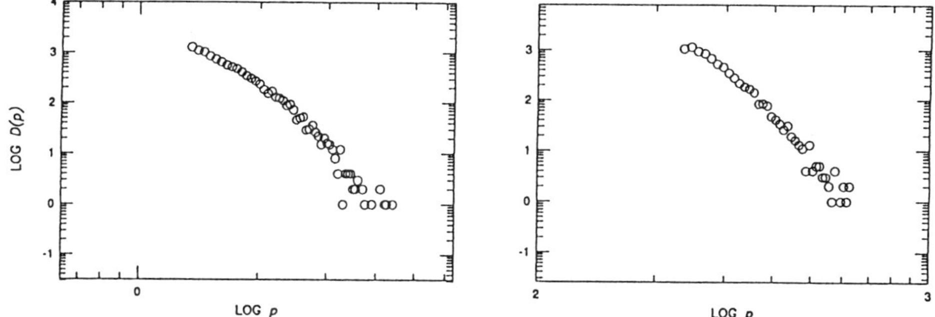


Fig. 2. Theoretical shot interval distributions, the frequency normalized by the frequency of the Poisson distribution, for shots with $p \geq 0.50\langle p \rangle$ (left) and $1.25\langle p \rangle$ (right). Here $\langle p \rangle$ is the time avarage of $p$.

Thrid, we illustrate the distribution of the peak value of shots, $D(p)$, in figure 3, using the theoretical model with ($m' = m/10$) in the left panel, and using the observational data in the right panel. Theoretically, we find $D(p) \propto p^{-7.7}$, in good agreement with the observational one, $\propto p^{-7.1}$.

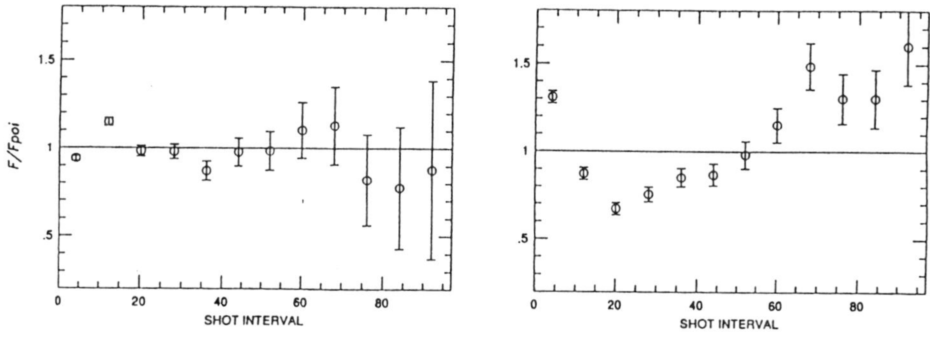


Fig. 3. Theoretical ($m' = m/10$ ; left) and observational (right) peak intensity distribution of shots.

We also investigate where and how often avalanches are triggered, which is shown in figure 4. Obviously, avalanches tend to occur at any radius when a gradual mass flow is not so large, while they tend to occur predominantly at outer radii , when a gradual mass flow increases.

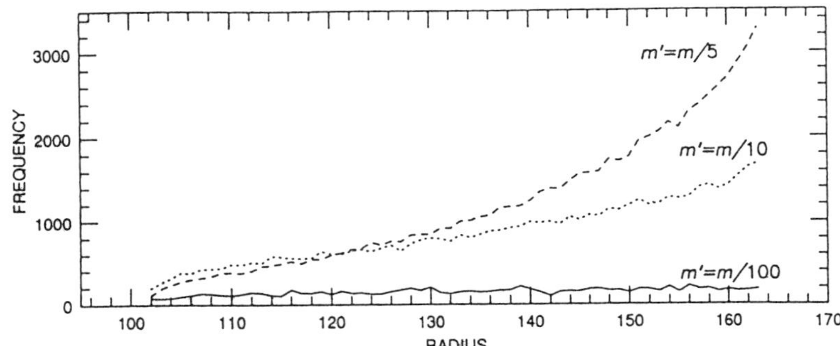


Fig. 4. Spatial distribution of the starting radii of avalanches. Solid, dotted, and short dash curves represent the cases with $m' = m/100$, $m/10$, and $m/5$, respectively.

## 3. Discussion

From a series of analysis using SOC model, we can get following three necessary conditions for X-ray fluctuations:

(1) Gas flow is two-dimensional. In the SOC model, it is essential that mass accretion takes place both in the radial and azimuthal directions [cf. equations (1)-(4)] to produce X-ray fluctuations as are observed.

(2) Physical quantities do not vary smoothly in the accretion disk. We assume that there are a lot of blobs on accretion disk surface.

(3) There should be mass reservoirs in the disk to explain the tendency seen in figure 2. We must, therefore, assume some critical conditions.

It is, however, difficult for the standard-type accretion disk by Shakura & Sunyaev (1973) to satisfy these requirements, since then gas flow is one-dimensional with $v_r \ll v_\phi$, where $v_r$ and $v_\phi$ are the radial and azimuthal component of the gas velocity. The advection-dominated accretion disk model is more favored to produce X-ray fluctuations than the standard-type disk, since the former is locally thermally unstable (see Manmoto et al. in this volume for time evolution of such disks). More detailed studies concerning the advection-dominated disk are needed as our future work.

## References

Bak P., Tang C., Wiesenfeld K. 1988, PhysRev A38, 364
Mineshige S., Ouchi N. B., Nishimori H. 1994a, PASJ 46, 97
Mineshige S., Takeuchi M., Nishimori H. 1994b, ApJL 435, L125
Negoro H., Kitamoto S., Takeuchi M., Mineshige S. 1995, ApJL 452, L49
Shakura N. I., Sunyaev R. A. 1973, A&A 24, 337
Takeuchi M., Miheshige S., Negoro H. 1995, PASJ 47, 617

# The Luminosity and Variability Amplitude Anticorrelation due to Accretion Disk Fluctuations

Gang BAO[1] and Marek ABRAMOWICZ[2]
*1. INFN Section of Padova; Department of Physics University of Padova, I-35131 Padova, Italy*
*2. Department of Astronomy and Astrophysics, Göteborg University and Chalmers University of Technology, S-412 96 Göteborg, Sweden*

**Abstract**

In this paper, a summary is given of our recent work (Bao & Abramowicz 1995) on the observed anticorrelations of the variability amplitude with the X-ray luminosity of active galactic nuclei (AGN). It is demonstrated that the bright spot model for the short term X-ray variability of AGN predicts that, statistically, sources with larger luminosities should have smaller variability amplitudes.

## 1. Introduction

The observed chaotic X-ray and UV lightcurves of both active galactic nuclei and galactic black hole candidates as well as the theoretical investigations suggest that the innermost parts of accretion disks around black holes are patchy and noisy. The accretion disk fluctuations can be well approximated as a group of "spots" orbiting the central black hole. The observed variability is then due mostly to the intrinsic "flickering" of the spots when the accretion disk is seen nearly face-on, and due to the strong periodic modulation of Doppler and general relativistic effects when the disk is seen at some high inclination angle. Research on the bright spot model was initiated by Abramowicz et al. (1989, 1991) and independently by Wiita et al. (1991).

Lawrence & Papadakis (1993) have analysed 12 high quality long–look observations of AGN from *EXOSAT* database. They have found that, statistically, slopes of the variability power spectra of all objects in the sample are close to a universal number $\alpha = 1.55$. They conclude that this is inconsistent with either the standard shot-noise process or the traditional 1/f noise,

*S. Kato et al. (eds.), Physics of Accretion Disks, 163–166.*
© 1996 OPA (Overseas Publishers Association) Amsterdam B.V.

but remarkably close to the predictions of the rotating bright spot model of Abramowicz et al. Lawrence & Papadakis have shown that the power spectrum amplitude shows large scatter from object to object, but varies systematically with the observed X-ray luminosity as $A \sim L_{\mathrm{x}}^{-0.5}$. This confirms earlier qualitative claims by many authors (e.g., McHardy 1988) that on average less luminous AGN vary more strongly.

Our present paper demonstrates that, apart from the universal power spectrum $p \sim f^{-1.55}$, the rotating hot spot model predicts the the amplitude-luminosity anticorrelation, which is in agreement with what was found in the *EXOSAT* high-quality data by Lawrence & Papadakis.

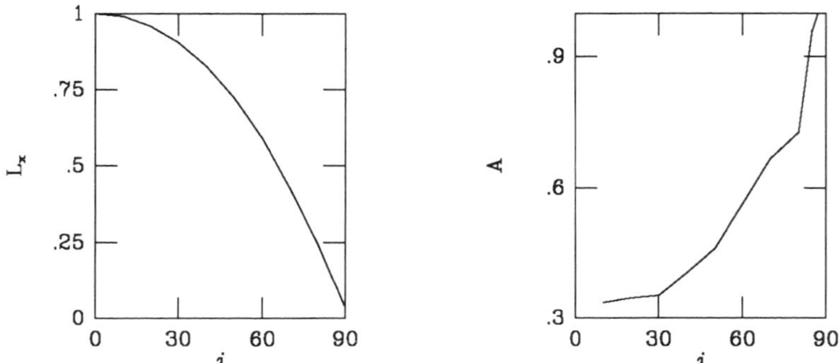


Fig. 1. Left: the apparent luminosity versus inclination of the disk. The intensity is normalized to unity, and the unit for inclination $i$ is in degree. Right: the variability amplitudes of a given source seen at different inclinations. The amplitude is normalized to unity, and the inclination $i$ is expressed in degrees.

## 2. Amplitude-Luminosity Anticorrelation

For simplicity, we assume that X-ray emission comes from the innermost part of an accretion disk around a Schwarzschild black hole. The energy flux observed in the rest frame of the disk at radial coordinate $r$ (or $R$, $R = r/M$) is

$$F_d = f(M; \dot{M}) R^{-3} a^{-1/2} b, \tag{1}$$

with

$$a = 1 - \frac{3}{R} \tag{2}$$

and

$$b = \frac{1}{\sqrt{Ra}} \left[ \sqrt{R} - \sqrt{6} - \frac{\sqrt{3}}{2} \left( \ln \frac{\sqrt{R} - \sqrt{3}}{\sqrt{6} - \sqrt{3}} - \ln \frac{\sqrt{R} + \sqrt{3}}{\sqrt{6} + \sqrt{3}} \right) \right] \tag{3}$$

where $f$ is a function of the mass of the black hole $M$ and the accretion rate $\dot{M}$. The observed X-ray luminosity of the disk at infinity is

$$L_{\rm d} = \int^{\rm disk} F_{\rm d} g^4 ds', \tag{4}$$

where $ds'$ is the area element observed at infinity, $g$ is the red shift factor containing both Doppler and gravitational effects. The luminosity from $N$ spots on the disk is

$$L_{\rm s} = \sum^{N} \int^{\rm spot} I(R) g^4 ds', \tag{5}$$

where $I(R)$ is the flux of a spot measured in its rest frame. So, the total X-ray luminosity is the summation of the disk and the spots (see figure 1, left), i.e.,

$$L = L_{\rm d} + L_{\rm s}. \tag{6}$$

Figure 1 (right) shows the variability amplitude $A$ for a given source seen at different inclinations. The amplitude $A$ is defined as the power at variability frequency $10^{-4}$ Hz for a black hole with mass $M = 10^8 M_\odot$. We can clearly see from the figure that for a disk with a given intrinsic luminosity, the variability amplitude is strongly anticorrelated with the observed luminosity, i.e., the larger the observed luminosity, the smaller is the variability amplitude.

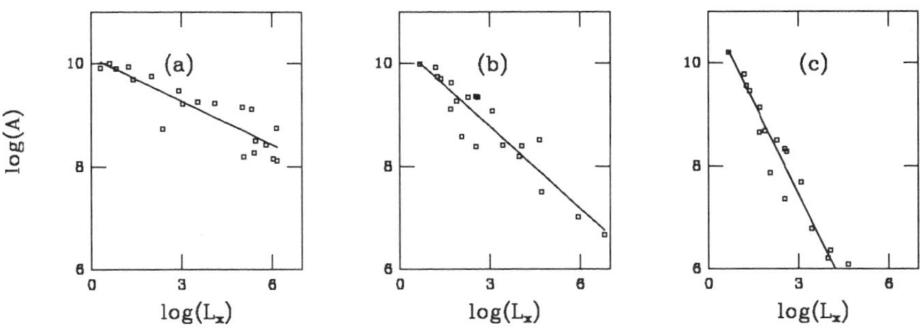


Fig. 2. The variability amplitudes versus the apparent luminosities of 20 sources. The (a), (b), and (c) show the cases for $\beta$ = 0.1, 0.2, and 0.5 respectively.

Figure 2 exhibits a simulation containing 20 sources, with their intrinsic luminosities spanning six decades. In the simulation we assume that there are 100 spots on the disk whose lifetime is eight times that of their local Keplerian periods. They are distributed randomly in their phases but follow a uniform distribution in radius. The inclination $i$ ranges from 85° to 5°, and a noise of 0.1% of the maximum signal is included. Both the intrinsic luminosity and the inclination are randomly chosen. We introduce a parameter contrast $x$, which is defined as the ratio of the intensity of spots to the intensity of disk, and the relation of contrast $x$ with the luminosity $L$ satisfies a power-law, $x \sim L^{-\beta}$. This parameter measures the difference in emission between the spots and the disk. Figure 2a, 2b, and 2c show the cases for $\beta = 0.1$, 0.2, and 0.5. Lawrence & Papadakis (1993) have shown that the variability amplitude varies systematically with the observed luminosity as $A \sim L_{\mathbf{x}}^{-0.5}$, and for this reason the results of our simulations are fitted to a power-law. The best fit gives the slope 0.30 ($\beta = 0.1$), 0.52 ($\beta = 0.2$), 1.19 ($\beta = 0.5$), respectively. If 0.5 is indeed the universal slope, then our fittings indicate that contrast index should be slightly smaller than 0.2.

## 3. Conclusions

The bright spot model very naturally explains the amplitude-luminosity correlation. According to the model, sources with smaller luminosities statistically show larger variability amplitudes. This is a robust prediction of the model and agrees very well with the observations.

## References

Abramowicz M. A., Bao G., Lanza A., Zhang X.-H. 1989, in Proceedings of the 23rd ESLAB Symposium, eds J. Hunt, B. Battrick (Noordvijk, ESA)

Abramowicz M. A. et al. 1991, A&A 245, 454

Bao G., Abramowicz M. A. 1995, ApJ submitted

Lawrence A., Papadakis I. 1993, ApJL 414, L85

McHardy I. M. 1989, in Proceedings of the 23rd ESLAB Symposium, eds J. Hunt, B. Battrick (Noordvijk, ESA)

Wiita P. J. et al. 1991, in Structure and Emission Properties of Accretion Disks, 6th I.A.P. Meeting/IAU Colloq. 129, eds C. Bertout et al. (Gif-sur-Yvette: Editions Frontières), p557

# Disk-Coronal Accretion Disk Model for Black Hole Candidates

Kaori NAKAMURA and Yoji OSAKI
*Department of Astronomy, School of Science, The University of Tokyo, Bunkyo-ku, Tokyo 113, Japan*

**Abstract**

A disk-coronal accretion disk model is considered for the black hole candidates to explain their X-ray spectra. In this model an optically thick disk is surrounded by a hot optically-thin corona. The disk emits the blackbody radiation which appears in the soft X-ray band while the corona produces the hard power-law X-ray as the results of inverse Compton scattering of the soft photons from the disk. Two cases are considered for the electron distribution in the corona: (1) thermal Maxwellian distribution and (2) a nonthermal power-law distribution. It is found that our model can reproduce X-ray spectra of the "high" state in the case of the nonthermal electron distribution.

## 1. Introduction

The black hole candidates are those galactic binary X-ray sources which are supposed to have a black hole as a compact star and X-rays are supposed to be produced in an accretion disk around the black hole.

In this paper we study X-ray spectra of black hole candidates. Their X-ray spectral features are characterized by two components: the soft and hard components. The soft component is thought to be produced by a superposition of blackbodies from accretion disk whose maximum color temperature rages from 0.4 keV to 1.2 keV. On the other hand, the hard component is approximated by the power-law spectrum whose energy index ranges from 0.7 to 1.5. Black hole candidates often show time variability and they often show two brightness levels called "high state" and "low state". There exists some correlation between the brightness level and the spectral features. In the high state the soft component is dominant and the spectral slope of the

*S. Kato et al. (eds.), Physics of Accretion Disks, 167–170.*
© 1996 OPA (Overseas Publishers Association) Amsterdam B.V.

hard component is steeper. In the low state the hard component is dominant and its spectral slope is flatter (Ebisawa et al. 1994).

To explain the X-ray spectral features of black hole candidates, we need a model that can produce both the thermal soft X-ray and the power-law hard tail. The thermal soft X-ray is usually supposed to be produced by the optically-thick accretion disk. As for the hard X-ray tail, Shapiro et al. (1976) was the first to propose the optically-thin hot disk for a model of Cyg X-1. In their model, the hot disk is situated inside the optically-thick cool disk in the radial direction. However, another model is possible in which a hot optically-thin region is separated in the vertical direction from the optically thick disk: a disk-coronal model. Here we use the disk-coronal model. The reason why we adopt the disk-coronal model is that there is observational evidence (Ebisawa et al. 1993) that the optically thick disk seem to extend to the very inner edge of the black hole accretion disk. Furthermore the detailed X-ray spectra of the hard power-law component indicate the existence of the so-called reflection component by the cool optically-thick material by which the hard X-ray seem to be intercepted with solid angle larger than $2\pi$. Such a configuration is not possible with such a model by Shapiro et al. (1976) in which the optically-thin hot disk is separated radially from the cool optically-thick disk but the disk-coronal model can satisfy this condition.

## 2. Disk-Coronal Model

Our model is an accretion disk model in which an optically thick disk is surrounded by an optically-thin, hot corona. Such a disk-coronal model was investigated for active galactic nuclei by Haardt & Maraschi (1991) and Nakamura & Osaki (1993). In our disk-coronal model, the disk is assumed to be optically thick to emit mainly soft X-rays in the form of blackbody radiation. On the other hand, the corona is assumed to be optically thin and most of the soft X-rays from the disk therefore pass through the corona without being scattered. However, a small part of the soft X-rays are inverse Compton scattered in the corona to produce the power-law spectrum at hard X-rays. A part of these hard X-rays may be used for producing electron-positron pairs. The half of the remaining hard X-rays are returned back to the disk and they are partly reflected and partly absorbed in the disk to heat up the disk.

The present model is basically a continuation of the previous model of Nakamura & Osaki (1993) in which the disk and coronal structures were solved self-consistently and emitted spectra were then calculated. However, some modifications are made to the previous model in the treatment of the energy release and the pressure balance between the disk and the corona.

As for the energy release in the disk and the corona is concerned, we assume that a fraction $f$ of energy is released in the corona and the remaining fraction of $1-f$ is released in the disk where $f$ is a model parameter in our model (see Svensson & Zdziarski 1994). As for the energy release in the disk, we use $\alpha$ prescription for the viscosity, where we assume that the viscous dissipation is proportional to the gas pressure, not to the total pressure.

The pressure of the corona is assumed to be equal to the surface pressure of the disk and the surface pressure is assumed to be $1/\tau$ times as large as the central pressure of the disk. Here, $\tau$ is the total optical depth of the disk.

As for the electron distribution in the corona, two cases are considered: one case of the thermal distribution and the other case of the non-thermal power-law distribution . In the case of the non-thermal power-law distribution, the power-law index is assumed to be 3 and 2.5, and the cut off energy is assumed to be $100m_ec^2$ and $10m_ec^2$.

## 3. Results and Conclusion

Once the mass of the black hole, the mass accretion rate, viscous parameter $\alpha$, and parameter $f$ are specified, all physical quantities such as surface densities of the disk and of the corona, temperatures of the disk at midplane and of the corona, are solved in our model and they are expressed as functions of radial distance from the black hole. We have also examined the effects of pairs. In what follows, we summarize our results of calculations.

In the model including the pairs, it is found that there is no pair dominant solution in the case of thermal electron distribution. On the other hand, in the case of power-law electron distribution corresponding to a given accretion rate, two solutions are found; one solution is pair dominant and the other is the normal solution with negligible pairs. There exists a maximum accretion rate in our model in the case of power-law electron distribution. This maximum accretion rate is larger when the cutoff energy of the power-law electron distribution is smaller. The pair dominant solution obtained in our model is found not to be realistic because the ion temperature is very high and the scale-height of the corona is unrealistically large.

Figure 1 shows X-ray spectra of our model for two different $f$ values and for two cases of electron energy distributions. As seen in figure 1, the distinction of the soft component from the hard component is more apparent in the case of the power-law electron distribution than in the case of the thermal electron distribution. Since the clear break from the soft thermal spectrum to the hard power-law spectrum is evident observationally, the X-ray spectrum in the high state of the black hole candidates is reproduced better in our model by the case of the non-thermal power-law distribution rather than that of the thermal distribution. It is also apparent from figure 1 that the smaller the parameter $f$ becomes, the more the soft component is produced than the hard component.

The cut-off energy of the hard X-ray spectrum is affected by the electron temperature of the corona or the cut-off energy of the power-law electron distribution. Observations of $\gamma$-ray are important in determining which models are better in this respect.

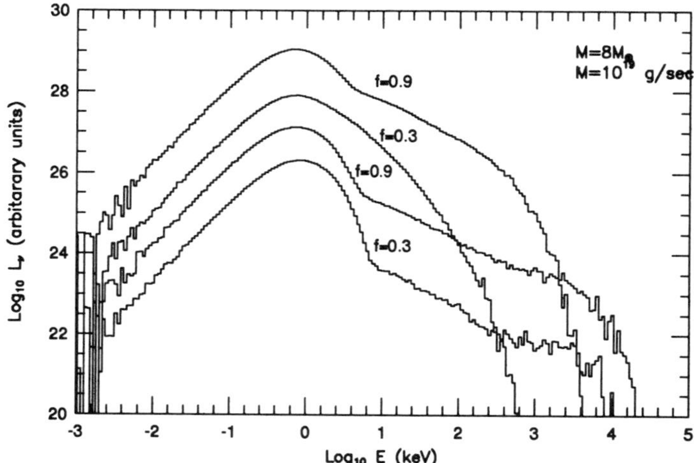


Fig. 1. X-ray spectra of our models. Two spectra above are those with two different $f$ values for the thermal electron distribution while the two spectra below are the same as above except for the power-law electron distribution in which the power-law index is 3 and the cut off energy is $100m_ec^2$.

In our model for the non-thermal power-law electron distribution, an unrealistic solution is sometimes obtained, for instance, the ion temperature becomes negative at large radii. This is because the power-law index and the cut off energy are fixed beforehand in our model. A more careful treatment is necessary and this and other improvements are a future work.

This research was partially supported by the Grant-in-Aid for Scientific Research of the Ministry of Education, Science and Culture (No. 4012).

## References

Ebisawa K. et al. 1994, PASJ 46, 375
Ebisawa K. et al. 1993, ApJ 403, 684
Haardt F., Maraschi L. 1991, ApJL 380, L51
Nakamura K., Osaki Y. 1993, PASJ 45, 775
Shapiro S. L., Lightman A. P., Eardley D. M. 1976 ApJ 204, 187
Svensson R., Zdziarski A. A. 1994 ApJ 436, 599

# Irradiation and Dwarf-Nova Type Instability in Back Hole X-Ray Novae

Soon-Wook KIM[1], J. Craig WHEELER[1], and Shin MINESHIGE[3]
*1. Department of Astronomy, University of Texas at Austin, RLM 15.308, Austin, TX 78712-1083, USA*
*2. Department of Astronomy, Kyoto University, Sakyo-ku, Kyoto 606-01, Japan*

**Abstract**

We study the effect of irradiation of accretion disks in black hole X-ray novae. The irradiated-disk-instability-model qualitatively displays the observed primary rise and, especially, the secondary reflare in X-ray nova outbursts. This implies that, unlike previously proposed models, the sudden enhancement of the mass transfer rate from the companion is not necessary to explain the phenomenon of the secondary reflare.

## 1. Introduction

The accretion-disk thermal-instability (or disk instability) model was proposed to explain outburst phenomena in dwarf novae. In this model, the accreted matter from a secondary star in a binary system is accumulated mainly in the outer portions of an accretion disk around the primary star during quiescence. As the disk density and temperature increase, a thermal instability associated with the ionization of hydrogen and helium results in a sudden heating and consequent brightening ("outburst") of the disk. The disk instability model has been very successful in explaining observations in dwarf nova outbursts (for reviews, see Osaki 1993 and references therein). It is inevitable that such an ionization instability should cause "dwarf nova-like" outbursts in other astrophysical accreting systems. For example, so-called "X-ray novae", black hole binary transients, have displayed similar outburst phenomena to dwarf nova outbursts (e.g., van Paradijs & Verbunt 1984). In this paper, we concentrate on the outburst phenomena in the black hole X-ray novae.

*S. Kato et al. (eds.), Physics of Accretion Disks, 171–174.*
© 1996 OPA (Overseas Publishers Association) Amsterdam B.V.

## 2. Observational Properties of Black Hole X-Ray Nova Outbursts

After the discovery of A0620-00 (Nova Mon 1975), the prototype of black hole binaries which show dwarf nova-like outbursts, many black hole X-ray nova have been discovered in recent years: Nova Vul 1988 (GS 2000+25), Nova Cyg 1989 (V404 Cyg; GS 2023+338), Nova Mus 1991 (GS 1124-683), Nova Per 1992 (GRO J0422+32) and Nova Vela 1993 (GRS 1009-45) (for review, see Tanaka & Lewin 1995; Kim et al. 1996b). These systems are also called "soft X-ray transients" because of the ultra-soft X-ray flux ($\lesssim$ keV). Typical black hole X-ray novae show three maxima in both X-ray and optical outbursts: (1) primary outburst maximum, (2) a secondary maximum (or "reflare"), $<$ 100 days after the primary maximum, and (3) a third broad maximum (or "bump"), a few hundred days later.

## 3. Irradiated Disk Instability in X-Ray Novae

In the model we are investigating (Kim et al. 1996b, c), two types of irradiation are considered: direct irradiation from the innermost hot disk and reflected irradiation by a corona or chromosphere above the disk. The indirect irradiation flux is given by

$$F_i(t) = C_X\Big(\frac{L_X(t)}{4\pi R^2}\Big), \tag{1}$$

where $C_X$ is a constant, $R$ is disk radius, and $L_X(t)$ is X-ray luminosity of the irradiation, which is proportional to the inner accretion rate of the disk. The direct irradiation flux is given by

$$F_d(t) = (1-A)\Big(\frac{L_X(t)}{2\pi R}\Big)\frac{d}{dR}\Big(\frac{H(t)}{R}\Big)^2, \tag{2}$$

where $A$ is X-ray albedo and $H$ is disk height. Note that hydrostatic balance implies $H/R \propto \sqrt{T}$, $T$ being disk temperature. The irradiation is, therefore, a function of the time-dependent mass accretion rate, temperature and location in the disk.

The most important aspect of this model is the "shadow effect": the dynamic effects of the disk height profile as the disk is irradiated. As a result, the outer disk is blocked from the irradiation produced in the hot disk region by the inner (in quiescence) and middle (in outburst) portions of the disk. After the outburst maximum, the irradiated disk instability causes additional heating in the shadowed, neutral outer disk, and hence the disk height in the outer disk increases. This causes a sudden exposure to irradiation in the outer disk and successfully produces the observed optical secondary reflare (Kim et al. 1996a, b).

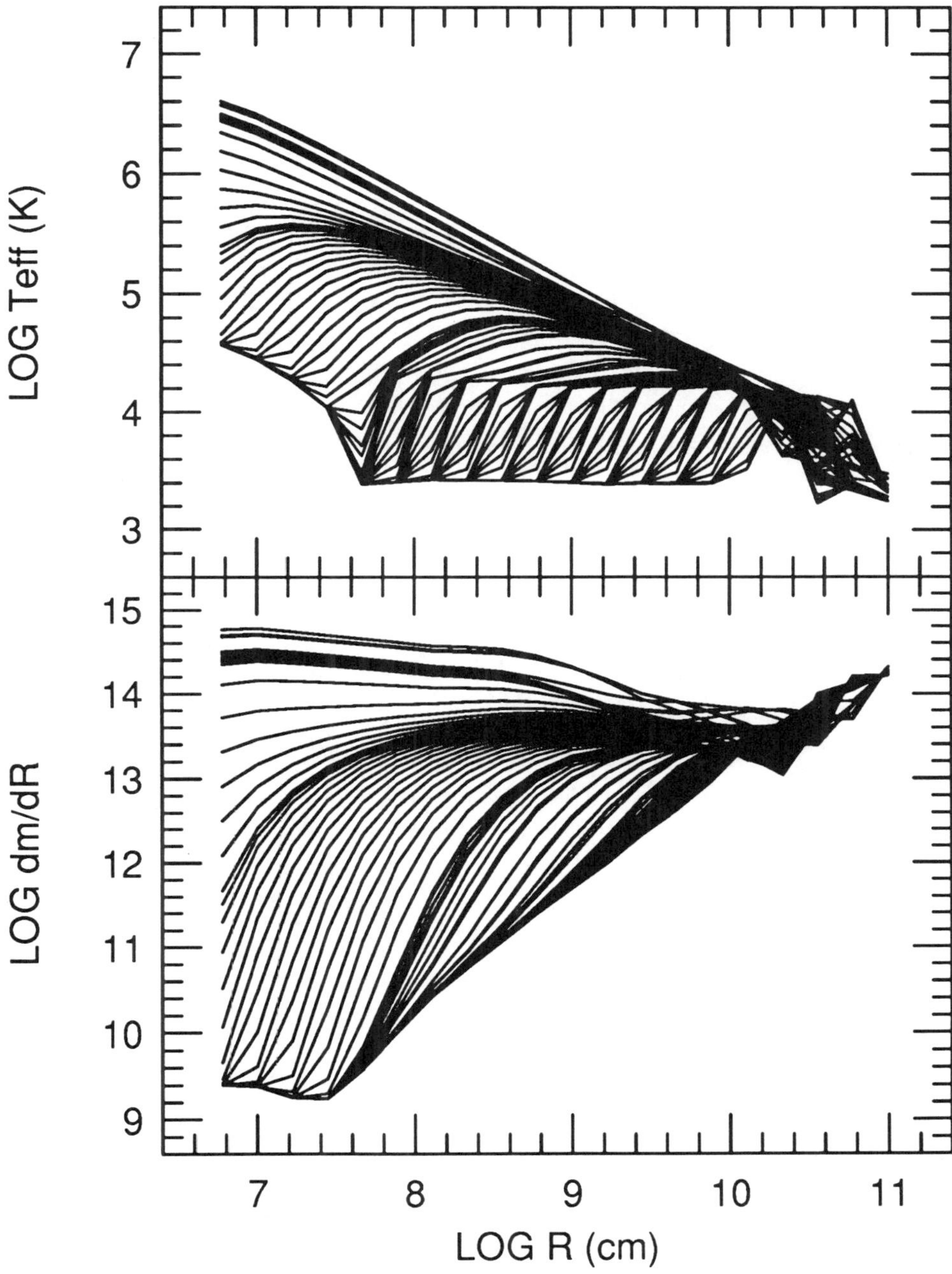


Fig. 1. The time-dependent evolution of the disk effective temperature (upper panel) and gradient of disk mass (lower panel) as a function of the disk radius for the evolution from the onset of the disk instability to the maximum light.

Figure 1 represents the time-dependent evolution of the disk from initiation to the maximum of an outburst in the model we described above. The outburst starts from the outer disk and propagates inward. Note that there are three distinctive temperature regions of the disk (see upper panel of figure 1): the inner disk is in a permanently hot, fully ionized state, with effective temperature ($T_{\rm eff}$) greater than $10^{3.85}$ K; the outermost disk is always in the cold neutral state; in between, the disk is thermally unstable. The outer disk temperature in quiescence, 1,700–3,000 K, is consistent with observations of A0620-00 and V404 Cyg (Haswell et al. 1993; Sanwal et al. 1996). In the lower panel of figure 1, we present the evolution of mass gradient in each disk ring, which is given by $dm/dR = 2\pi R\ \Sigma(R,t)$, where $\Sigma$ is the disk surface density. As an outburst starts, the mass accumulated in the outer disk is transferred inward, which results in a huge "avalanche" as shown.

The irradiated disk instability model (Kim et al. 1996b, c) successfully reproduces the observed optical primary and secondary maxima. This strongly indicates that the secondary reflare is a self-regulated disk phenomenon. This model does not require a sudden enhanced mass transfer rate from the companion star to explain the secondary reflare as in previous studies (Augusteijn et al. 1993; Chen et al. 1993).

## References

Augusteijn T., Kuulkers E., Shaham J. 1993, A&A 279, L13

Chen W., Livio M., Gehrels N. 1993, ApJL 408, L5

Haswell C. A., Robinson E. L., Horne K., Stiening R. F., Abbott T. M. C. 1993, ApJ 411, 802

Kim S.-W., Wheeler J. C., Mineshige S. 1996a, in Cataclysmic Variables and Related Objects, IAU Colloquium No. 158, in press

Kim S.-W., Wheeler J. C., Mineshige S. 1996b, in preparation

Kim S.-W., Wheeler J. C., Mineshige S. 1996c, in preparation

Osaki Y. 1993, in Cataclysmic Variables and Related Physics, ed O. Regev, G. Shaviv (Institute of Physics Publishing, Bristol), p152

Sanwal D., Robinson E. L., Zhang E., Colomé C., Harvey P. M., Ramseyer T. F., Hellier C., Wood J. H. 1996, ApJ in press

Tanaka Y., Lewin W. H. G. 1995, in X-Ray Binaries, ed W. H. G. Lewin, J. van Paradijs, E. P. J. van den Heuvel (Cambridge University Press, Cambridge), p126

van Paradijs J, Verbunt F. 1984, in High Energy Transient in Astrophysics, ed S. E. Woosley (AIP, New York), p49

# Viscous Evolution of Eccentric Disks around a Black Hole

Yasufumi KOJIMA
*Department of Physics, Hiroshima University, Higashi-Hiroshima 739, Japan*

**Abstract**

Evolution of eccentric disks is numerically calculated to study the combined effect of viscosity and post-Newtonian gravity. The viscosity is effective for smoothing the local density distribution, but ineffective for the global circularization of the flow. The initial eccentricity vanishes by the mixing of the flows originated from the relativistic differential precession.

## 1. Introduction

Flow in the standard accretion disk theory is a priori assumed to be circular. The deformation of the disk is expected to gradually vanish as the energy and angular momentum are transported under the viscosity, even if exits. Syer & Clarke (1992, 1993) critically examined the circularization process. They showed that the initial eccentricity of the disk is long lived under the Newtonian gravity, unless there is no other stirring force. Therefore, the eccentric disk may be possible. Lyubarskij et al. (1994) modeled the eccentric accretion disk in the Newtonian gravity. Eracleous et al. (1995) compared the line profiles emitted from the relativistic eccentric disk with observational data of some AGN, which can not be fitted with axisymmetric circular disk. The emission line profile from the rotating disk is double-peaked structure in general, the red-shifted and blue-shifted peaks corresponding to receding and approaching components. The blue peak is stronger than the red one in the emission from the circular relativistic disks due to Doppler boost. Some observed broad lines reveal the opposite characteristic. For this reason, Eracleous et al. (1995) studied the possibility of the eccentric disk.

The relativistic eccentric disk is, however, transient unlike the Newtonian disk, because the pericentre of the elliptical orbit is no longer constant. It

*S. Kato et al. (eds.), Physics of Accretion Disks, 175–178.*
© 1996 OPA (Overseas Publishers Association) Amsterdam B.V.

is important to examine how long the eccentric accretion disk survives in the relativistic system such as in the central part of AGN, where gas may be supplied by the debris of tidally disruptive stars around a massive black hole. The initial configuration is assumed to be elliptical and the subsequent viscous evolution is numerically simulated by the smoothed particle technique. The flow is mainly controlled by the gravity from the central black hole. The self-gravity of the disk and the thermal pressure are assumed to be negligible. The viscous force is also small in the magnitude, but taken into account because of the importance in the long term evolution. Associated with these forces, there are two characteristic time-scales, dynamical time-scale $t_{\rm d} = 2\pi[R_o^3/(GM)]^{1/2}$ and viscous time-scale $t_{\rm v} = R_o^2/\nu$, where $R_o$ is the typical size of the system and $\nu$ is the kinematical viscosity. We simply assume constant $\nu$, and choose the ratio of these time-scales as $t_{\rm v} = 10^3 t_{\rm d}$. The relativistic factor is chosen as $GM/R_oc^2 \sim 10^{-2}$ in the calculation.

## 2. Numerical Results

We have set up nested elliptical annuli with the eccentricity $e = 2/3$ and common semi major axis $a$, at $t = 0$. The each fluid element initially moves along the elliptical annulus with the Keplerian speed. The density inside the disk is set up to be uniform along each annulus, but distributed in the Gaussian form around the central annulus of $a/R_o = 1$. This configuration is stationary if the viscosity and the general relativistic effects are absent. The subsequent viscous evolution is numerically calculated. We have confirmed the following points (Kojima 1995).

(1)The elliptical shape is almost unchanged in the Newtonian gravity, although the density distribution inside the disk is smoothed away by the viscosity until $t \sim t_{\rm v}$. The viscosity affects the smoothing of the local density irregularity, but little the global shape. Smoothing time depends on the density distribution. Larger density contrast is much quickly smoothed away. In this way, elliptical shape is long lived. The eccentric accretion disk may be possible in the Newtonian gravity.

(2)The post-Newtonian gravity causes violent differential precession of the disk, as shown in figure 1. The figure corresponds to the snap shot at $t/t_{\rm d} = 160$ i.e., $t/t_{\rm v} = 0.16$. The result of non-viscous evolution at the same time is also shown in the right panel of figure 1 for comparison. Both global structure are almost the identical, although the local density distribution is much smooth in the viscous evolution. This means the kinematical circularization. The time-scale of the circularization is estimated as $t_1 \sim (2\pi/\chi)(a/\Delta a)t_{\rm d}$, where $\chi$ is the pericentre advance angle per revolution, $\chi = 6\pi GM/[a(1-e^2)c^2]$, and $\Delta a$ is the extent of $a$. The time-scale corresponds to $\sim (ac^2/GM)t_{\rm d}$ in our problem.

(3)When the central black hole is rotating and the disk is mis-aligned with the spin axis, then the disk is stirred in three dimensional space and settles into an axisymmetric geometrically thick disk. The effect is the post$^{1.5}$-Newtonian correction, so that the time-scale $t_2$ is longer than $t_1$. The value of $t_2$ depends on the configuration, but approximately $\sim [ac^2/(GM)]^{3/2} t_{\rm d}$. The effect is significant, if the disk is formed near the black hole. The process to the torus is shown in figure 2. The left and right panels show the projection onto x-y and x-z planes, respectively. The figures correspond to the snap shot at $10^{-3} t/t_{\rm d} = t/t_{\rm v} = 0.25$.

## 3. Discussion

Eccentric disk may be formed in consequence of tidal disruption of a star around a massive black hole. We have simulated subsequent fate of the elliptical disk under the post-Newtonian gravity. The viscosity smoothes the local density irregularity. Steep density gradient causes strong diffusion and leads to the prompt smoothing. The resultant flow is not necessary circular. The initial shape is gradually erased with a longer time-scale, as the matter accretes or expands. In this way, the viscosity plays little role on the global circularization of the flow. The initial memory such as the eccentric disk may be preserved for a long time in the distant region from the black hole. General relativistic effects are significant near the black hole. The differential precession stirs the whole disk and eventually circularizes the flow. The time-scale of the circularization becomes smaller than the viscous one in the disk near the black hole. Lens-Thirring effect is also important in the inner disk. If the angular momentum of the disk is mis-aligned with the spin axis of the rotating black hole, then the disk settles into an axisymmetric torus.

Emission line profiles may be useful as the diagnosis of the disk structure. The broad line width is ascribed to the Doppler effect with velocity $v/c \sim \Delta\lambda/\lambda$. The kinematical velocity field on the disk can be inferred from the line profile, if the local thermal velocity and other effects are small. The broad lines with $v > 10^3$ km s$^{-1}$ from AGN are the candidate. The theoretical models except Eracleous et al. (1995) are based on assumption of the axisymmetric circular flow. The eccentric disk is possible, but survives temporarily in the relativistic system. The life time of the eccentricity at the orbital radius $r$ is $\sim [rc^2/(GM)] t_{\rm d} \sim GMc^2/v^5$, where Kepler's law is used. This means that non-axisymmetry of the line profile at the high velocity component, say, $v = 10^4$ km s$^{-1}$ can be seen only at $\sim 10(M/10^6 M_\odot)$ $(v/10^4 \text{km s}^{-1})^{-5}$ yr, while that of the lower velocity component survives for a longer time.

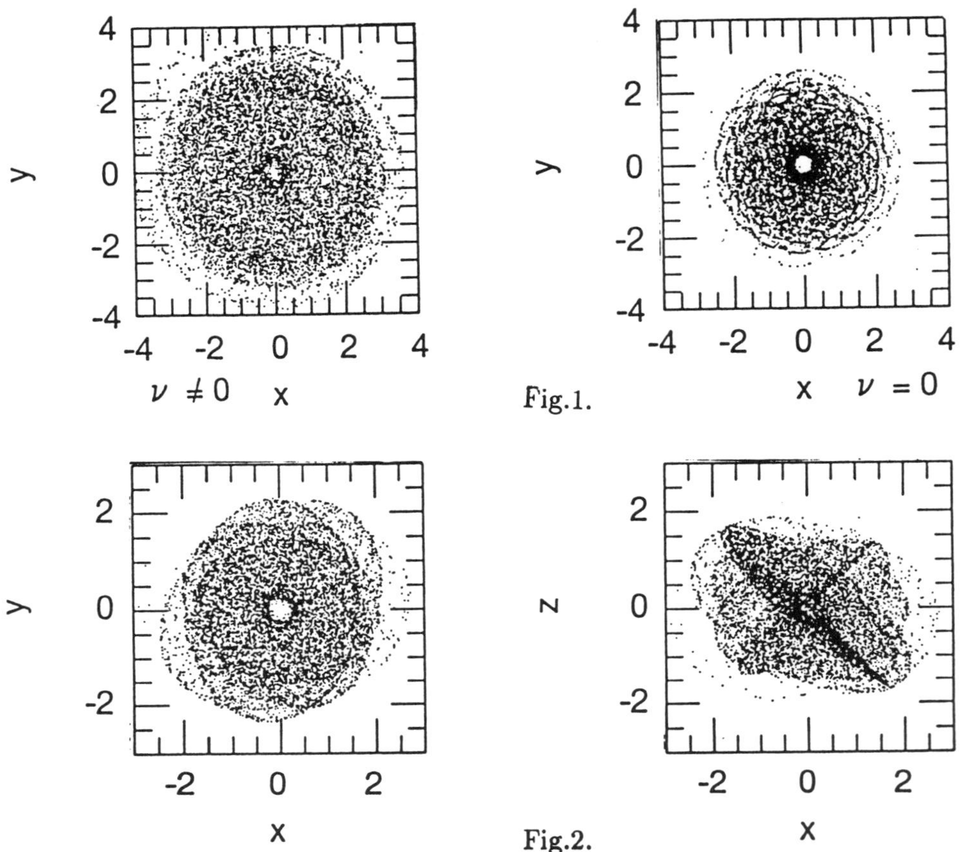


Fig.1.

Fig.2.

## References


Eracleous M., Livio M., Halpern J. P., Storchi-Bergmann T. 1995, ApJ 438, 610

Kojima Y. 1995, in preparation

Lyubarskij Yu. E., Postnov K. A., Prokhorov M. E. 1994, MNRAS 266, 583

Syer D., Clarke C.J. 1992, MNRAS 255, 92

Syer D., Clarke C.J. 1993, MNRAS 260, 463

# Interaction of Accretion Disks with Magnetic Stars

Frederick K. LAMB[1,2] and Pascal R. DAUMERIE[2]
*1. University of Illinois at Urbana-Champaign, Department of Astronomy, USA*
*2. University of Illinois at Urbana-Champaign, Department of Physics, 1110 W. Green St., Urbana, IL 61801-3080, USA*

**Abstract**

Many young stellar objects, as well as neutron stars and white dwarfs in binary systems, are thought to be magnetic stars accreting from Keplerian disks. We summarize the results of recent numerical calculations of the interaction of Keplerian accretion disks with magnetic stars based on the physical picture developed by Ghosh & Lamb. We also describe briefly the quite different picture of disk-magnetosphere interaction proposed recently by Shu and co-workers, calling attention to several theoretical and observational difficulties with their model.

## 1. Introduction

The interaction of Keplerian accretion disks with magnetic stars is an important and interesting problem with applications to young stellar objects (YSOs) as well as magnetic white dwarfs and neutron stars in binary systems. Evidence is accumulating that classical T Tauri stars (CTTS) are accreting from Keplerian disks and have dynamically important magnetic fields (see Königl 1991; Bodenheimer 1995). By now, the evidence that many white dwarfs in binary systems have dynamically important magnetic fields and are fed by Keplerian disks is compelling (see Córdova 1995; Warner 1995). So far, the most detailed information on the interaction of Keplerian accretion disks with magnetic stars has come from studies of accretion-powered pulsars in binary systems (see Ghosh & Lamb 1991, hereafter GL91; White et al. 1995).

*S. Kato et al. (eds.), Physics of Accretion Disks, 179–184.*
© 1996 OPA (Overseas Publishers Association) Amsterdam B.V.

## 2. Model of Daumerie et al.

While differing in details, Keplerian accretion flows onto magnetic white dwarfs, neutron stars, and YSOs share important basic features. Near the star, accreting matter is sufficiently ionized that the electrical conductivity is high. Because the electrical conductivity of the accreting plasma is high, the stellar field interacts with and couples to it largely via *macroscopic* processes—such as reconnection to small-scale magnetic fields generated in the disk by magnetoturbulence—that are closely related to the processes that transport angular momentum outward through the disk (Ghosh & Lamb 1978, 1979a,b, 1991). Even a very tenuous plasma ($n_e \sim 10^{-4}$–$10^{-1}$ cm$^{-3}$ around YSOs and $\sim 10^{2}$–$10^{10}$ cm$^{-3}$ around neutron stars) is adequate to support field-aligned currents between the disk and star (see Zylstra 1988; GL91).

Inside the inner corotation radius $\varpi_{\rm c,i}$ the stellar magnetic field is strong enough to force accreting plasma to corotate with the star. Between $\varpi_{\rm c,i}$ and $\varpi_0$ is a boundary layer where the magnetic couple to the star is strong enough to remove the angular momentum of the disk plasma in a radial distance $\Delta\varpi$ ($< \varpi_0$) and to support plasma inflow along field lines to the magnetic poles. Outside $\varpi_0$ is an outer transition zone where the angular momentum flux carried by the twisted stellar magnetic field diminishes rapidly with increasing radius and the viscous stress is increasingly important.

In the boundary layer and outer transition zone, the motion of the disk plasma relative to the star twists the magnetic field. Studies of twisting in this geometry show that it produces current sheets that are unstable to tearing modes and reconnection. If it occurs rapidly enough, reconnection through the (inhomogeneous and clumpy) disk will limit the twist. If not, the magnetic field above and below the disk will become increasingly twisted, causing the magnetic energy to grow and the poloidal field $B_p$ to expand (Zylstra 1988; GL91). This expanded and twisted field configuration cannot persist, because there are field configurations with reduced twist that have much lower energies and are accessible via reconnection within the magnetosphere. As a result of these processes, we expect the twist to grow and relax episodically.

Outflow of disk plasma along stellar magnetic field lines that episodically thread the outer transition zone may produce an erratic wind from the disk. A steady wind is very unlikely because outflow from the disk along stellar field lines that initially thread the disk quickly disconnects these field lines from the star, after which magnetoturbulence within the disk will rapidly tangle them into a small-scale, disordered field, terminating the outflow (Keplerian disks do not by themselves produce ordered, large-scale magnetic fields; see section 3). Any outflow from the Keplerian disk along magnetic field lines that do not connect to the star cannot affect significantly the torque on the star. Unsteady reconnection of the twisted stellar field in the magnetosphere above and below the disk may also produce flaring and episodic mass ejection.

We have incorporated these physical processes into a consistent set of non-ideal MHD equations for steady plasma flow in a slim disk and have solved these equations numerically, using a relaxation method (Daumerie et al. 1996). This model includes for the first time the effect of the star's rotation on co- and counter-rotating disk flows (its effect on the magnetic couple was already included in previous work), the effect of radial pressure gradients within the disk, and an effective viscous stress throughout the flow. We model the time-averaged consequences of reconnection in an electrodynamically consistent way by introducing an effective electrical conductivity $\sigma_{\text{eff}}$ in the disk and adjusting its local value so that the electrical currents in the disk give a local azimuthal magnetic pitch consistent with magnetic tearing and reconnection; the resulting currents and magnetic fields are then used in the MHD equations. The model is completed by the radial and azimuthal momentum equations and the equations of continuity, vertical hydrostatic equilibrium, and radiative transfer, and by the equation of state. When the model equations are written in nondimensional form, the characteristic length that emerges is the Alfvén radius $r_A \equiv \mu^{4/7}(GM)^{-1/7}\dot{M}^{-2/7}$, where $M$ and $\mu$ are the mass and magnetic dipole moment of the neutron star.

We solve the model equations while requiring the azimuthal velocity to approach the Keplerian velocity far from the star. In all our solutions, the location of the inner edge of the Keplerian flow is given approximately by the standard azimuthal stress balance condition $(B_p B_\phi/4\pi)\, 4\pi\varpi^2\Delta\varpi = \dot{M}_{\text{BL}}\varpi^2\Omega_{\text{K}}$, where $\dot{M}_{\text{BL}}$ is the mass flux through the boundary layer (see GL91). We compute the accretion torque from the material, viscous, and magnetic stresses given by the solution. The accreting matter and the magnetic couple to the plasma in the disk inside the outer corotation radius $\varpi_{\text{c,o}} \equiv (GM/\Omega_{\text{s}}^2)^{1/3}$ produce a spin-up torque whereas the magnetic couple to the plasma in the disk outside $\varpi_{\text{c,o}}$ produces a spin-down torque. As a result of the competition between these two torques, the net torque produced by a disk flow circulating in the same sense as the stellar spin produces a strong spin-up torque $\sim \dot{M}(GMr_A)^{1/2}$ if the star is spinning slowly or—equivalently—the accretion rate is high, but brakes the star's spin if it is rapid or the accretion rate is low, in agreement with the observed behavior of accretion-powered pulsars.

## 3. Model of Shu & Co-workers

Recently, Shu and co-workers (Shu et al. 1994a,b, 1995; Najita & Shu 1994; Ostriker & Shu 1995) have proposed a model of disk-magnetosphere interaction that differs significantly from the model described in section 2. According to Shu et al., their proposal is motivated by two concerns about the model described in section 2. First, they are concerned that the twisting of the stellar magnetic field caused by its interaction with the plasma in the inner disk and its subsequent relaxation make detailed predictions difficult. They attempt to sidestep this problem by simply *assuming* that the plasma in that part of the disk that couples to the stellar magnetic field rotates uniformly at exactly the same rate as the star. As we explain below, this assumption is not physically consistent. The work described in section 2, in contrast, seeks to model what is expected to happen, rather than what would be convenient.

Second, Shu et al. (1994a) are concerned that there may be too few charged particles around the inner disks of CTTS to support the field-aligned currents required in the model of section 2, although they do not estimate the density required. As noted in section 2, for stellar magnetic fields of 1–$10^3$ G the electron density needed is only $10^{-4}$–$10^{-1}$ $cm^{-3}$, orders of magnitude lower than the expected density.

In their model, Shu et al. assume that the inner edge of the Keplerian flow, which they accept is at $r_A$, is always located exactly at the outer corotation radius, which they denote $r_x$, and that the disk is completely depleted of plasma at a radius $r_t$ somewhat smaller than $r_x$. They call the radial interval between $r_x$ and $r_t$ the "T region" and assume that the stellar magnetic field couples to the disk only in this region. In order to avoid having to consider twisting of the stellar magnetic field, Shu et al. postulate that the angular velocity $\Omega$ of the accreting plasma is constant throughout the T region and equal to the angular velocity $\Omega_s$ of the star (in fact, in computing the gravitational potential near $r_x$ they assume that $\Omega$ is also equal to $\Omega_s$ above and below the disk and outside $r_x$). They assume further that plasma flows from the disk toward the star along the field lines that thread the disk in the region just outside $r_t$ and treat the flow in this region using the equations of ideal MHD. Shu et al. also assume that the torque on the star is zero. Outside $r_x$, Shu et al. postulate a large-scale, ordered magnetic field that threads the disk and has the shape and strength required to drive a magnetocentrifugal wind. They assume that the wind carries sufficient angular momentum and mass to terminate the Keplerian flow at $r_x$; hence $r_x$ is necesarily $\approx r_A$.

There are several observational and theoretical difficulties with this model. The assumption that $r_x$ is always equal to $r_A$ requires that $\dot{M}$ satisfy a unique relation involving $\Omega_s$, $\mu$, and $M$ which, if it can be satisfied at all, limits the applicability of the model to a specific accretion rate. Observations of accreting white dwarfs and neutron stars show that in these systems the accretion rate often varies on time scales much shorter than the spin evolution time scale, so a model that makes this assumption cannot be applied to these systems. Nor is there any reason to expect the accretion rates in YSOs to satisfy this relation at all times.

Shu et al. (1994a) argue that the torque on a magnetic star accreting from a Keplerian disk is always essentially zero, since the magnetic field near the stellar surface is very strong and therefore cannot be twisted very much. This argument is fallacious. The quantity that determines the magnetic torque is the magnetic stress $B_\phi B_p$, not the magnetic pitch $B_\phi / B_p$. If $B_p$ is large, as Shu et al. assume, the magnetic torque can be large even if $B_\phi$ is small. Ostriker & Shu (1995) argue that the torque on a magnetic star should be essentially zero if the spin evolution time scale is comparable to the time over which accretion occurs. In reality, this argument is valid only if the accretion rate changes on a time scale that is much longer than the spin evolution time scale. This is not the case for accretion-powered pulsars, which show significant variations in their accretion rates on time scales $\sim$ days to years, which is shorter than the shortest expected spin evolution time scale ($\sim$

$10^2$) yrs and much shorter than the duration of the accretion phase ($\sim 10^6$–$10^8$ yrs). There is no reason to expect the situation to be completely different for accreting white dwarfs or YSOs. The assumption that the accretion torque is always essentially zero is also inconsistent with the observed behaviors of accretion-powered pulsars, many of which show spin-up and spin-down rates that imply torques comparable to the maximum theoretical torque $\sim \dot{M}(GM\varpi_0)^{1/2}$. There is some evidence that the spin rates of some YSOs are evolving.

The assumption that the torque on the star is zero requires the field lines that connect the star and the T region of the disk to have a negative azimuthal pitch, *i.e.*, to form a trailing spiral, in order for the magnetic torque to cancel the spin-up torque produced by the infall of orbiting matter. There is no way for accreting matter to produce such a field geometry in ideal MHD. In non-ideal MHD, maintenance of a trailing spiral is possible only if the plasma at the disk footpoints has an orbital frequency less than the spin frequency of the star, since only then will there be an EMF of the appropriate sign to maintain the electrical currents in the magnetosphere required by the trailing azimuthal field. Thus, the assumption of negative azimuthal pitch (and zero torque) is physically inconsistent with the assumption that the orbital frequency of the plasma in the T region of the disk is the same as the spin frequency of the star.

The assumption that the torque on the star is zero requires that angular momentum flow radially outward through the T region at a rate $\sim \dot{M}(GMr_x)^{1/2}$. This angular momentum flux cannot be carried by the ordered magnetic field that threads the T region, since this field exerts no relevant stress. Shu et al. assume there is some source of anomalous viscosity within the disk in the T region—they mention magnetoturbulence—that provides the needed stress. However, this would require an angular velocity gradient $\partial\Omega/\partial\varpi$ in the T region $\sim \dot{M}(GMr_x)^{1/2}/(2\pi\varpi^3\nu_{\rm eff}\Sigma) \sim (\partial\Omega_K/\partial\varpi)_x$. This is inconsistent with the assumption that plasma in the T region rotates uniformly. Even a small departure from uniform rotation at angular frequency $\Omega_s$ will twist the magnetic field threading the T region, forcing consideration of relaxation processes such as those discussed in section 2.

The gravitational force dominates the centrifugal force in the T region, since $\Omega < \Omega_K$ there, and the magnetic field shape assumed by Shu et al. (1994a) and Ostriker & Shu (1995) in this region also produces a radially inward electromagnetic force. The resulting net inward force cannot be balanced by a thermal pressure gradient, because the disk is assumed to be completely depleted of plasma at $r_t$. Thus, the assumed shape of the magnetic field is inconsistent with the assumption that the plasma in the T region of the disk is in dynamical equilibrium.

The magnetic field along which the wind is assumed to leave the disk is disconnected from the star and therefore cannot be organized by the star or its magnetic field. This field must therefore be generated within the disk. Studies of $\alpha$-$\Omega$ dynamo action in geometrically thin disks show that the quadrupolar field component is the fastest-growing; it produces a magnetic field with a $z$-

component that reverses sign at the midplane of the disk. This is not the field geometry assumed by Shu et al. The strongest components of the magnetic field produced by magnetoturbulence in a geometrically thin disk of height $h$ have length scales $\lesssim h$. Thus, the assumption by Shu et al. of a stationary magnetic field that threads the disk, is disconnected from the star, and is strongest on a scale $r$ is not supported by these calculations. Without such a field, the wind mechanism proposed by Shu et al. is not possible.

## Acknowledgments

F.K.L. thanks Professor Kato and the other organizers of the workshop for inviting him to participate and for providing reimbursement for travel and living expenses. This research was supported in part by NASA grant NAG 5-2925 and NSF grant AST93-15133.

## References

Bodenheimer P. 1995, ARAA 33, 199
Córdova F. 1995, in X-Ray Binaries, ed W. H. G. Lewin, J. van Paradijs, E. P. J. van den Heuvel (Cambridge), p331
Daumerie P. R., Lamb F. K., Ghosh P. 1996, in preparation
Ghosh P., Lamb F. K. 1978, ApJ 223, L83
Ghosh P., Lamb F. K. 1979a, ApJ 232, 259
Ghosh P., Lamb F. K. 1979b, ApJ 234, 296
Ghosh P., Lamb F. K. 1991, in Neutron Stars: Theory and Observation, ed J Ventura, D. Pines (Kluwer), p364
Königl A. 1991, ApJL 370, L31
Najita J., Shu F. 1994, ApJ 429, 809
Ostriker E., Shu F. 1995, ApJ 447, 813
Shu F., Najita J., Ostriker E., Wilkin F. 1994a, ApJ 429, 781
Shu F., Najita J., Ruden S., Lizano S. 1994b, ApJ 429, 797
Shu F., Najita J., Ostriker E., Shang H. 1995, ApJL 455, L155
Warner, B. 1995, Cataclysmic Variable Stars (Cambridge)
White N. E., Nagase F., Parmar A. N. 1995, in X-Ray Binaries, ed W. H. G. Lewin, J. van Paradijs, E. P. J. van den Heuvel (Cambridge), p331
Zylstra G. 1988, Ph.D. thesis, University of Illinois at Urbana-Champaign

# The Anomalous and Preeclipse Dips in Her X-1 with Respect to the Coronal Wind Model

Susanne SCHANDL
*Max-Planck Institut für Astrophysik, Karl Schwarzschildstr. 1, D-85748 Garching, Germany*

## Abstract

The interaction between the accretion stream and the thin, tilted and twisted accretion disk can explain the observed complex dip structure in the X-ray lightcurve of Her X-1. Based on the disk shape solutions of the coronal wind model, the stream falls into the warped disk mostly from above or below where the hydrodynamic interaction leads to a cold clumpy spray and a thickening of the involved accretion disk rings. Crossing the line of sight to the central X-ray source the spray causes the observed preeclipse dips and the thickening generates the anomalous dips. We show that the observed column densities, spectra and orbital phases of occurrence of both dip features can be explained.

## 1. The Coronal Wind Model

The X-ray binary HZ Her/Her X-1 consists of a $2.2 M_{\odot}$ Roche lobe filling companion and an accreting magnetic neutron star. Besides the 1.24 s spin period of Her X-1 and the 1.7 day orbit the binary shows a long periodic behaviour on a timescale of 35 days.

This behaviour is understood in terms of a tilted and twisted accretion disk which the observer on earth sees nearly edge on, because the inclination of the system is very high ($> 80°$) (Gerend & Boynton 1976). A warped disk precesses in the tidal field of the companion (Katz 1973) where the calculated precession period agrees with the observed 35 day period.

The region of the disk which rises above the orbital plane by more than the inclination therefore covers the central X-ray source temporarily during the precession. Additionally, there exists an X-ray induced corona which reduces the intensity at the end of the main-on where its optically thick part covers Her X-1.

Our coronal wind model (Schandl & Meyer 1994) reproduces this corona which is optically thick at small radii ($r \leq 10^{10}$ cm) and which temporarily covers the central neutron star because the inner region of the disk is tilted.

*S. Kato et al. (eds.), Physics of Accretion Disks, 185–190.*
© 1996 OPA (Overseas Publishers Association) Amsterdam B.V.

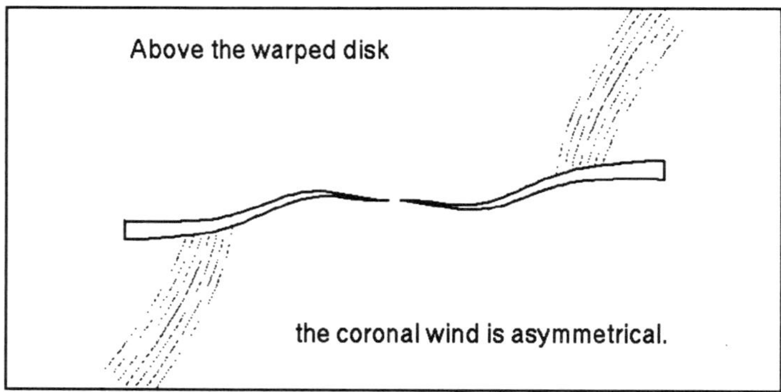


Fig. 1. The vertical cut through the warped disk with its X-ray driven coronal wind above the asymmetrical illuminated surface shows schematically the mechanism which produces and maintains the tilt and twist of the accretion disk.

Because the Compton effect is the main heating and cooling process in the upper layer of the corona and the spectrum is hard enough that the inner energy exeeds the gravitational potential at the outer regions, coronal gas leaves the system forming a coronal wind (Begelman et al. 1983). The angular momentum of the gas flow gets lost from the system, too. Integration over the whole disk rings yields the repulsive wind torque acting on the disk at each radius. Above a warped and therefore asymmetrical illuminated disk with an asymmetrical wind loss the wind torque is finite (see figure 1). We obtained the stationary disk shape by solving the torque equation where the acting torque equals the torque by the coronal wind, by the viscosity and by the tidal forces of the companion star. For the boundary conditions we excluded any torque from outside. The resulting disk shape is shown in figures 2 and 3 and the corresponding lightcurve is plotted in figure 4.

## 2. The Interaction between Accretion Stream and Disk

Additionally to the long periodic variation and the eclipses by the companion on the orbital period of 1.7 days the X-ray intensity undergoes two types of dips (Giacconi et al. 1973). The *preeclipse dips* occur just before the eclipses and march backwards in orbital phase as the 35 day cycle proceeds. They last for 2 to 5 hours and show strong variations on short timescales of the intensity, the spectrum and the pulse shape.

The other type of dips are the *anomalous dips* which occur only just after turn-on at orbital phases around 0.2 or 0.7. These are the same phases where the turn-on itself takes place (Giacconi et al. 1973). The anomalous dips are somewhat shorter, 1 to 2 hours, but show cold matter absorption features in the spectrum, too (Reynolds & Parmar 1995).

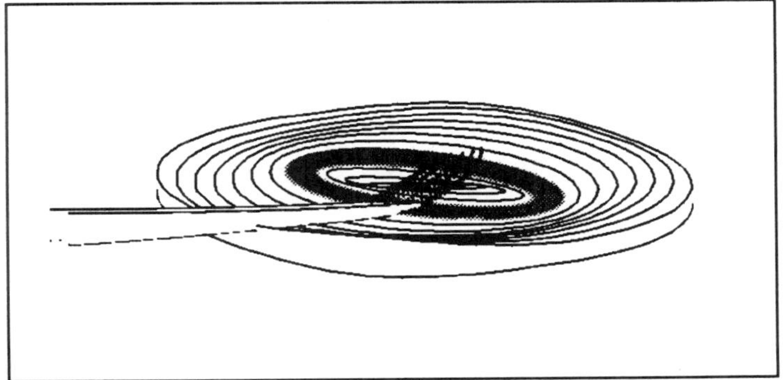

Fig. 2. The impact of the accretion stream on the disk surface causes a spray of matter and a thickening of the disk which is carried through the disk along Keplerian orbits (marked grey). Both effects reproduce the observed preeclipse and anomalous dips.

Considering the interaction of the accretion stream and the thin, warped disk we find that the stream which falls in the orbital plane hits the disk mostly from above or below and not at the rim. Estimation of the surface densities shows that the contribution of the stream to the disk is small compared to the disk matter itself. The supersonic motion of disk and stream matter causes strong shocks. Because of the small angle the pressure of the stream matter after it has run through the shock is a bit smaller than the disk pressure and thus the stream does not go through the disk. Otherwise there is more disk matter involved into the shocks than stream matter because of the larger surface density. The details of the estimations are published in Schandl (1996).

### *2.1 The Preeclipse Dips*

The matter running through the shock fronts gets refracted because only the velocity component perpendicular to the front gets reduced to subsonic values. In particular, some of the matter at the surface of the accretion disk that impacts the penetrating obstacle will get refracted upwards resulting in a cold and clumpy spray spreading above the disk surface. When this runs in front of the X-ray source (see figure 2) it will reduce the signal in the way of the preeclipse dips showing cold matter absorption features in the spectrum and intensity variations on small timescales. From the observed column densities of the order of $10^{23}\,\mathrm{cm}^{-2}$ (Giacconi et al. 1973; Choi et al. 1994; Reynolds & Parmar 1995) and the duration of the dips one can calculate the mass flow in the spray. Compared to the model estimations this spray corresponds to only a few percent of the matter involved into the shock.

Another important consistency of our model and the observations are the times and phases where the dips occur. For this we assume that the spray flows in a direction as a stream refracted at the disk surface would move. When this spray crosses the line of sight between observer and central X-ray source the dip occurs. The simulated dips are shown in the X-ray light curve in figure 4a.

The simulated period of the dips of about 1.66 days calculated from figure 4b reproduces the observed preeclipse dip occurrence (Giacconi et al. 1973) very well. This includes the preeclipse dip Jones & Forman (1976) observed during the low state (at $\phi_{35} = 0.6$) at orbital phase 0.56.

Note that this period actually compares to the synodic orbital period but is caused by a quite different mechanism in our model.

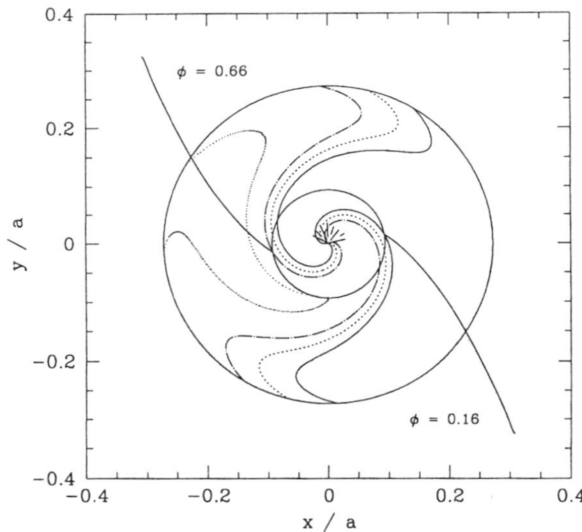


Fig. 3. The warped accretion disk from the top in units of the separation a. The spirals indicate the intersection between the disk and the orbital plane (solid line - upper surface; dashed dotted line - lower surface; dotted line - disk center). Here the stream impinges the disk. The stripes at the inner disk show the phases where the optically thick part of the corona covers the source. The region inside the dotted area rises by more than 8.5° indicating that it covers the central X-ray source if it intercepts the line of sight to the earth. The precession is clockwise and thus the reader sees the disk at turn-on. The disk ring doing the turn-on is also plotted. The accretion stream starting at the inner Lagrangian point impacts the disk at this radius at two phases, once from the upper, once from the lower surface. The resulting thickening of the ring causes the anomalous dips.

*2.2 The Anomalous Dips*

Otherwise the penetration and break-up of the accretion stream results in heating, turbulent motion and an addition of matter to the disk. All this produces a thickening of the disk. It gets gradually smoothed out by viscosity and thermal conductivity on a timescale which is comparable to the Keplerian orbital period. Therefore the thickening settles down until the disk matter runs again through the turbulent impact region creating a saw-tooth profile of the disk thickness over orbital phase.

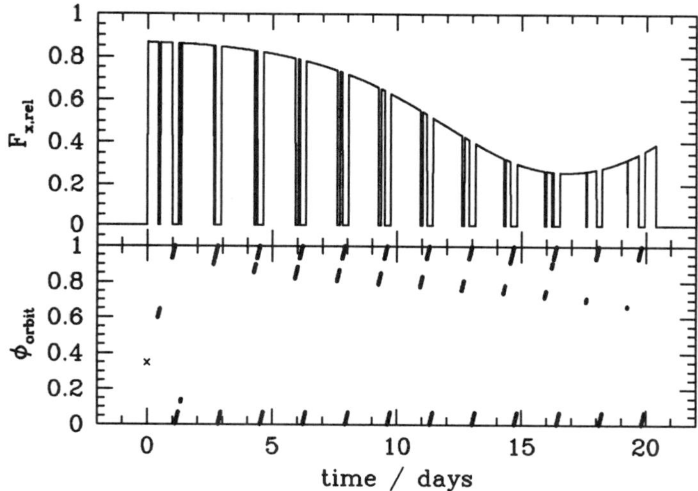


Fig. 4. In the upper figure the calculated light curve of the relative X-ray intensity $F_{x,rel} = e^{-\tau}$, where $\tau$ is the optical depth in the corona, including the eclipses and the anomalous and preeclipse dips, is plotted. One can see that our coronal model yields a density which is still too small compared to the observations though of correct order of magnitude. A density (and optical depth) which is 2 to 3 times higher would result in a total drop of the intensity at the end of the main-on around day 15. The increase later in the cycle corresponds to the observed short-on which ends due to the disk crossing the line of sight. The turn-on is marked by a cross in the lower picture where the orbital phases of the dips and eclipses are shown. This kind of representation clarifies the marching back of the preeclipse dips in orbital phase.

At turn-on the disk opens the line of sight to the central X-ray source. This means the observer looks marginally over the disk surface. More precisely it is the surface at a special disk radius doing the turn-on ($r_{\text{turn-on}} = 6 \times 10^{10}$ cm, drawn in figure 3), because of the twist of the disk. And it is this disk ring which we look marginally over and on which our view depends very sensitively. If this disk ring undergoes the thickening it will cover the X-ray source doing an anomalous dip. This occurs only at turn-on, because we look afterwards far above the disk and its thickened region. And it may occur at two distinct orbital phases because the special disk ring gets impacted twice per orbit, once from above, once from below with a synodic orbital phase difference of about 0.5 (figure 3). The phases $\phi = 0.16$ and $\phi = 0.66$ (figure 4) agree with the observed times. To reproduce the anomalous dips about 10 percent of the energy contributed by the stream has to heat up the disk und thus enlarging the disk scale height by 1/8. The spectrum will be the same as at turn-on, because it is the same disk matter covering the source.

## 3. Conclusions

The interaction between the accretion stream and the thin warped disk has been found to be the origin of the complex dip structure in the X-ray signal. The spray explains the spectral and time variable intensity structure of the preeclipse dips. The mass flow in the spray calculated from the column densities of cold matter observed during the dips can be generated by a few percent of the disk matter involved into the stream impact scenario. The disk shape resulting from the coronal wind model reproduces the occurrence times of these dips.

The anomalous dips are modelled by a temporal thickening of the disk region we look marginally over at turn-on. Considering the energy input by the stream we estimated a thickening of the correct order. The simulated orbital phases of the stream impact at the corresponding critical radius are the same as the observed phases.

## Acknowledgements

I am grateful to Friedrich Meyer for checking the concept of this model and for the precious ideas for the estimates. I thank R. Staubert for calling my attention to the dip feature problem and M. Ruffert for helpful discussions about shocks.

## References

Begelman M. C., McKee C. F. 1983, ApJ 271, 89
Choi C. S., Nagase F., Makino F., Dotani T., Min K. W. 1994, ApJ 422, 799
Gerend D., Boynton P. E. 1976, ApJ 209, 562
Giacconi R. et al. 1973, ApJ 184, 227
Jones C., Forman W. 1976, ApJL 209, L131
Katz J. I. 1973, Nature Phys. Sci. 246, 87
Reynolds A. P., Parmar A.N. 1995, A&A 297, 747
Schandl S. 1996, A&A in press
Schandl S., Meyer F. 1994, A&A 289, 149

# Corona Formation from Disk Evaporation

Fridrich MEYER
*Max-Planck-Institute for Astrophysics, Karl-Schwarzschild-Str. 1*
*D -85740 Garching, Germany*

**Abstract**

We discuss consequences of energy and mass exchange between accretion disk coronae and their underlying disk and show that this can uniquely determine the coronal accretion flow. We further suggest that in AGN the hot coronal ions contribute to disk irradiation and allow stronger reprocessed light from the accretion disk surface than obtained by the X-ray irradiation alone.

## 1. Introduction

Hot coronae above cool accretion disks leave observational signatures by the appearance of hard X-rays (dwarf novae in quiescence), hard power-law tails of frequency spectra, caused by Comptonization (galactic black hole candidates), and X-ray heating and Fe-line fluorescence of disk surfaces (active galactic nuclei).

While in some cases a corona is formed and sustained through irradiation of the disk surface by a central source, as in low mass X-ray binaries, in many cases the coronae appear self-sustained, presumably by the same frictional release of gravitational energy that heats standard accretion disks. This situation will in particular be found near the accretion centers where energies released are already comparable to the total accretion energy. The physical situation is thus that of a normal accretion disk, sandwiched between more extended coronal layers that themselves form two other, thick accretion disk halves. This configuration has a number of interesting aspects many of which have found extensive consideration in the literature. Here attention is given to the interaction of disk and corona at there common interface. By the example of dwarf novae accretion disks in quiescence it is shown how mass exchange and formation of a uniquely determined corona is the consequence of the proper boundary condition. We then discuss a few general features of this situation and show how some puzzles in the observations may be resolved.

*S. Kato et al. (eds.), Physics of Accretion Disks, 191–195.*
© 1996 OPA (Overseas Publishers Association) Amsterdam B.V.

## 2. Dwarf Nova Accretion Disk in Quiescence

Here the compact center is a white dwarf. It is surrounded by a cool post-outburst accretion disk with low internal mass flow rate, typically below ten to the minus twelve solar masses per year. Observations show that in this stage characteristic X-rays in the keV-range appear. These, it is claimed, point to a coronal accretion flow: The X-ray luminosity observed requires an order of magnitude higher accretion rate than what a cool disk can carry inward. The existence of a hot corona above such a cool and, in this case, nearly inert disk is puzzling.

In treating the equations of dynamical equilibrium and of mass and energy conservations that govern this corona two approximations simplify the complex 2-dimensional physical situation. Firstly, one treats the coronal gas like a thick accretion disk in which radial gravitational attraction by the central white dwarf and radial pressure gradients are compensated by centrifugal forces of near-Keplerian rotation. The remaining uncompensated vertical downward gravity then leads to the usual barometric decrease of pressure and density with height. One may then also treat non-molecular friction in the usual alpha-parametrization and obtain the inward coronal mass flow from the frictional outward angular momentum transport caused by the differential Kepler-like rotation.

Secondly, as it turns out, coronal density, temperature, and mass flow decrease strongly with radial distance from the white dwarf. One can then approximate the innermost, all-important region as a radially homogenous zone, whose vertical structure is determined by gravity, frictional release of heat, inward mass loss and in our case also mass and energy loss into a wind leaving the corona at height. In addition, and crucial for the formation and balance of the corona, one has downward conductive heat flow and upward mass evaporation from the cool underlying disk surface.

We describe now the mechanism which fills gas into the corona. The high temperature of the coronal gas leads to a high thermal (electron) conductivity and a high downward heat flow. The conductivity however becomes increasingly poorer as the lower temperatures of the cool disk surface are approached. The stationary solution thus requires that the heat flow arising from the temperature difference between corona and disk can be radiated away. Since the relevant layers are optically thin this requires an appropiate density to be established. Should the density be too low, the gas will be heated up and become part of the corona. The conduction and its temperature profile will then dig deeper into denser surface layers of the cool disk until the required equilibrium has been reached.

By pressure equilibrium this also establishes the pressure and the density in the corona and these are kept up as long as the corona is hot. Any disk corona, however, looses mass inwards by frictional removal of angular momentum. This loss is continuously replaced by matter evaporating from the cool disk. Thus a steady coronal accretion flow is established. Our numerical solution (Liu et al. 1995) yields values for the coronal accretion rate in agreement with those required by the observations. It can further be noted that these results depend only on the gravitational potential gradients and on elementary radiative and conductive properties of the gas.

Though in these solutions the wind carries away only about one fifth of the mass, the main part being accreted, it still acts as a sensitive thermostat of the corona. This is due to the essentially exponential dependance of the density in the sonic point on temperature. Thus the energy carried away with the wind increases rapidly with increasing temperature. As a result the temperature remains limited to significantly below the virial value. If wind losses are prevented (e.g., by placing a lid on top of the corona) much higher densities and accretion rates are obtained (Narayan, private communication). In this case the coronal temperature and density rise until radiation from the bulk of the corona prevents further growth.

Since the continuing accretion is kept up by disk evaporation the inner disk matter may be completely used up. Then an inner hole will form in the cold accretion disk which slowly grows in size. In the dwarf nova case this has consequences for the following dwarf nova outburst. In this outburst the innermost and potentially hottest disk regions are at first missing and will only be filled in on the radial diffusion time of the hot disk. We think that this phenomenon is related to the so-called UV-lag observed in a number of dwarf novae.

## 3. A Few General Features of Disk Coronae

When one only requires pressure equilibrium at the coronal-disk interface fairly arbitrary accretion rates in corona and underlying disk would be compatible. But we have seen that the possible evaporation and condensation requires an additional, thermal, boundary condition which strictly determines the coronal flow. We note that the coronal energy liberation by irradiation and conduction also affects the temperature of the disk underneath and thus possibly its own accretion rate. The physical situation is thus intricate but also rich in perhaps fruitful possibilities.

The rather strong radial dependance of the corona, mainly due to the radial decrease of gravity that supplies the heat source, has the effect that significant coronal flow only occurs sufficiently close in. Now imagine an accretion disk with a mass accretion rate provided from outside. As one approaches the inner regions more and more of this mass flows through its corona. Finally, at a critical radius, depending on the mass accretion rate all is carried in form of the hot coronal gas and further in the underlying disk has disappeared. From then on the accretion might take the form of an "advective flow" as, e.g., investigated by Narayan and collaborators. They have shown that central black holes can swallow the major part of the accretion energy in such flows.

Particular interest rests on the radii where accretion disk and corona still coexist. Especially in active galactic nuclei this seems to hold still close to the central black hole. Then the coronal ions reach MeV temperatures. The electrons receive their energy through collisions from the ions. They are kept cooler by Compton collisions with cool photons from the disk surface or possibly also synchrotron photons. The Comptonised photons then yield the observed X-rays. They irradiate the underlying disk which shows the resulting Fe fluorescence lines. Further evidence of this irradiation is coherent time variability of X-rays and the UV light of the disk. Since no time lag between the fluctuation of the two components is observed they must be essentially co-spatial, and since the time scale of variation is too short to be explainable by changes in the disk mass accretion rate one has concluded that disk irradiation is the source of the disk UV light.

One problem has however made this conclusion questionable: The observed X-ray luminosity is consistently only a fraction of the disk UV luminosity. This has led to speculations of special beaming of the X-rays towards the disk and away from the observer. We point out here that this problem might be resolved.

The coronal MeV protons have a long collisional mean free path, which would allow them to leave the system altogether. This indicates that magnetic fields are present that turn the free ion motion into gyration around the field lines. Along the field lines the charged ions may travel freely and can reach footpoints of magnetic flux tubes in the disk if they are scattered into the so-called "loss cone", i.e., if the angle between their velocity vector and the downward field has a sine smaller than the ratio of the magnetic field strengths in the corona to that at the footpoint. This provides a high energy particle irradiation of the disk surface that derives from the primary energy source and might well be larger than that by the secondary X-rays. All three compnents, energetic protons, X-rays, and the irradiated disk will of course vary coherently as required.

A final remark concerns the galactic black hole candidate sources. In quiescence one may expect a similar disk evaporation situation as for the dwarf novae. One might then expect a coronal evaporation and accretion flow much stronger than in the underlying disk. Such a disk where it provides the strongest coronal flow will however also show irradiation by energetic ions and X-rays from electron-photon collisions. This can easily raise its surface temperature to values higher than those that result only from the disks own very low quiescent accretion rate. Even if these irradiation temperatures are higher than those at which a self-heated disk would go into outburst this would not occur here: The heating from the outside quenches the dwarf nova type instability since a small increase in midplane temperature does not lead to the required increase in pressure and friction, as both are controlled from the outside irradiation.

## 4. Conclusions

The above discussion shows some of the remarkable aspects of self-sustained coronae above accretion disks. In particular it is shown that the proper boundary conditions at the interface of disk and corona lead to evaporation and condensation flows between the two components and determine the coronal properties locally, as a function of the distance from the accreting center. Features in this corona-disk interaction might explain some interesting observational puzzles.

## Acknowledgements

I am grateful to Professor Shoji Kato and the organizers of the conference for providing support for living expenses during the meeting and acknowledge the Deutsche Forschungsgemeinschaft for partial travel support.

## References

Liu F. K., Meyer F., Meyer-Hofmeister E. 1995, A&A 300, 823

Photographs at the banquet. Mrs Narayan, Dr Meyer, Dr Narayan (upper) and Drs Lasota, Abramowicz and Duschl (lower).

# Effects of Radiation Forces on Disk Accretion

Frederick K. LAMB[1] and M. Coleman MILLER[2]
*1. Department of Physics, University of Illinois at Urbana-Champaign 1110 W.Green Street, Urbana, IL 61801-2030, USA*
*2. University of Chicago, Department of Astronomy and Astrophysics 5640 S. Ellis, Chicago, IL 60637, USA*

**Abstract**

We discuss the nature of radiation forces. Near relativistic stars, special and general relativistic effects and the finite angular size of the radiation source can increase the angular momentum loss rate to radiation by a factor $\sim 5$ and the total loss by a factor $\sim 2$ compared to the losses near a Newtonian point source. Near rotating sources, the radiation tries to force matter to orbit the star with a characteristic frequency that depends on radius. In sources with luminosities $\gtrsim 0.01$ times the Eddington luminosity, radiation forces can dramatically affect the accretion flow near the source, with important consequences for the temporal and spectral properties of the source and the spin evolution of the star.

## 1. Introduction

Many accreting X-ray sources in binary systems have luminosities $L \gtrsim 0.01\, L_{\mathrm{E}}$, where $L_{\mathrm{E}} \equiv 1.3 \times 10^{38}\,(M/M_{\odot})$ erg $\mathrm{s}^{-1}$ is the Eddington critical luminosity. A subset that includes some atoll sources and some other type I X-ray burst sources are thought to be neutron stars with magnetic fields so weak that they are dynamically unimportant. In these systems, radiation forces can dramatically alter the character of the accretion flow near the neutron star, terminating Keplerian motion above the stellar surface, masking the effects of general relativity, altering the spectrum of the emitted radiation, and affecting the spin evolution of the star (Miller & Lamb 1993, hereafter ML93; Lamb & Miller 1995, hereafter LM95; Miller & Lamb 1996, hereafter ML96). The effects of radiation forces may make it possible to distinguish accreting black holes from accreting neutron stars that have very weak magnetic fields.

*S. Kato et al. (eds.), Physics of Accretion Disks, 197–202.*
© 1996 OPA (Overseas Publishers Association) Amsterdam B.V.

In section 2 we give a general expression for the radiation force and point out that it is not the same as the radiation pressure or the radiation pressure gradient, which are frequently but mistakenly used as synonyms for the radiation force. In section 3 we discuss the effects of radiation forces on test particles orbiting around nonrotating radiation sources and gravitating masses and show that special and general relativistic effects can greatly increase the radiation drag force near a relativistic star. We discuss rotational effects in section 4. In section 5 we describe some of the qualitative changes in accretion disks that can be produced by radiation forces. Our conclusions are summarized in section 6.

## 2. Radiation Force

The force on a particle at position $x^\gamma$ produced by scattering of radiation in the frequency range $\nu$ to $\nu + d\nu$ in a solid angle $d\Omega$ around the direction $\mathbf{n}$ as seen in the rest frame of the particle is

$$\frac{1}{c} I(\boldsymbol{n}, \nu; x^\mu) \int_0^\infty d\nu' \int_{4\pi} d\Omega' \, \frac{d\sigma}{d\Omega'}(\boldsymbol{n}, \nu; \boldsymbol{n}', \nu') \left( \boldsymbol{n} - \frac{\nu'}{\nu} \boldsymbol{n}' \right) d\nu \, d\Omega \,, \qquad (1)$$

where $I(\boldsymbol{n}, \nu; x^\gamma)$ is the specific intensity of the incident radiation, $d\sigma/d\Omega'$ is the differential scattering cross section, and $\nu'$ is the frequency of the radiation scattered in the direction $\boldsymbol{n}'$. In writing this expression we have assumed that the radiation is unpolarized and that the scattering interaction is linear, i.e., that processes such as induced scattering can be neglected. The force produced by interaction with the full radiation field may be obtained by integrating expression (1) over $\boldsymbol{n}$ and $\nu$ (see LM95).

Acceleration of particles by radiation is often attributed to "radiation pressure". This usage is inaccurate and potentially misleading, as can be seen by considering scattering by a particle in an isotropic radiation field. The radiation pressure is nonzero, since it would collapse an evacuated reflecting balloon, but if (as in Thomson scattering) the momentum of the scattered radiation is zero as seen in the rest frame of the particle, the radiation exerts no force on the particle. Radiation pressure is not the same as radiation force. When we speak of the radiation force in the following discussion, we shall mean the force given by expression (1).

In general, the force on a particle is not proportional to any components of the radiation stress-energy tensor. If the momentum transfer cross section is independent of angle and frequency, the radiation force on a particle is in the direction of and proportional to the radiation flux in the comoving frame, but if the cross section is angle- or frequency-dependent, this is not generally true. Consider, for example, the force on a disk with one perfectly reflecting and one perfectly absorbing face, in an isotropic radiation field. Although the radiation flux is zero, the radiation force is not (the disk will be accelerated in the direction of the normal to the absorbing face). The radiation force is proportional to the radiation pressure *gradient* only if the momentum transfer cross section is independent of both angle and frequency, and the mean free paths of all photons are much smaller than all other length scales of interest (see LM95).

The radiation force on particles moving at nonrelativistic speeds relative to a radiation source can be divided into two components: a velocity-independent, "outward" force, and a velocity-dependent "drag" force that opposes the motion. Both components affect matter accreting onto a neutron star or black hole. Because of the substantial complications introduced by angle- or frequency-dependence of the momentum transfer cross section, in the following discussion we shall assume that, as in Thomson scattering, the cross section is independent of angle and frequency, so that the radiation force is proportional to and in the direction of the radiation flux.

## 3. Angular Momentum Loss

Under some circumstances, interaction with radiation can remove a substantial fraction of the angular momentum of accreting matter. Radial transport of angular momentum within an optically thick accretion disk is relatively unimportant. More interesting is the loss of angular momentum from a star or orbiting matter via emission of radiation, which can reduce the spin-up rate of the star. The most dramatic effects occur when radiation from a central source, such a star or a boundary layer around a star or black hole, interacts with matter orbiting in an accretion disk. For this reason, we shall focus here on the latter situation. Two distinct but related questions are of interest, namely, what is the rate at which angular momentum is transferred to the radiation field, and what is the total fraction of the angular momentum of the accreting matter that is eventually transferred to the radiation? The answer to the first question determines the time scale on which orbiting matter loses its angular momentum whereas the answer to the second determines the magnitude of the effect on the accretion flow.

To explore these two questions, consider the effect of radiation forces on matter orbiting a nonrotating star of luminosity $L$ that produces a spherically symmetric radiation field. For simplicity, let us assume that Newtonian gravity is a good approximation, that the motion is nonrelativistic, that the accretion flow is optically thin, that momentum transfer is via Thomson scattering, that the electrons are tied to the ions by electric and magnetic fields or collisions, and that the radiation is streaming radially outward from the star. Then $T^{tr} \approx T^{rr} \approx U$, where $U$ is the energy density of the radiation, and all other components of the radiation stress energy tensor are negligible.

The radiation flux produces a radially outward component of the radiation force that opposes the radially inward force of gravity. As a result, the orbital period of a particle in a circular orbit ($u_r = u_\theta = 0$) at radius $r$ is

$$t_\phi \equiv \frac{2\pi r}{u_\phi} = 2\pi \left( \frac{r^3}{\epsilon GM} \right)^{1/2} , \tag{2}$$

which is a factor $\epsilon^{-1/2}$ greater than in the absence of radiation. Here $\epsilon \equiv (L_\mathrm{E} - L)/L_\mathrm{E}$.

A particle with $u_\phi \neq 0$ also experiences an azimuthal radiation drag force, which transfers angular momentum from the particle to the radiation field and decreases the azimuthal velocity at a rate

$$\frac{du_\phi}{dt} = -\frac{\sigma_\mathrm{T}}{mc} u_\phi U \; . \tag{3}$$

The angular momentum of the particle is therefore lost in the characteristic time

$$t_{\mathrm{drag},\phi} = \frac{mc}{\sigma_\mathrm{T} U} \; . \tag{4}$$

The inspiral of the particle will be rapid if $t_{\mathrm{drag},\phi} \leq t_\phi$, which occurs at and inside the critical radius

$$R_\mathrm{crit} = (2\pi)^2 \frac{(1-\epsilon)^2}{\epsilon} \frac{GM}{c^2} \; . \tag{5}$$

The critical radius increases rapidly as $L$ approaches $L_\mathrm{E}$. Thus, if $L \approx 0.5\, L_\mathrm{E}$, $R_\mathrm{crit} \approx 10$ km, showing that radiation drag is important only near the star. If instead $L \approx L_\mathrm{E}$, $R_\mathrm{crit} \approx 250\,(0.05/\epsilon)$ km, and radiation drag is important even far away (Fortner et al. 1989).

The fraction of the angular momentum of the accreting matter that can be removed from the flow onto an accretion-powered star in the Newtonian point source approximation is at most equal to the efficiency $\eta \equiv L/\dot{M}c^2 \sim 0.2$ of the accretion process (Fortner et al. 1989). If the star is powered by some other process, such as a nuclear flash, the fraction that is lost to radiation can approach unity (ML93; see also Walker & Mészáros 1992).

Near a neutron star, the radiation drag force is increased by special and general relativistic effects and the finite angular size of the star (ML96). Table 1 summarizes these effects for a nonrotating, uniformly radiating star of radius $6M_\odot$. The angular momentum loss *rate* is increased by the larger photon density, Doppler effects, relativistic beaming of the scattered radiation, and the increase (by two factors of the redshift) of the locally measured luminosity, producing an overall factor $\sim 5$ increase in the loss rate over the rate in the Newtonian point-source approximation. The *total* angular momentum loss is increased by the Doppler effects, beaming, and one factor of the redshift, producing an overall increase by a factor $\sim 2$. Thus, up to $\sim 50\%$ of the angular momentum and energy of the matter accreting onto an accretion-powered neutron star can be removed by radiation. The loss rates and fractional losses near black holes are likely to be smaller, but may still be substantial. These losses can cause qualitative changes in the accretion flow (ML93; ML96).

Table 1. Increases in angular momentum loss rate and total loss.

| | Photon density | Doppler | Beaming | GR | Total |
|---|---|---|---|---|---|
| Loss rate | 2 | 4/3 | 4/3 | 3/2 | ~5 |
| Total loss | — | 4/3 | 4/3 | $(3/2)^{1/2}$ | ~2 |

The effect of radiation forces on the motion of accreting matter is altered by the changes in the exterior spacetime geometry caused by rotation of the central gravitating mass and by the changes in the radiation field produced by motion of the radiating matter (ML96). In general, the latter effect is much more important than the former. For all neutron stars with measured spin rates, retaining only terms that are first-order in the angular momentum of the star is an excellent approximation (ML96). This greatly simplifies calculations, since the spacetime exterior to a slowly rotating massive object is unique to first order in the angular momentum of the object (Hartle & Thorne 1968). To this order, spacetime around a slowly rotating star is the same as that around a black hole with the same mass and angular momentum. We emphasize that this is *not* true for rapidly rotating stars. Rotation of the source increases the inspiral time for particles in prograde orbits, causing the total angular momentum and energy loss to be greater near a rotating source, if the radiation energy density is low (ML96).

## 4. Effects on Accretion Disks

Radiation drag can remove a significant fraction of the angular momentum of matter orbiting near a neutron star or black hole (ML93; Fukue & Umemura 1995; ML96), increasing the inward radial velocity (especially near the radius of the marginally stable orbit, where the specific angular momentum is a minimum) and decreasing the optical depth of the flow. As a result of this "induced transparency", a flow may be optically thin in the vertical and radial directions even if it would be optically thick in the absence of radiation forces. We find that when radiation drag is included, the radial optical depth of the inner disk around a neutron star is less than unity for accretion rates $\dot{M} < 2 \times 10^{17}\,(h/r)\,\mathrm{g\,s^{-1}}$, where $h$ is the half-thickness of the disk (ML96).

Creation of an optically thin region can change completely the radial and vertical structure of the inner disk and the spectral and temporal properties of the radiation produced there. For luminosities $L \gtrsim 0.01\,L_{\mathrm{E}}$, the steady state inward radial velocity near the star is $v_r \approx 0.1$–$0.2\,\mathrm{c}$, and the azimuthal velocity near the central source is significantly less than the Keplerian angular velocity there (ML96).

Near rotating sources, the radiation force tries to force the accreting matter to orbit with the "equilibrium" angular momentum $u_{\phi 0} \sim v_0 R$, where $R$ is the radius and $v_0$ is the equatorial rotational velocity of the radiation source (ML96). At radii $r > R$, the equilibrium angular frequency $\Omega_0$ scales as $r^{-2}$. This effect may introduce new characteristic frequencies in the motion of accreting matter near black holes and nonmagnetic neutron stars.

## 5. Conclusions

Radiation forces can have important effects on disk accretion by neutron stars and black holes. If the luminosity is $\gtrsim 0.01\, L_{\rm E}$, radiation forces will generally terminate Keplerian flow at a radius greater than the stellar radius, masking the existence of a marginally stable orbit. Radiation forces can also reduce (or even eliminate) the velocity boundary layer at the stellar surface. Because they can produce large radial and vertical velocity shears within the accretion disk and an optically thin region at the inner edge, radiation forces can profoundly alter the radial and vertical structure of the disk, significantly affecting the spectral and temporal properties of the radiation from the disk. Radiation forces can also change significantly the profiles and spectra of thermonuclear bursts. Removal of angular momentum from accreting matter by radiation forces can reduce neutron star spin-up rates by factors $\sim$2–3, even for luminosities $L \approx 0.2\, L_{\rm E}$, possibly limiting the spin frequencies that can be produced by accretion and preventing neutron stars from being spun up to the spin rates at which gravitational wave instabilities are important. Finally, radiation forces may introduce new, potentially observable, periodicities in the motion of matter around neutron stars and black holes.

## Acknowledgments

F.K.L. thanks Professor Kato and the other organizers of the workshop for inviting him to participate and for providing reimbursement for travel and living expenses. M.C.M. acknowledges the support of a Compton Observatory Fellowship. This research was supported in part by NASA grant NAG 5-2925 and NSF grant AST93-15133 at the University of Illinois, and NASA grants NAG 6-2868 and 5-28543 at the University of Chicago.

## References

Fortner B. A., Lamb F. K., Miller G. S. 1989, Nature 342, 775
Fukue J., Umemura M. 1995, PASJ 47, 429
Hartle J. B., Thorne K. S. 1968, ApJ 153, 807
Lamb F. K., & Miller M. C. 1995, ApJ 439, 828
Miller M. C., Lamb F. K. 1993, ApJL 413, L43
Miller M. C., Lamb F. K. 1996, ApJ in press
Walker M. A., Mészáros P. 1989, ApJ 346, 844

# Cosmological Accretion Disks Driven by Radiation Drag

Masayuki UMEMURA

*Center for Computational Physics, University of Tsukuba, Ibaraki 305, Japan*

## Abstract

The accretion disk driven by the radiation drag which is exerted by the isotropic cosmic background radiation is analyzed. From a viewpoint of timescales, the drag force seems more efficient than the $\alpha$-viscosity to transfer angular momenta in high redshifts greater than 200. Steady accretion due to the radiation drag is possible both for a point-mass and dark-matter potentials if the drag coefficient is constant. In the former case, the solution can connect to the center. Also, the non-steady evolution is numerically investigated. The cosmological collapse of a density fluctuation leads to a rigidly rotating accretion disk in a dark-matter potential. The disk shrinks through the angular momentum loss due to the drag. Consequently, the disk evolves into an optically-thick, self-gravitating disk, which is gravitationally unstable for non-axisymmetric modes.

## 1. Radiation Drag

### 1.1. Equation of Radiation Hydrodynamics

The equation of radiation hydrodynamics where the radiation force is dominated by the Thomson scattering is given by

$$\frac{\partial \boldsymbol{v}}{\partial t} + \boldsymbol{v}\nabla\boldsymbol{v} = -\nabla\Phi - \frac{1}{\rho}\nabla p + \frac{\sigma_T \chi_e}{\mu m_p c}[\boldsymbol{F}_\gamma - (E_\gamma + \boldsymbol{P}_\gamma)\boldsymbol{v}] \tag{1}$$

in the small $\boldsymbol{v}/c$ limit, where $\sigma_T$ is the Thomson scattering cross section, $E_\gamma$ is the radiation energy density, $\boldsymbol{F}_\gamma$ is the radiation flux, $\boldsymbol{P}_\gamma$ is the radiation

*S. Kato et al. (eds.), Physics of Accretion Disks, 203–208.*
© 1996 OPA (Overseas Publishers Association) Amsterdam B.V.

stress tensor, $\chi_e$ is the ionization fraction, $\mu$ is mean molecular weight, and the others have usual meanings. For the cosmic microwave background radiation, $E_\gamma = aT_{\gamma 0}(1+z)^4$, where $z$ is the cosmological redshift, and $|\boldsymbol{P}_\gamma| = E_\gamma/3$ because the radiation fields are isotropic. Furthermore, if the medium is optically thin against the Thomson scattering opacity, then $\boldsymbol{F}_\gamma = 0$. Thus, the radiation drag force is given by $-\beta \boldsymbol{v}$ with

$$\beta(z) = \frac{4}{3}\frac{\sigma_T \chi_e a T_{\gamma 0}^4}{\mu m_p c}(1+z)^4. \tag{2}$$

### *1.2. Efficiency of Angular Momentum Transfer*

The azimuthal component of equation (1) is

$$\frac{dj}{dt} = -\beta(z) j, \tag{3}$$

where $j$ is the specific angular momentum. An estimate on the magnitude of angular momentum is made based on arguments of the cosmological collapse of a primordial density fluctuation. A density fluctuation gains angular momentum due to the tidal torque exerted by ambient fluctuations primarily in the linear evolution stage (Peebles 1969; Heavens & Peacock 1988), and the final spin parameter is typically

$$\lambda \equiv \frac{J_{\rm T}|E_{\rm T}|^{1/2}}{GM_{\rm T}^{5/2}} \sim 0.05, \tag{4}$$

where $J_{\rm T}$ is the total angular momentum, $E_{\rm T}$ is the total energy, and $M_{\rm T}$ is the total mass of the fluctuation. After the fluctuation is virialized, the spin parameter $\lambda$ can be translated into the specific angular momentum as

$$j_b \simeq R_{\rm max} v_\phi \simeq R_{\rm max}\sigma\lambda = (2GM_T R_{\rm max})^{1/2}\lambda, \tag{5}$$

where $R_{\rm max}$ is the radius of the maximum expansion. If considering a spherical fluctuation, the maximum expansion radius is given by

$$R_{\rm max} = \left(\frac{4M_b}{3\pi\rho_c\Omega_b}\right)^{1/3}(1+z)^{-1}, \tag{6}$$

where $M_b$ is the baryonic mass, and $\Omega_b$ is the baryon density parameter.

The timescale of the angular momentum transfer due to the radiation drag is

$$\tau_\beta = \frac{1}{\beta(z)} = 3.4\times 10^6 {\rm yr}\ \chi_e^{-1}\left(\frac{T_{\gamma 0}}{2.735K}\right)^{-4}\left(\frac{1+z}{200}\right)^{-4}. \tag{7}$$

On the other hand, the Hubble time at $z$ is

$$\tau_{\rm H} = 4.6 \times 10^6 {\rm yr} \left(\frac{0.5}{h}\right)^{-1} \left(\frac{1+z}{200}\right)^{-3/2}, \tag{8}$$

where $h$ is the Hubble constant in the units of 50 km s$^{-1}$ Mpc$^{-1}$. $\tau_\beta$ becomes shorter than $\tau_{\rm H}$ at high redshift epochs of $z > 200$ if the medium is ionized ($\chi_e \sim 1$). Thus the radiation drag could be an effective mechanism which transfers angular momenta at such high redshift epochs (Loeb 1993).

### *1.3. Comparison with α-viscosity*

The azimuthal equation of motion where the $\alpha$-viscosity works is

$$\frac{dj}{dt} = -\frac{1}{2}\alpha c_s^2, \tag{9}$$

if the rotation is Keplerian and the gas is isothermal with the sound velocity $c_s$. Using equations (5) and (6) for $j$, the timescale of angular momentum transfer due to $\alpha$-viscosity is given by

$$\begin{aligned} \tau_\alpha &= 1.1 \times 10^7 {\rm yr}\ \alpha^{-1} \left(\frac{M_b}{10^6 M_\odot}\right)^{2/3} \left(\frac{\Omega_b}{0.05}\right)^{-2/3} \\ &\times \left(\frac{\lambda}{0.05}\right) \left(\frac{h}{0.5}\right)^{-1/3} \left(\frac{1+z}{200}\right)^{-1/2} \left(\frac{c_s}{10^3 {\rm K}}\right)^{-1}. \end{aligned} \tag{10}$$

If this is compared with $\tau_\beta$, the radiation drag could be more efficient at high redshifts even if $\alpha$ is unity. [It should be noted that in a primordial disk the $\alpha$ may be much less than unity, because the rotation is not Keplerian but nearly rigid when a central massive object does not exist (Umemura et al. 1993).]

## 2. Steady β Accretion Disk

First we consider the steady axisymmetric accretion driven by radiation drag (Fukue & Umemura 1994). The basic equations for a geometrically thin disk are described as follows:

The continuity equation is

$$2\pi r \Sigma v_r = -\dot{M}, \tag{11}$$

where $\Sigma$ is the surface density, $v_r$ the radial infalling velocity, and $\dot{M}$ the constant accretion rate.

The equation of motion in the radial direction is

$$v_r \frac{dv_r}{dr} - \frac{v_\varphi^2}{r} = -\frac{1}{\Sigma}\frac{d\Pi}{dr} - \frac{d\Phi}{dr} - \beta v_r, \tag{12}$$

where $v_\varphi$ is the azimuthal rotation velocity, $\Pi$ is the pressure integrated over the vertical direction, and $-d\Phi/dr$ is the gravitational force exerted by the external gravitational potential (ignoring self-gravity). Here the coefficient of the drag, $\beta$, is assumed to be constant, *temporally* and *spatially*. In a point-mass case, the gravitational force is $-d\Phi/dr = -GM/r^2$, where $M$ is the mass of a central object, while in a dark matter case, $-d\Phi/dr = -r\Omega_{\rm DM}^2$, where $\Omega_{\rm DM}$ is constant.

The equation of motion in the azimuthal direction is expressed as

$$v_r \frac{dv_\varphi}{dr} + \frac{v_r v_\varphi}{r} = -\beta v_\varphi. \tag{13}$$

We assume also that the disk gas is isothermal for simplicity.

In the point-mass case, introducing the radial Mach number $\mathcal{M}_r$ $(= v_r/c_s)$ and azimuthal Mach number $\mathcal{M}_\varphi$ $(= v_\varphi/c_s)$ and measuring the radius in units of $GM/c_s^2$, the basic equations become

$$\frac{d\mathcal{M}_r}{dr} = \frac{\mathcal{M}_r \left( \frac{\mathcal{M}_\varphi^2}{r} + \frac{1}{r} - \frac{1}{r^2} - \hat{\beta}\mathcal{M}_r \right)}{\mathcal{M}_r^2 - 1}, \tag{14}$$

$$\frac{d\mathcal{M}_\varphi}{dr} = -\left( \frac{\mathcal{M}_\varphi}{r} + \hat{\beta}\frac{\mathcal{M}_\varphi}{\mathcal{M}_r} \right), \tag{15}$$

where $\hat{\beta} = GM\beta/c_s^3$. These equations are so-called *wind equations* for the present problem. For a given $\beta$, we find transonic solutions which connect to the asymptotic solutions near the center which are

$$v_r = -2\beta r - 4\beta c_s^2 \frac{r^2}{GM}, \quad (r \to 0) \tag{16}$$

$$v_\varphi = \sqrt{\frac{GM}{r}} - c_s^2 \sqrt{\frac{r}{GM}}. \quad (r \to 0) \tag{17}$$

These imply that the centrifugal force is almost in balance with the gravitational force near the center (Keplerian rotating) and the steady accretion is possible where the radial velocity is proportional to the radius at the very central region.

On the other hand, the wind equations in a dark-matter potential are, measuring the radius in units of $c_s/\Omega_{\rm DM}$,

$$\frac{d\mathcal{M}_r}{dr} = \frac{\mathcal{M}_r \left( \frac{\mathcal{M}_\varphi^2}{r} + \frac{1}{r} - r - \hat{\beta}\mathcal{M}_r \right)}{\mathcal{M}_r^2 - 1}, \tag{18}$$

$$\frac{d\mathcal{M}_\varphi}{dr} = -\left( \frac{\mathcal{M}_\varphi}{r} + \hat{\beta}\frac{\mathcal{M}_\varphi}{\mathcal{M}_r} \right), \tag{19}$$

where $\hat{\beta} = \beta/\Omega_{\rm DM}$. We find transonic solutions which cross the critical curve at one point for a given $\beta$. In a dark-matter potential the gravitational acceleration becomes negligible near the center, while thermal pressure does not. Thus, any solutions cannot connect to the center. Nonetheless, these solutions are applicable to the early phases of the accretion from large radii.

## 3. Non-steady Evolution of Cosmological Accretion Disk

In the previous section, the *steady* disk accretion is analyzed, ignoring self-gravity and assuming isothermality. Here, 3D numerical analyses are presented for the non-steady disk accretion, including self-gravity and the thermal history (Umemura et al. 1993). The equation of motion is given by

$$\frac{d\boldsymbol{v}}{dt} = -\boldsymbol{\nabla}\Phi - \frac{1}{\rho}\boldsymbol{\nabla}p + \mathcal{T} \times \boldsymbol{r} - \beta(z)[\boldsymbol{v} - H(z)\boldsymbol{r}], \tag{20}$$

where $\mathcal{T} \times \boldsymbol{r}$ is the tidal force which produces angular momenta, and $\Phi$ is the potential of self-gravity including dark matter. The amplitude of the tidal force is determined so that it increases the spin parameter of the dark matter up to a value of $\lambda = 0.05$ at maximum expansion. The thermal history of the gas is described by the energy equation,

$$\frac{\rho}{\gamma - 1}\frac{d}{dt}\left(\frac{P}{\rho}\right) - \frac{P}{\rho}\frac{d\rho}{dt} = -\Lambda \ , \tag{21}$$

where $\gamma = 5/3$ is the adiabatic exponent, and $\Lambda$ is the cooling function combining Compton, Bremsstrahlung, recombination, and line cooling for a primordial abundance with hydrogen and helium (Umemura 1993).

During the evolution of the gas cloud we calculate its radial optical depth around the center of mass (averaged over angles), $\tau \equiv \int \langle n_e \rangle_{d\Omega} \sigma_{\rm T} dr$, where $n_e$ is the electron density. If $\tau$ exceeds unity in a sphere of radius $r$ the radiation drag of the gas enclosed by this sphere is turned–off, a crude but conservative approximation for the self-shielding which eventually releases the central regions from the radiation drag.

The nonlinear dynamics of the gas is followed by a Smooth Particle Hydrodynamics (SPH) code that was optimized for the present calculation. The SPH method is combined with an $N$-body scheme in which the gravitational interactions are calculated by direct summations.

The numerical results show the four-step non-steady evolution. First, a top-hat density fluctuation collapses with the gain of angular momentum by the tidal torque. Second, a rigidly-rotating disk forms in centrifugal balance within a dark matter potential. Third, the disk shrinks *homologously* through angular momentum transfer due to the radiation drag. Finally, an optically-thick, self-gravitating disk forms, which is similar to a Maclaurin disk.

## 4. Discussion

The fate of the optically-thick, self-gravitating disk is an open question. However, we give a speculation. The Toomre's $Q$-value estimated using the properties of the disk is

$$\begin{aligned} Q &= \frac{2\Omega c_s}{\pi G \Sigma} \\ &= 0.06 \left(\frac{\Omega}{10^{-4}\mathrm{yr}^{-1}}\right)\left(\frac{\Sigma}{500 \mathrm{~g~cm}^{-2}}\right)\left(\frac{c_s}{10 \mathrm{~km~s}^{-1}}\right) \ll 1. \end{aligned} \tag{22}$$

Thus, the disk is expected to be gravitationally unstable against tightly winding modes. Actually, we can see the appearance of non-axisymmetric modes in the numerical results. However, they do not seem to be tightly winding waves, but global spiral modes. This implies that the local analyses are not sufficient, but the global analyses are required on the stability. Anyhow, these non-axisymmetric modes may play an important role for the further evolution of the disk in the sense that they may also transfer angular momenta effectively (Kikuchi 1995).

If the non-axisymmetric modes transfer angular momenta outwards, the inner parts of the disk infalls further and the outer parts expands. After a dense core forms at the center, the disk would be stabilized around the center where the self-gravity is much weaker than the gravity of the central object. On the other hand, the outer parts might become optically thin again by the expansion, and the radiation drag would work. Then, the gas infalls through the angular momentum loss. Again, the disk could be unstable against non-axisymmetric modes because of the enhancement of the self-gravity. These processes might proceed recurrently. Finally, a central massive object may form with a surrounding small disk. Elaborate analyses on such evolution are presented elsewhere.

## References

Fukue J., Umemura, M. 1994, PASJ 46, 87
Heavens A., Peacock J. 1988, MNRAS 232, 339
Kikuchi N. 1995, this volume
Loeb A. 1993, ApJ 403, 542
Peebles P. J. E. 1969, ApJ 155, 393
Umemura M. 1993, ApJ 406, 361
Umemura M., Loeb A., Turner, E. L., 1993, ApJ 419, 459

# Beta Accretion Disks Driven by External Radiation Drag

Jun FUKUE
*Astronomical Institute, Osaka-Kyoiku University, Asahigaoka, Kashiwara, Osaka 582, Japan*

## Abstract

Accretion disks driven by external radiation drag exerted by a central luminous source are presented under the steady and sub-relativistic approximations. Due to the effect of external radiation drag, the angular momentum of the gas is removed. As a result, a nearly Keplerian rotating disk far from the center turns to a radially infalling one near the center. The relativistic case is briefly discussed.

## 1. Radiation Drag in Accretion Phenomena

The interaction between the accreting gas and the radiation field, in particular *Compton drag*, attracts many researcher's interest in recent years (see table 1). Under these circumstances, we have been investigating the gaseous accretion disk driven by radiation drag – $\beta$-disks (Fukue & Umemura 1994, 1995). In this paper I explain the basic properties of the $\beta$-disk, and briefly show the preliminary results on the extension to the relativistic regime.

Before we proceed further, it should be noted that, in these $\beta$ processes, the radiation drag coeffecient $\beta$ is *not a parameter* (like $\alpha$ in the standard disk) *but a function* of the radiation energy density $E$ if we express the drag force, which is proportional to the velocity, as $-\beta \boldsymbol{v}$. That is, for cosmological $\beta$-disks (Fukue & Umemura 1994), $\beta = \frac{\chi_e \sigma_T}{mc}\frac{4}{3}E = \frac{\chi_e \sigma_T}{mc}\frac{4}{3}\varepsilon_{\gamma 0}(1+z)^4$, where $\chi_e$ is the ionization fraction, $\sigma_T$ the Thomson cross section, $m$ the particle mass, $c$ the light speed, $\varepsilon_{\gamma 0}$ the present energy density of the cosmic background radiation, and $z$ the redshift. For binary $\beta$-disks (Fukue & Umemura 1995), it is expressed as $\beta = \frac{\chi_e \sigma_T}{mc}2E = \Gamma \frac{r_g c}{r^2}$, where $r$ is the radial distance, $r_g$ the Schwarzschild radius of the central object, and $\Gamma$ the normalized luminosity defined below.

*S. Kato et al. (eds.), Physics of Accretion Disks, 209–214.*
© 1996 OPA (Overseas Publishers Association) Amsterdam B.V.

Table 1. The Importance of the Compton (Radiation) Drag around an Accretion Disk.

| **Cosmological Case** $z \sim 400$ | | **Binary Case** $z \sim 0$ |
|---|---|---|
| | OBJECT | |
| Protoquasars | | X-ray stars<br>Cataclysmic variables |
| | SOURCE | |
| "1000K" background | | Central star<br>Accretion disk itself |
| | EFFECTS | |
| Massive black hole formation<br>*Loeb (1993)*<br>*Umemura et al. (1993)* | | QPOs in X-ray stars<br>*Fortner et al. (1989)*<br>*Lamb (1989, 1991)*<br>Siphon process in CVs<br>*Meyer and Meyer-H (1994)* |
| Cosmological $\beta$ disks<br>*Fukue and Umemura (1994)*<br>*Tsuribe et al. (1994)*<br>*Takahashi et al. (1995)* | | $\beta$ accretion disks<br>*Fukue and Umemura (1995)* |

## 2. Subrelativistic Regime

Let us suppose an axisymmetric gaseous disk rotating around a central object of mass $M$. The disk is assumed to be geometrically thin, and therefore, the physical quantities are integrated over the vertical direction. It is also assumed to be effectively optically thin. We use cylindrical coordinates $(r, \varphi, z)$ with the $z$-axis along the rotation axis of the disk, and adopt the Newtonian formalism for the gas.

Then, the continuity equation is

$$2\pi r \Sigma v_r = -\dot{M}, \tag{1}$$

where $\Sigma$ is the surface density, $v_r$ the radial infalling velocity, and $\dot{M}$ the constant accretion rate. Using the density $\rho$ of the disk gas and the half-thickness $H$ of the disk, the surface density is expressed as $\Sigma = 2\rho H$.

To the lowest order of $\boldsymbol{v}/c$, the equation of motion in the radial direction under the influence of radiation fields is

$$v_r \frac{dv_r}{dr} - \frac{v_\varphi^2}{r} = -\frac{1}{\Sigma}\frac{d\Pi}{dr} - \frac{GM}{r^2} + \frac{1}{\rho}\frac{n_e \sigma_T}{c}\left(F - E v_r - v_j P^{rj}\right), \tag{2}$$

where $v_\varphi$ is the azimuthal rotation velocity, $\Pi$ $(= 2pH)$ the gas pressure integrated over the vertical direction, $n_e$ the electron number density, $E$ the radiation energy density, $F$ the radiation flux, and $P^{ij}$ the radiation stress tensor.

The equation of motion in the azimuthal direction under the influence of radiation fields is

$$v_r \frac{dv_\varphi}{dr} + \frac{v_r v_\varphi}{r} = -\frac{1}{\rho} \frac{n_e \sigma_T}{c} \left( E v_\varphi + v_j P^{\varphi j} \right). \tag{3}$$

Here, the right-hand side comes from external radiation drag, which plays an essential role in the present model.

For radiation fields produced by a central source with luminosity $L$, the radiation flux $F$ is expressed as $F = L/4\pi r^2$ and $E$ becomes $F/c$ at large $r$. Furthermore, $P^{rr} = E$ and other components of $P^{ij}$ vanish at large $r$.

The equations of motion are then rewritten as

$$v_r \frac{dv_r}{dr} - \frac{v_\varphi^2}{r} = -\frac{1}{\Sigma}\frac{d\Pi}{dr} - \frac{GM}{r^2}\left(1 - \Gamma + 2\Gamma\frac{v_r}{c}\right), \tag{4}$$

$$v_r \frac{dv_\varphi}{dr} + \frac{v_r v_\varphi}{r} = -\frac{GM}{r^2}\Gamma\frac{v_\varphi}{c}. \tag{5}$$

Here, we introduce the parameter $\Gamma$:

$$\Gamma = \chi_e \frac{m_p}{m}\frac{L}{L_E}, \tag{6}$$

where $\chi_e$ is the ionization rate, $m_p$ the proton mass, $m$ the particle mass, and $L_E$ ($= 4\pi c GM m_p/\sigma_T$) the Eddington luminosity. For fully ionized electron-proton plasmas, this parameter $\Gamma$ is just the central luminosity normalized by the Eddington luminosity, and therefore, should be less than or equal to unity (for electron-positron plasmas $\Gamma$ could become of the order of $m_p/m_e$).

## 3. Drag-Driven-Disks

In order to understand the basic properties of the present drag-driven-disk (DDD), here I pick up a cold case, where the pressure-gradient force is ignored (cf. Fukue & Umemura 1995 for a warm case).

In this case, equations (4) and (5) have the asymptotic solutions at infinity:

$$v_r = -\frac{2GM\Gamma}{cr}, \tag{7}$$

$$v_\varphi = \sqrt{\frac{GM(1-\Gamma)}{r}}, \tag{8}$$

whereas they have the asymptotic solutions near to the center such as

$$v_r = v_\infty, \tag{9}$$

$$v_\varphi \propto \frac{1}{r} e^{-r_0/r}. \tag{10}$$

Here, the characteristic velocity $v_\infty$ and the characteristic radius $r_0$ are respectively expressed as

$$v_\infty = -\frac{1-\Gamma}{2\Gamma}c, \tag{11}$$

$$r_0 = \frac{GM(1-\Gamma)}{2v_\infty^2} = \frac{\Gamma^2}{1-\Gamma}r_g, \tag{12}$$

where $r_g = 2GM/c^2$ is the Schwarzschild radius of the central object. Inside this characteristic radius $r_0$ the angular momentum is quickly lost and the rotating flow turns to radially infalling flow with *terminal speed* $v_\infty$, where the effective gravitational force is balanced by the radiation drag force.

In figure 1 several examples of the numerical solutions, which satisfy boundary conditions above, are shown.

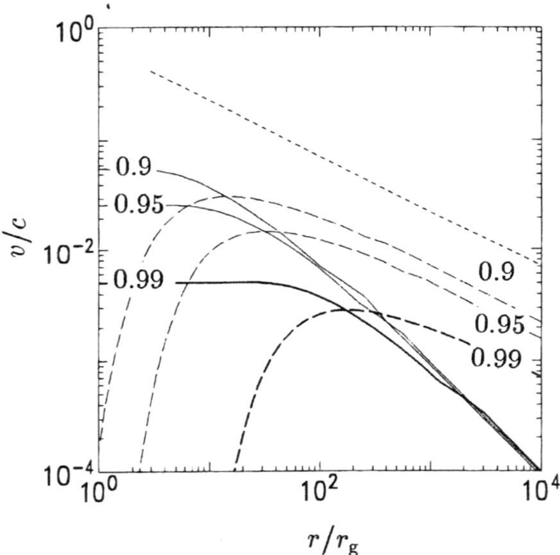


Fig. 1. Examples of numerical solutions in the cold regime. The abscissa is the radius in units of $r_g$ and the ordinate is the velocity in units of $c$. Solid curves denote the infall velocity $v_r$ while dashed ones represent the rotation velocity $v_\varphi$. The dotted line expresses the Keplerian velocity. The values of $\Gamma$ are attached on each curve.

As well known, no accretion take place for $\Gamma = 0$ under the Newtonian approximation. [It should be noted that in the relativistic case accretion is possible, since the height of the angular-momentum barrier becomes finite (see Fukue 1987 and references therein).] However, when there is an interaction between the disk gas and radiation fields ($\Gamma \neq 0$), the angular momentum is removed and accretion may become possible even in the Newtonian case.

Such accretion disks, where the angular momentum is removed via the external drag of radiation fields from the central source, would be important in several astrophysical contexts. For example, in the case of an X-ray burster the radiation density at the burst phases is very high in the inner region of the accretion disk, and therefore, the gas-accretion processes are remarkably enhanced due to the external radiation drag. In proto-quasars radiation drag may be the most dominant mechanism to remove angular momentum of the gas. Also, in a protoplanetary disk the external radiation drag may play an important role if a sufficient amount of dust exists in the disk.

## 4. Relativistic Regime

Now I briefly examine the relativistic case (Fukue 1996). The full set of basic equations for photohydrodynamics is found in several literatures (e.g., Lindquist 1966; Hsieh & Spiegel 1976; Fukue et al. 1985).

For simplicity, I use the general relativity for gas, but the special relativity for radiation; the general relativistic effects such as the gravitational bending of light are ignored (cf. Abramowicz et al. 1990; Yamada & Fukue 1993). Furthermore, the flow is supposed to be steady and axisymmetric.

In figure 2 examples of the numerical solutions, which satisfy appropriate boundary conditions, are shown.

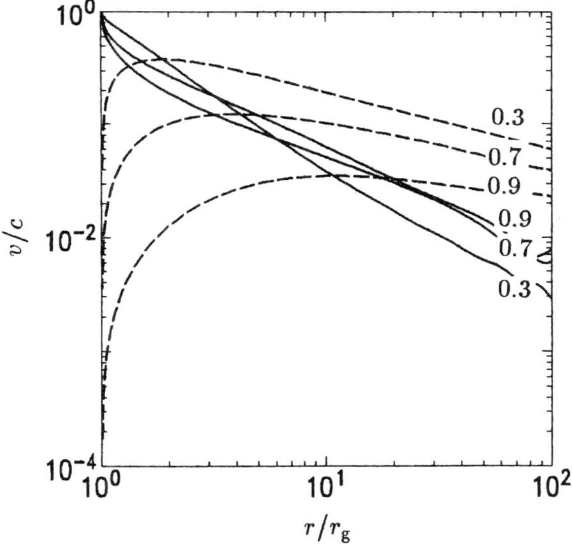


Fig. 2. Examples of numerical solutions in the relativistic regime. The abscissa is the radius in units of $r_g$ and the ordinate is the velocity in units of $c$. Solid curves denote the three-velocity $v$ for the corotating observer, while dashed ones represent the azimuthal velocity $v_\varphi$ measured by the static observer. The values of $\Gamma$ are attached on each curve.

In the relativistic case the terminal velocity disappears very close to the Schwarzschild radius and the radial infall velocity approaches the light speed, although the angular momentum is also removed similar to the subrelativistic case.

## 5. Concluding Remarks

Due to radiation drag, the angular momentum is removed and gas accretion takes place. In the subrelativistic regime, the angular momentum is quickly lost inside the characteristic radius $r_0 = \frac{\Gamma^2}{1-\Gamma} r_g$, where $\Gamma$ is the effective normalized luminosity. In the lowest order of $v/c$, there appears the terminal velocity $v_\infty = -\frac{1-\Gamma}{2\Gamma} c$, at which the effective gravity is balanced by radiation drag. Very close to the Schwarzschild radius, the terminal velocity disappears due to the relativistic effect.

In addition to a drag-driven-disk — $\beta$ *disk* (Fukue & Umemura 1995) discussed here, radiation drag may play an important role in several situations; e.g., in an accretion disk corona — $\beta$ *corona* (see Watanabe 1996, and also in this volume), and in an accretion disk wind — $\beta$ *wind* (see Tajima 1996, and also in this volume).

Moreover, we should consider the relativistic case more carefully (cf. Abramowicz et al. 1990). The transition case from an optically-thin regime to an optically-thick regime is also important, in particular, for stability problems (see Tsuribe in this volume). Finally, we should combine the present $\beta$-disk with the traditional $\alpha$-disk into an '$\alpha\beta$-disk'.

## References

Abramowicz M. A., Ellis G. F. R., Lanza A. 1990, ApJ 361, 470
Fortner B. A., Lamb F. K., Miller G. S. 1989, Nature 342, 775
Fukue J. 1987, PASJ 39, 309
Fukue J. 1996, in preparation
Fukue J., Umemura M. 1994, PASJ 46, 87
Fukue J., Umemura M. 1995, PASJ 47, 429
Fukue J., Kato S., Matsumoto R. 1985, PASJ 37, 383
Hsieh S.-H., Spiegel E. A. 1976, ApJ 207, 244
Lamb F. K. 1989, in 23rd ESLAB Symposium, p215
Lamb F. K. 1991, in Neutron Stars: Theory and Observations, ed J. Ventura, D. Pines (Kluwer Academic Publishers, Dordrecht) p.445
Lindquist R. W. 1966, Ann. Phys. 37, 487
Loeb A. 1993, ApJ 403, 542
Meyer F., Meyer-Hofmeister E. 1994, A&A 288, 175
Tajima Y., Fukue J. 1996, PASJ submitted
Takahashi A., Fukue J., Sanbuichi K., Umemura M. 1995, PASJ 47, 425
Tsuribe T., Umemura M., Fukue J. 1994, PASJ 46, 597
Umemura M., Fukue J. 1994, PASJ 46, 567
Umemura M., Loeb A., Turner E. L. 1993, ApJ 419, 459
Watanabe Y., Fukue J. 1996, PASJ submitted
Yamada T., Fukue J. 1993, PASJ 45, 97

# Rapid Mass Accretion by Background Radiation Force

Tohru TSURIBE[1,2] and Masayuki UMEMURA[1]
1. *Research Center for Computational Physics, University of Tsukuba, Tsukuba, Ibaraki 305, Japan*
2. *Department of Astronomy, School of Science, The University of Tokyo, Bunkyo-ku, Tokyo 113, Japan*

## Abstract

The proto-AGN disks are investigated. The dynamical effect of the cosmic background radiation (CBR) is especially studied in the view point of the angular momentum transport. The full-angle-dependent plane-parallel radiation transfer in the moving media is performed. It is found that the efficiency of the angular momentum transport from disk to the background radiation decreases exponentially with optical depth independent of the flow velocity. An example of the evolution and the probability distribution of the angular momentum of the disk is also displayed. If spin-up disks which have large angular momentum $\lambda > 0.05$ forms at $1 + z > 200$, and the gas can be reionized at this epoch, these disks shrink rapidly and tend to have the typical angular momentum and typical Thomson scattering optical depth of the order of unity. Thus AGN seed black holes can form at $1+z > 10$ by CBR force and $\alpha$-viscosity.

## 1. Introduction

In the large redshift epoch, the energy density of CBR is larger than that of the present day $[E_{\rm R} = 10^8((1+z)/400)^4 aT_{\gamma 0}^4]$. Thus, the interaction between disk and CBR is effective in the dynamical evolution of the disks. The main obstacle against to form supermassive QSO black hole at high redshift is the angular momentum. One interesting idea is that the angular momentum is effectively removed due to the interaction between the gas and CBR (Loeb

*S. Kato et al. (eds.), Physics of Accretion Disks, 215–218.*
© 1996 OPA (Overseas Publishers Association) Amsterdam B.V.

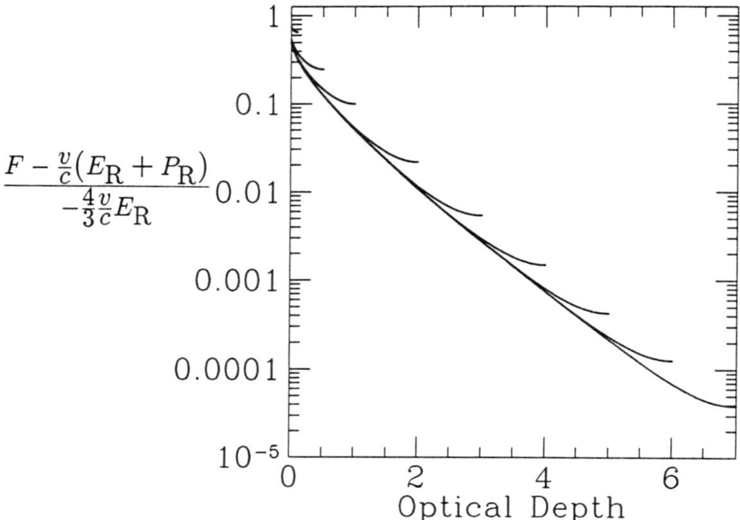


Fig. 1. Net radiation resistant force to the matter of the disk normalized by the constant drag force. Resistant force decreases with optical depth exponentially.

1993). Considering the interaction of matter and radiation, cosmological accretion disks can be divided into three different regions. In extremely optically-thin region ($\tau \ll 1$), the matter feels the radiation drag force. In the extremely optically-thick region ($v_{\mathrm{diffusion}} < v_{\mathrm{flow}}$), photons are trapped inside the matter and no momentum is transported from matter to CBR fields. Although the treatment of first two regions is rather simple (Umemura et al. 1993, Fukue & Umemura 1994), it is not simple in the intermediate region ($\tau > 1, v_{\mathrm{diffusion}} > v_{\mathrm{flow}}$). In this region CBR photons can transport into the inner part of the disk and interact with matter. But it has not been well known the importance of this transport effect. In order to understand this effect quantitatively, the calculation of the radiation transfer is performed (section 2). Some applications and cosmological consequence are discussed in section 3.

## 2. Background Radiation Field in Rotating Disks

Suppose a rotating disk embedded by the isotropic background radiation. At the first stage of proto-AGN disk evolution, there should be no central black hole. Thomson scattering is the main interaction between matter and CBR. In this paper, only Thomson scattering is considered for simplicity. To calculate the frequency-integrated transfer equations, we adopt following assumptions: 1) The disk is geometrically thin to use plane-parallel geometry.

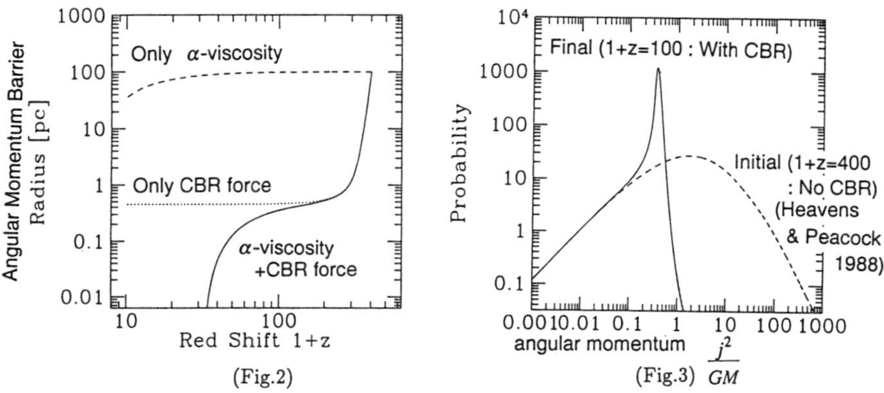


Fig. 2. Evolution of the angular momentum of the disks. Kepler rotation and isothermal perfect ionized gas is assumed. In this example, $T_{\rm CBR} = 2.74$ K at present, $M_{\rm disk} = 10^5 M_\odot$, $c_s = 10$ km s$^{-1}$, $\alpha = 0.001$ are assumed.

Fig. 3. Probability function of disk angular momentum. The optical depth of the disk at the peak is of the order of unity almost independent of initial distribution.

2) Only the diffusion regime ($v_{\rm dif} > v_{\rm flow}$) is considered to use steady radiation in hydrodynamics. 3) Rotation velocity is a function of only radius and time. 4) Outer boundary radiation field is isotropic. Using both the variable Eddington factor method (Miharas & Miharas 1984) and the accelerated lambda iteration method, consistent results were obtained. The net radiation force to the matter is shown in figure 1 normalized by a constant drag force. The radiation force to the matter is the sum of the drag force and the induced flux force. In optically thin disks, the radiation force is large because the induced flux is small. The induced flux increases with optical depth. In the optically thick disk, only near the disk surface is optically thin and inner part of the disk is optically thick. In the central optically-thick region, drag force and induced flux force cancel and net radiation force is nearly zero. Important result is that this net force decreases almost exponentially with optical depth. This result can be understood easily, but this was not obvious until this calculation was done. The *r*-*z* shear accretion flow will arise by this radiation effect.

## 3. Cosmological Evolution of the Disks

In previous section, we found that the angular-momentum transport efficiency decreases exponentially with optical depth. In this section the dynamical evolution of the disk is discussed, assuming that this efficiency is the mean value of $e^{\tau_1}$ and $e^{\tau_2}$, where $\tau_1$ and $\tau_2$ are the optical depths from disk surface of top and bottom, respectively. We should note that this simple exponential approximation overestimates the efficiency of the angular momentum transport. The result of the previous section is smaller than this approximation because the photon that is not perpendicular to the disk feels a larger optical depth than that is measured perpendicular to the disk.

First, the 1D axisymmetric hydrodynamical calculation is done. By this calculation, it is found that if the pressure gradient is negligible, initial power-law type surface-density distributions change into final top-hat-like distributions. This is because gas accretes more rapidly in the outer optically-thin region than the inner optically-thick region. The gas accretion stops at the final radius, where the optical depth is about 3. After that, disks can no more shrink in cosmological time scale.

Next, the cosmological evolution of the angular momentum of disks is discussed. An example of redshift evolution is shown in figure 2. Large initial angular-momentum disk can shrink only factor of three by $\alpha$-viscosity. When only radiation force acts, disk shrinks rapidly at first and becomes optically thick. After that disk can not shrink any longer. If $\alpha$-viscosity and radiation force are included, disks can shrink over the fourth of magnitude. This is delightful for the formation of the central black hole at the high redshift.

Finally, angular momentum probability function is shown in figure 3. Before the radiation force does not work, the density fluctuation obtains the angular momentum widely distributed by the tidal interaction (Heavens & Peacock 1988). After the radiation effect works, a very sharp peak appears in the distribution function. This shows that most of the disks become to have typical angular momentum ($\lambda \sim 0.05$) because disks which have large initial angular momentum are optically thin and accrete rapidly until disk optical depth is of the order of unity. This result depends mainly on the processes of scattering effect and almost does not depend on the initial distribution function.

## Acknowledgements

The authors would like to express their sincere thanks to Dr. J. Fukue, Dr. T. Nakamoto and Dr. F. Nakamura for their helpful discussions and comments.

## References

Fukue J., Umemura M. 1994, PASJ 46, 87

Heavens A., Peacock J. 1988 MNRAS 232, 339

Loeb A. 1993, ApJ 403, 542

Miharas D., Miharas B. W. 1984, Foundations of Radiation Hydrodynamics (Oxford Univ. Press, New York)

Umemura M., Loeb A., Turner E. L. 1993, ApJ 419, 459

# Radiative Winds from Accretion Disks under Radiation Drag

Yukiko TAJIMA and Jun FUKUE
*Astronomical Institute, Osaka-kyoiku University,
Asahigaoka, Kashiwara, Osaka 582, Japan*

## Abstract

We consider accretion disk winds, which are driven by disk radiation fields, taking account of radiation drag as well as radiation flux. The winds moving, in intense radiation fields, with a velocity close to the light speed are affected by radiation drag. We calculate, under the near-disk approximation, the trajectories of particles and obtain the conditions whether particles escape or not. Due to radiation drag, it becomes difficult for particle winds to escape, comparing with the case of non-drag. When the particles cannot escape, they float at some heights under the balance between the vertical gravitational force and the disk radiation flux. In such a case they will become an accretion disk conona spreading above/below the disk.

## 1. Radiative Disk-Winds

Let us suppose particles ejected from the surface of a geometrically thin disk around a central object of mass $M$. In addition to the gravitational force of the central object and the centrifugal force, the radiative force from the disk (and the central object) acts on particles (cf. Icke 1980). Furthermore, when the velocity of particles becomes of the order of the light speed, the particles are affected by radiation drag, which is in proportion to their velocities.

Here, for simplicity, we use the *near-disk approximation*, in which particles are assumed to suffer from the radiation fields of the disk just below them. We calculate the motion of the particles using the Newtonian formalism. We adopt cylindrical coordinates $(r, \varphi, z)$ with the $z$-axis along the rotation axis of the disk.

*S. Kato et al. (eds.), Physics of Accretion Disks, 219–222.*
© 1996 OPA (Overseas Publishers Association) Amsterdam B.V.

The equations of motions for particles become as follows (cf. Fukue & Umemura 1995):

$$\frac{dv_r}{dt} = \frac{v_\varphi^2}{r} - \frac{GMr}{R^3} + \frac{\sigma_{\rm T}}{mc}(F_r - 2Ev_r) - \frac{\sigma_{\rm T}}{mc}\frac{4}{3}E_{\rm d}v_r, \tag{1}$$

$$\frac{1}{r}\frac{d}{dt}(rv_\varphi) = -\frac{\sigma_{\rm T}}{mc}Ev_\varphi - \frac{\sigma_{\rm T}}{mc}\frac{4}{3}E_{\rm d}(v_\varphi - v_{\rm K}), \tag{2}$$

$$\frac{dv_z}{dt} = -\frac{GMz}{R^3} + \frac{\sigma_{\rm T}}{mc}(F_z - Ev_z) + \frac{\sigma_{\rm T}}{mc}\left(F_{\rm d} - \frac{4}{3}E_{\rm d}v_z\right), \tag{3}$$

where $R = \sqrt{r^2 + z^2}$, $\sigma_{\rm T}$ the Thomson scattering cross section, $m$ the particle mass, $(v_r, v_\varphi, v_z)$ and $v_{\rm K}$ the velocity components of the particles and the Keplerian rotating velocity of the disk, $(F_r, F_z)$ and $F_{\rm d}$ the components of the flux from the central object and the disk, $E$ and $E_{\rm d}$ the radiation energy densities of radiation fields from the central object and the disk, respectively.

In equation (2) the right-hand side expresses the radiation drag forces caused by radiation fields from the central object and the disk. It should be noted that the radiation drag due to the disk radiation field is proportional to $(v_\varphi - v_{\rm K})$. This is because the radiation field produced by the disk is corotating with the disk. We assume the near-disk approximation so that the disk radiation flux is only in the vertical direction.

In the neighborhood of the disk surface, the radiation flux $F_{\rm d}$ and the energy density $E_{\rm d}$ of radiation fields from the standard disk are expressed as $F_{\rm d} = \sigma T_{\rm d}^4$ and $E_{\rm d} = 2\sigma T_{\rm d}^4/c$, where $T_{\rm d}$ is the surface temperature of the standard disk.

Using the Eddington luminosity, $L_{\rm E}$, of the central object, we define the luminosity parameters as $\Gamma \equiv \chi_e L/L_{\rm E}$ and $\Gamma_{\rm d} \equiv \chi_e L_{\rm d}/L_{\rm E}$, where $\chi_e$ is the ionization rate, $L$ and $L_{\rm d}$ the luminosities of the central object and the disk, respectively.

Taking $\Gamma$ and $\Gamma_{\rm d}$ and measuring the radius and velocity in units of $r_{\rm g}$ and $c$, the equations of motion are normalized as follows:

$$\frac{dv_r}{dt} = \frac{v_\varphi^2}{r} - \frac{(1-\Gamma)r}{2R^3} - \frac{\Gamma}{R^2}v_r - \frac{4r_{\rm in}\Gamma_{\rm d}}{r^3}\left(1 - \sqrt{\frac{r_{\rm in}}{r}}\right)v_r, \tag{4}$$

$$\frac{1}{r}\frac{d}{dt}(rv_\varphi) = -\frac{\Gamma}{2R^2}v_\varphi - \frac{4r_{\rm in}\Gamma_{\rm d}}{r^3}\left(1 - \sqrt{\frac{r_{\rm in}}{r}}\right)(v_\varphi - v_{\rm K}), \tag{5}$$

$$\frac{dv_z}{dt} = -\frac{(1-\Gamma)z}{2R^3} - \frac{\Gamma}{2R^2}v_z + \frac{3r_{\rm in}\Gamma_{\rm d}}{2r^3}\left(1 - \sqrt{\frac{r_{\rm in}}{r}}\right)\left(1 - \frac{8}{3}v_z\right). \tag{6}$$

As initial conditions, we assume that, just after particles emanate from the disk at the initial radius $r_0$, they corotate with the disk, having with the Keplerian velocity of the disk; $v_r = 0$, $v_\varphi = \sqrt{GM/r_0}$, $v_z = 0$ at $r = r_0$ and $z = 0$. With these initial conditions, we calculate numerically the trajectories of particles.

## 2. Motions of Paticle Winds

Let us here examine the case where the central object is a *black hole* ($\Gamma = 0$). Several trajectories are shown in figure 1. As seen from figure 1, in a less luminous case the non-dragged winds can escape infinity, while the dragged winds are trapped by the gravitational force of the central object. The dragged winds blow outward and eventually float at some typical height, where the gravitational force in the vertical direction is balanced with the force due to the disk flux.

When the disk becomes luminous, the dragged winds as well as the non-dragged ones can escape infinity. Their trajectories, however, become wider in the radial direction for the dragged winds than for the non-dragged winds. This is because in the former case the particle winds get angular momentum from the disk radiation fields which corotate with the disk.

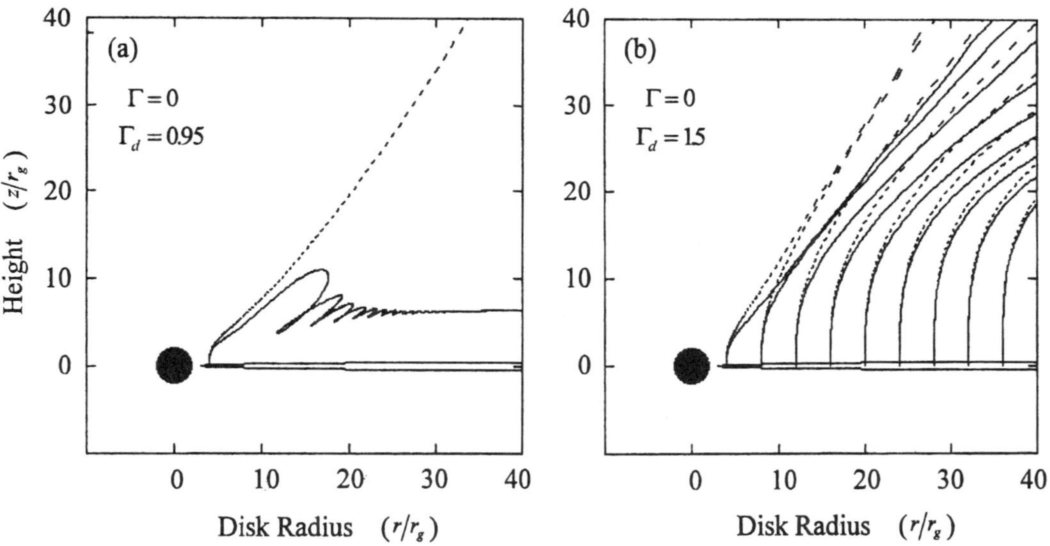


Fig. 1. Trajectories of particle winds. The solid curves and dotted ones represent the dragged and non-dragged winds, respectively. (a) $\Gamma_{\rm d} = 0.95$ and $r_0 = 4r_{\rm g}$. (b) $\Gamma_{\rm d} = 1.5$ and $r_0 = 4r_{\rm g}$ to $36r_{\rm g}$ in steps of $4r_{\rm g}$.

Figure 2 shows the escape condition of particle winds. It is easy for winds to blow from inner region of the disk. Radiation drag, however, makes winds difficult to blow, comparing with the case of non-dragged winds.

## 3. Concluding Remarks

We have examined the accretion disk wind which is driven by the disk *radiation field* under the influences of *radiation drag* as well as radiation flux.

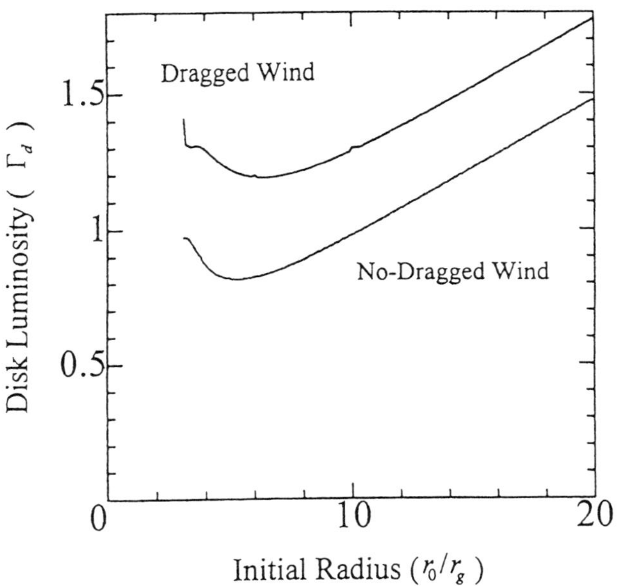


Fig. 2. The escape condition of particle winds. For the parameter space above the curves, radiative winds can blow, while winds cannot escape for the parameter space below the curves.

Generally, radiation drag acts so as to suppress winds to blow. In particular, in the present case, the vertical motion is mainly suppressed due to radiation drag, since we use the near-disk approximation and the disk radiation flux has only the $z$ component.

As for angular momentum gain/loss, the situation is not so simple, because the disk radiation fields have also angular momentum. After particles are ejeced from the disk surface, they turn outward direction because of the centrifugal force. As leaving from the disk, the angular momentum of particles drops quickly at $r > r_0$, while the angular momentum of the disk is Keplerian. Hence, the rotational velocity of particles becomes smaller than that of the disk just below particles. Due to this difference of the rotational velocities between the winds and the disk, the winds are dragged by the disk radiation field and *obtain* the angular momentum. This is the reason that the dragged winds widen in the radial direction, comparing with the non-dragged winds.

Finaly, the winds from less luminous disks cannot escape and eventually float at some typical height above the accretion disk, where the vertical gravitational force is balanced with the force of the radiation pressure; the winds may become the accretion disk corona.

## References

Fukue J., Umemura M. 1995, PASJ 47, 429
Icke V. 1980, AJ 85, 329

# Advection Corona Driven by Radiation Drag

Yoichi WATANABE and Jun FUKUE
*Astronomical Institute, Osaka-Kyoiku University,*
*Asahigaoka, Kashiwara, Osaka 582, Japan*

**Abstract**

We have examined the motion of an accretion disk corona ($\beta$-corona) which is dynamically driven via radiation drag exerted by a central luminous source and an accretion disk itself. We adopted the standard model ($\alpha$-model) as a disk model. When the central source is brighter, then the specific angular momentum of the coronal gas is lost by radiation drag of the radiation field from the central source. The rotational velocity becomes less than a Keplerian one and falls toward the central object. On the other hand, when the accretion disk is brighter, the infall is supressed by radiation drag of the radiation field from the disk.

## 1. Introduction

In the current picture of an accretion disk, the angular momentum of the disk gas is transferred outward via *internal viscosity*. Very recently, a new mechanism in which angular momentum is removed via *radiation drag* due to the external radiation fields, such as cosmological background radiation, or radiation from neutron stars or central luminous sources, has been proposed and investigated.

We here study how the accretion disk corona is suffered from *external radiation drag*. We consider radiation drag due to radiation fields of the central luminous source and the accretion disk itself.

## 2. Basic Equations

Let us suppose an axisymmetric disk corona which is steadily rotating around a central object of mass $M$. We ignored the self-gravity of the disk gas. The disk corona is assumed to be geometrically thin, and therefore, the physical quantities are integrated over the vertical direction. It is also assumed to be effectively optically thin. We use cylindrical coodinates $(r, \varphi, z)$ with

*S. Kato et al. (eds.), Physics of Accretion Disks, 223–226.*
© 1996 OPA (Overseas Publishers Association) Amsterdam B.V.

the $z$-axis along the rotation axis of disk coronae, and adopt the Newtonian fomalism for simplicity. Under these assumptions, the basic equations governing the structure of the disk corona, which is immersed in radiation fields produced by the central luminous object and the accretion disk, are described as follows.

The continuity equation is

$$2\pi r \Sigma v_r = -\dot{M}, \tag{1}$$

where $\Sigma$ is the surface density of the coronal gas, $v_r$ the radial infalling velocity of the corona and $\dot{M}$ the constant accretion rate in the corona. Using the density $\rho$ of the coronal gas and the half-thickness $H$ of the corona, the surface density is expressed as $\Sigma = 2\rho H$.

The equations for radiation hydrodynamycs in a moving plasma are found in, e.g., Hsieh and Spiegel (1976) and Fukue et al. (1985). To the lowest order of $\boldsymbol{v}/c$, where $\boldsymbol{v}$ is the bulk velocity of the plasma and $c$ is the light speed, the equation of motion in the radial direction under the influence of radiation fields is

$$v_r \frac{dv_r}{dr} - \frac{v_\varphi^2}{r} = -\frac{1}{\Sigma}\frac{d\Pi}{dr} - \frac{GM}{r^2} + \frac{1}{\rho}\frac{n_e \sigma_T}{c}\left(F - v_j P^{rj} - E v_r\right), \tag{2}$$

where $v_\varphi$ is the azimuthal rotation velocity of the corona, $\Pi(= 2pH)$ the corona gas pressure integrated over the vertical direction ($p$ being the pressure of the coronal gas), $n_e$ the electron number density, $\sigma_T$ the Thomson scattering cross section, $E$ the radiation energy density, $F$ the radiation flux and $P^{ij}$ the radiation stress tensor. The third term on the right-hand side of equation (2) is the radiation force and the radiation drag force.

The equation of motion in the azimuthal direction under the influence of radiation fields is

$$v_r \frac{dv_\varphi}{dr} + \frac{v_r v_\varphi}{r} = -\frac{1}{\rho}\frac{n_e \sigma_T}{c}\left(v_j P^{\varphi j} + E v_\varphi\right). \tag{3}$$

Here, the right-hand side comes from the external radiation drag. The hydrostatic balance in the vertical direction is integrated as

$$\frac{\Pi}{\Sigma} = \frac{GM}{r^3}\frac{H^2}{2}. \tag{4}$$

Finaly, the equation of state is

$$p = \frac{\mathcal{R}}{\mu}\rho T, \tag{5}$$

where $\mathcal{R}$ is the gas constant, $\mu$ the mean molecular weight and $T$ the temperature of the coronal gas.

For a radiation field produced by a central source with luminosity $L$, the radiation flux $F$ of central source is expressed as $F = L/4\pi r^2$ and the radiation energy density $E$ of central source becomes $F/c$ at large $r$. Furthmore, the $rr$- component of the radiation stress tensor $P^{ij}$ of the central source is expressed as given as $P^{rr} = E$, and the other components vanishes. The radiation energy density of an $\alpha$-disk becomes $E_\mathrm{d} = 2\sigma T_\mathrm{d}^4/c$, where $T_\mathrm{d}$ is the surface temperature of the $\alpha$-disk and $\sigma$ the Stefan-Boltzmann constant. In addition, the $rr$-, $\varphi\varphi$-, and $zz$- components of the radiation stress tensor $P^{ij}$ of the $\alpha$-disk is expressed as $P^{rr} = P^{\varphi\varphi} = P^{zz} = \frac{1}{3}E$, and the other components are zero.

The equations of motion are then rewritten as

$$v_r\frac{dv_r}{dr} - \frac{v_\varphi^2}{r} = -\frac{1}{\Sigma}\frac{d\Pi}{dr} - \frac{GM}{r^2}\left(1 - \Gamma + 2\Gamma\frac{v_r}{c}\right) - \frac{8r_\mathrm{in}GM}{r^3}\Gamma_\mathrm{d}\left(1 - \sqrt{\frac{r_\mathrm{in}}{r}}\right)\frac{v_r}{c}, \tag{6}$$

$$v_r\frac{dv_\varphi}{dr} + \frac{v_r v_\varphi}{r} = -\frac{GM}{r^2}\Gamma\frac{v_\varphi}{c} - \frac{8r_\mathrm{in}GM}{r^3}\Gamma_\mathrm{d}\left(1 - \sqrt{\frac{r_\mathrm{in}}{r}}\right)\frac{v_\varphi - v_\mathrm{K}}{c}. \tag{7}$$

Here, we introduce the parameters $\Gamma = \chi_\mathrm{e}L/L_\mathrm{E}$ and $\Gamma_\mathrm{d} = \chi_\mathrm{e}L_\mathrm{d}/L_\mathrm{E}$, where $\chi_\mathrm{e}$ is the ionization rate, $L_\mathrm{E}$ is the Eddington luminosity, and $v_\mathrm{K} = \sqrt{GM/r}$.

## 3. Cold Case

Now we examine the cold case, where the pressure-gradiant force is ignored. We impose the boundary conditions such that the rotation of the corona is Keplerian at infinity. The asymptotic solutions are then as follows:

$$|v_r| \sim \frac{\Gamma}{r} - \frac{\Gamma^3}{\sqrt{1-\Gamma}}\frac{1}{r^2}, \tag{8}$$

$$v_\varphi \sim \sqrt{\frac{1-\Gamma}{2r}} - \frac{3}{\sqrt{2}}\frac{\Gamma^2}{\sqrt{1-\Gamma}}\frac{1}{r\sqrt{r}}, \tag{9}$$

measuring the radius in units of the Schwartzschild radius $r_\mathrm{g}$ of the central object and the velocity in units of the light speed $c$.

## 4. Results

Examples of numerical solutions in the cold case are shown in figure 1. When $\Gamma$ is larger, then $v_\varphi$ is smaller and $|v_r|$ is larger. When $\Gamma_\mathrm{d}$ is larger, then $v_\varphi$ is larger and $|v_r|$ is smaller. In other words, when the central luminous source is brighter, the angular momemtum of the coronal gas is removed due to radiation drag exerted by the central source radiation, and the infall of the coronal gas is promoted. On the other hand, when the accretion disk is brighter, the infall of the corona gas is supressed.

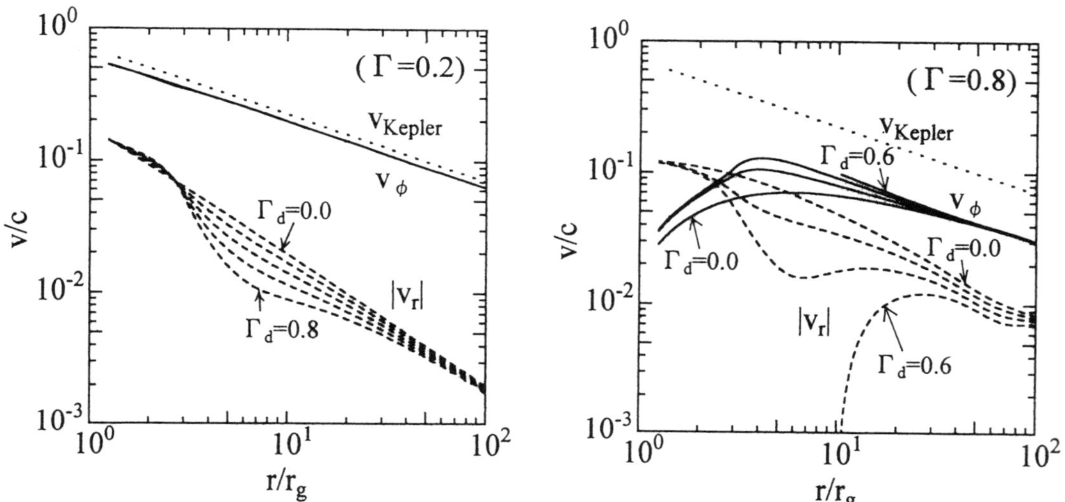


Fig. 1. Examples of numerical solutions in the cold case. The abscissa is the radius $r$ in units of $r_g$ and the ordinate is the velocity in units of $c$. Solid curves express the infall velocity while dashed curves represent the rotating velocity. Dotted curves denote the Keplerian rotating velocity. The parameter are (a) $\Gamma = 0.2$ and (b) $\Gamma = 0.8$ and $\Gamma_d = 0.0, 0.2, 0.4, 0.6, 0.8$.

## References

Fukue J., Umemura M. 1995, PASJ 47, 429

Fukue J., Kato S., Matsumoto R. 1985, PASJ 37, 383

Hsieh S.-H., Spiegel E. A. 1976, ApJ 207, 244

# Spiral Structure and the Stability of Stellar Disks

J. A. Sellwood
*Rutgers University, Department of Physics and Astronomy, PO Box 849, Piscataway, NJ 08855, USA*

## Abstract

The generation of transient spiral waves by swing-amplification in self-gravitating disks is now regarded as the underlying reason for the predominance of spiral patterns in galaxies. The physics of this process is well understood and should be readily applicable to gaseous accretion disks. There is less agreement on a mechanism for the maintenance of long-lived spirals, or the continuous regeneration of fresh spirals, in galaxies, but most current ideas in this area do not seem likely to carry over to gaseous disks.

## 1. Introduction

In some respects, disk galaxies can themselves be regarded as gigantic accretion disks in which most of the effective viscosity comes from the gravitational stresses associated with spiral waves. Thus my main focus here will be on the controversial topic of the origin of spirals in galaxies. In the limited space available, I am able to do no more than summarize my own view of the current state of the theory.

## 2. Gravity Torques

It is easy to state what is required to maximize the gravitational stresses that transport the angular momentum. A formula for the couple exerted on the inner disk by the outer disk seems to have been first written down by Lynden-Bell & Kalnajs (1972):

$$C_z = \frac{1}{4\pi G} \int_S f_R f_\phi dS.$$

*S. Kato et al. (eds.), Physics of Accretion Disks, 227–232.*
© 1996 OPA (Overseas Publishers Association) Amsterdam B.V.

Here, the integration surface is a right cylinder parallel to the spin axis and $f_R$ and $f_\phi$ are the radial and azimuthal parts of the perturbation forces that arise from the spiral density variations. Since the couple depends on their product, its magnitude rises directly as the square of the amplitude of a perturbation of constant shape. For a fixed amplitude, the couple also peaks for open spiral waves with pitch angles around 45°, dropping to zero both for rings and for bars. Strong, open spiral patterns transport the most angular momentum, therefore. Devising a theory for the generation of such disturbances seen in galaxies has proved to be surprisingly difficult, however.

## 3. Swing Amplification

It is now clear that the so-called "swing amplification" process is the fundamental reason for the existence of spiral patterns in shearing disks. It was first recognized by Goldreich & Lynden-Bell (1965) and by Julian & Toomre (1966), and has been reviewed by Toomre (1981). Any disturbance in the disk induces a large-scale spiral response which looks very promising for a short period. The typical crest-to-crest wavelength at which amplification peaks is

$$\lambda_{\rm crit} = \frac{4\pi^2 G\Sigma}{\kappa^2},$$

where $\Sigma$ is the disk surface density and $\kappa$ is the Lindblad epicycle frequency. Because the spirals are strongest at some finite inclination angle, the preferred azimuthal wavelength, $2\pi R/m$, will be greater than $\lambda_{\rm crit}$. Since the inclination angle for peak amplification increases with the rate of shear, the ratio $2\pi R/m$ to $\lambda_{\rm crit}$ is 1-2 for best amplification in a flat rotation curve, while it is 2-4 in a Keplerian disk. Thus for open patterns to produce a large torque, we require a high surface density disk, since $\lambda_{\rm crit}$ is larger and the shear is reduced.

Unfortunately, a swing-amplified spiral winds up and fades rapidly as the remorseless shear continues, and patterns can therefore be long-lived only if the seed disturbance is constantly regenerated or if there is some sort of feed back from trailing to leading waves.

## 4. Stellar vs Gaseous Disks

The process of swing amplification works equally well in both stellar and gaseous disks, but other ideas for the maintenance of strong, open spiral waves in galaxies are specific to particle disks. In particular, the wave-mechanics of feed-back loops or other processes needed to maintain spiral activity over long periods differ in fundamental respects between gaseous and stellar disks.

First, stellar disks cannot support compressive (sound) disturbances and non-axisymmetric waves are therefore confined to the region between the Lindblad resonances, straddling co-rotation. Second, strong wave-particle interactions occur at Lindblad resonances in stellar disks, where incident, small-amplitude waves are absorbed. Third, the strong, open spirals that exert considerable gravity torques, must extract and deposit angular momentum from the resonant particles at the ends of the spiral arms (Lynden-Bell & Kalnajs 1972). We therefore expect strong spirals in galaxies to be short-lived since narrow resonances will saturate quickly. In fact, resonant scattering by large-amplitude waves may play a pivotal role in the continuous regeneration of fresh spirals in particle disks (see §6.2). None of these processes occurs in a gaseous disk, where no Lindblad resonances occur.

## 5. Global Modes

Ideally, one might hope that spiral instabilities would emerge as solutions of the global eigen-mode problem for small perturbations to some equilibrium model. The basic set of equations to be solved were written down long ago (Kalnajs 1971) but solutions have been obtained in just a few highly idealized cases. Much more has been learned from local approximations and $N$-body simulations, and relating these less elegant approaches can be still more informative.

A whole zoo of global instabilities has been found so far. Jeans-type instabilities, in which gravity is the destabilizing agent, include axisymmetric modes (Toomre 1964), bar modes (Hohl 1971; Kalnajs 1972), edge modes (Toomre 1989), lop-sided modes (Zang & Hohl 1978; Sellwood 1985), *etc.* Bending modes (Toomre 1966; Fridman & Polyachenko 1984; Merritt & Sellwood 1994), on the other hand, are driven by anisotropic velocity dispersion and gravity exerts a stabilizing influence. No galaxy model exhibits all these instabilities simultaneously, of course, and we now have some understanding of the regions of parameter space where each can be expected (Sellwood 1994).

Most of these modes seem irrelevant to accretion disks. Lop-sided modes, for example, which might have the most interest for accretion disk dynamics, come in two kinds in stellar disks. The simplest are found in rather bizarre disks with roughly equal counter-rotating streams of stars (Merritt & Stiavelli 1990; Sellwood & Merritt 1994), an impossible situation in gaseous disks. The second are the $m = 1$ analogue of the well-known bar instability (Zang 1976; Sellwood 1985) which requires an extremely massive disk. The so-called sling-amplified modes described by Shu et al. (1990) are not seen in stellar systems because their feed-back is via sound waves reflecting off the edge.

None of these instabilities leads to mild spiral density waves. Most initially smooth models possess either a violently disruptive instability, such as a bar, or nothing at all. A rare example of a completely stable model is the Mestel disk used by Toomre (1981) to illustrate swing-amplification; the last few frames of his most revealing figure 7 illustrates what was already known from Lynden-Bell & Kalnajs (1972) and Mark (1974) that that even a vigorous transient disturbance is damped as the wave "rolls onto the beach" at the Lindblad resonance. In such a situation, the wave action is absorbed at the resonance through wave-particle interaction, in a manner very analogous to Landau damping.

## 6. Theories of Spiral Waves

Since Lindblad resonances were recognized to damp waves, C. C. Lin and his collaborators have developed a theory of spiral wave generation in smooth disks in which the inner Lindblad resonance is shielded by a "Q-barrier". Bertin et al. (1989) have found slowly growing, tightly wrapped, mildly unstable spiral modes in cool disks with a dynamically hot inner region. They argue that such modes probably lead to low-amplitude, quasi-stationary spiral patterns. Even if these solutions behave as they argue, and the same models do not permit more vigorous evolution, such mild, tightly wrapped waves will transport so little angular momentum (Bertin 1983) that they are of little interest in the context of accretion disks.

It has gradually dawned on us, however, that it may be incorrect to assume that disk galaxies are smooth, well mixed systems, and this idea has led to two further theories for the origin of spiral patterns. Further ideas related to bar forcing and tidally induced patterns are clearly important for some galaxies, but are not of much interest in the context of accretion disks.

### *6.1. Swing-amplified noise*

Toomre (1990) and Toomre & Kalnajs (1991) emphasize that shot noise associated with a random distribution of particles will be swing amplified. As the inner particles drift by the outer ones, one observes an ever changing "kaleidoscope" of spiral arm fragments. The level of shot noise from stars alone is too low to make strong spirals by this mechanism, but Toomre envisages that fluctuations from star clusters and giant molecular clouds would be large enough excite what is observed. This essentially local theory does not attempt to offer an explanation for a "grand-design" pattern, although the most strongly amplified features will be quite large-scale, since $\lambda_{\rm crit}$, which determines the crest-to-crest spacing, is typically comparable to the local radius in many galaxies.

### *6.2. Groove modes*

Lovelace & Hohlfeld (1978) and Sellwood & Kahn (1991) showed that a local deficiency in the angular momentum density of stars in a disk drives an instability. The latter authors named it the groove instability, since a deficiency of particles with a particular angular momentum would, in the absence of random motion, yield a disk with a groove, but the instability persists even when random motions are large-enough to wash out a noticeable density feature. The basic instability originates through a process analogous to Landau excitation in a plasma with a double-peaked distribution function and produces a co-rotating wave-like distortion to the edges of the groove. This would be a mild disturbance, were it not for the enthusiastic support response of the surrounding disk, which since it is again due to the swing-amplifier, gives greatest encouragement to waves on scales close to $\lambda_{\mathrm{crit}}$.

The instability saturates when the groove is closed, leaving the disk with an azimuthal density variation at co-rotation which takes some time to disperse; the spiral response of the surrounding disk decays as the forcing perturbation disperses, and the angular momentum it transfers is deposited at the Lindblad resonances. The absorption of angular momentum and resonant scattering is readily observed in the simulations. It creates a deficiency of particles in the disk with the resonant angular momentum, which can lead to a new instability with co-rotation of the new wave at the radius of the Lindblad resonance of the old. Other processes, such as wave reflection of the disturbed region, may also contribute to new instabilities. Thus a self-perpetuating cycle of instabilities is set up in the simulations, which is ultimately limited by the increasing random motion in the disk.

Because the disk amplifies the disturbance, each new groove is deeper than the last, and the limiting amplitudes of successive instabilities increase. Moreover, particle noise can provide the initial seed. We therefore conclude that any system of particles, no matter how large, can always produce large amplitude spirals even when the equivalent smooth disk is stable, which is consistent with my experimental results with up to two million particles.

Interesting as I find all this, I once again have to conclude that a theory of spiral wave generation based on resonant scattering can be of little relevance to accretion disks.

## 7. Conclusions

A strongly shearing, massive disk is all that is required for the swing amplifier to produce large-scale, open, transient spiral patterns. Their amplitude is determined by the magnitude of the input signal, however, and will be very feeble in an initially quiescent disk. If a mechanism is available to produce seed density variations, or if some form of feed-back can occur, then the swing amplifier will give us the large-amplitude spirals we desire. But the mechanisms by which this could happen in gaseous accretion disks are unlikely to resemble any of the current ideas for generating spiral waves in galaxies.

More is known about the global modes of stellar disks than of gaseous disks, yet it also seems unlikely that this body of theory will prove to be much help for accretion disks. Lop-sided modes are most likely to carry over to the gaseous case since for these modes there is no possibility of a Lindblad resonance in the inner part of the disk; the instability grows vigorously in stellar systems when the disk is more massive than the central object.

**References**

Bertin G. 1983, A&A 127, 145
Bertin G., Lin C. C., Lowe S. A., Thurstans R. P. 1989, ApJ 338, 78
Fridman A. M., Polyachenko V. L. 1984, Physics of Gravitating Systems, (Springer-Verlag, New York)
Goldreich P., Lynden-Bell D. 1965, MNRAS 130, 97
Hohl F. 1971, ApJ 168, 343
Julian W. H., Toomre A. 1966, ApJ 146, 810
Kalnajs A. J. 1971, ApJ 166, 275
Kalnajs A. J. 1972, ApJ 175, 63
Lovelace R. V. E., Hohlfeld R. G. 1978, ApJ 221, 51
Lynden-Bell D., Kalnajs A. J. 1972, MNRAS 157, 1
Mark J. W-K. 1974, ApJ 193, 539
Merritt D., Sellwood J. A.1994, ApJ 425, 551
Merritt D., Stiavelli M. 1990, ApJ 358, 399
Sellwood, J. A. 1985, MNRAS 217, 127
Sellwood, J. A. 1994, in Numerical Simulations in Astrophysics, ed J. Franco, S. Lizano, L. Aguilar & E. Daltabuit (Cambridge University Press) p90
Sellwood J. A., Kahn F. D. 1991, MNRAS 250, 278
Sellwood J. A., Merritt D. 1994, ApJ 425, 530
Shu F. H., Tremaine, S., Adams, F., Ruden, S. P. 1990, ApJ 358, 395
Toomre A. 1964, ApJ 139, 1217
Toomre A. 1966, in Geophysical Fluid Dynamics, notes on the 1966 Summer Study Program at the Woods Hole Oceanographic Institution, ref. no. 66-46, p111
Toomre A. 1981, in Structure and Evolution of Normal Galaxies, ed S. M. Fall, D. Lynden-Bell (Cambridge University Press, Cambridge) p111
Toomre A. 1989, in Dynamics of Astrophysical Discs, ed J. A. Sellwood (Cambridge University Press) p153
Toomre, A., 1990. in Dynamics and Interactions of Galaxies, ed R. Wielen (Springer-Verlag:Berlin, Heidelberg) p292
Toomre A., Kalnajs A. J. 1991, in Dynamics of Disc Galaxies, ed B. Sundelius (Göteborgs University Press) p341
Zang T. A. 1976, PhD MIT
Zang T. A., Hohl F. 1978, ApJ 226, 521

# The Bar Formation by Lynden-Bell Mechanism

Natsumi FURUYA
*Department of Astronomy, Faculty of Science, Kyoto University, Sakyo-ku, Kyoto 606-01, Japan*

**Abstract**

The Lynden-Bell mechanism is regarded as one of the theories of the bar formation in barred spiral galaxies. In this article, we show by numerical integrations of orbits that the mechanism is actually effective in an weak nonaxisymmetric potential. If stars are in the abnormal region, they are expected to form a bar structure. They move on oval orbits whose major axes are aligned along the trough of the potential and oscillate around it. If the bar-like potential grows exponentially with time, orbits whose frequencies are initially different from ($\sim$ 10% times faster or slower than) that of the bar are trapped and the amplitudes of oval oscillations become smaller. These results suggest that the gravitational interaction among the stars may form a bar structure through this effect. The bar becomes more stronger, however, the amplitude of oval oscillation tends to increase again. The study of orbits under a self-consistent potential will make clear the relation of the Lynden-Bell mechanism to the strong bar formation.

## 1. Introduction

According to the observations, about half of the disk galaxies have bar-like structures. It is not yet clear how these bar structures are formed. Considering the interaction of a single stellar orbit with a weak bar-like potential, Lynden-Bell (1979) suggested that there are orbits whose elongations will be aligned along the potential valley and oscillate about it, and then a bar structure in the galaxy will be formed by the assembly of many such orbits.

Lynden-Bell called the regions in which orbits behave as is stated above the abnormal region, and discover this region analytically for the Isochrone potential (Hénon 1959):

*S. Kato et al. (eds.), Physics of Accretion Disks, 233–236.*
© 1996 OPA (Overseas Publishers Association) Amsterdam B.V.

$$U = (2r^2 + 1 - \sqrt{4r^2 + 1})/2r^2. \tag{1}$$

Though Lynden-Bell discussed the orbit using the axisymmetric potential, actually the star moves in the nonaxisymmetric potential. Therefore we suspect the motion of the star may be different from his expectation not only in a strong bar-like potential but also in an weak bar-like potential.

In this article, the orbits of test stars in the nonaxisymmetric potential are calculated by means of numerical integrations in order to study if the mechanism is actually effective in more realistic circumstances. The model and results are presented in section 2. In section 3 we make a brief consideration about the process through which bar structure is formed.

## 2. Numerical Integration of the Orbits

Let us take the Hamiltonian function of the system as follows,

$$H = \frac{1}{2}\dot{r}^2 + \frac{1}{2}\frac{J_\theta^2}{r^2} + U(r) + \psi_b. \tag{2}$$

We use the Isochrone potential, equation (1), as the axisymmetric potential $U(r)$ and expand the bar perturbation $\psi_b$ in polar harmonics as follows,

$$\psi_b = \sum_{m=-\infty}^{\infty} \psi_m(r,t) \exp[im(\theta - \int \Omega_p dt)]. \tag{3}$$

Since a bar-like perturbation is dominated by the quadrupole ($m = 2$), we may write

$$\psi_b = \psi_2(r,t) \cos(2\theta - 2\int \Omega_p dt). \tag{4}$$

We assume that the pattern speed is constant and the amplitude of the perturbation has the following form:

$$\psi_2 = a_0 \exp[\alpha t], \tag{5}$$

although this is not self-consistent. Here $a_0$ and $\alpha$ are constants.

We solve the canonical equations by means of the 4th-order Runge-Kutta method. Since the pattern speed of a bar formed by Lynden-Bell mechanism is the order of $\langle\Omega_i\rangle_{\rm typ}$, the mean precession rate of typical orbits contributing to the bar formation, we choose $\Omega_p = 0.08$, about a half of $(\Omega_i)_{max}$ for the Isochrone potential. The initial amplitude is given to be $a_0 = -10^{-5}$.

Figure 1 shows the behavior of stars which are near ILR orbits under the steady potential. Notice that the major axis of the orbit which are in the abnormal region in the unperturbed state oscillates around the bar as Lynden-Bell expected. On the other hand, if the initial orbits are in the normal region, the major axis oscillates around the axis perpendicular to the bar.

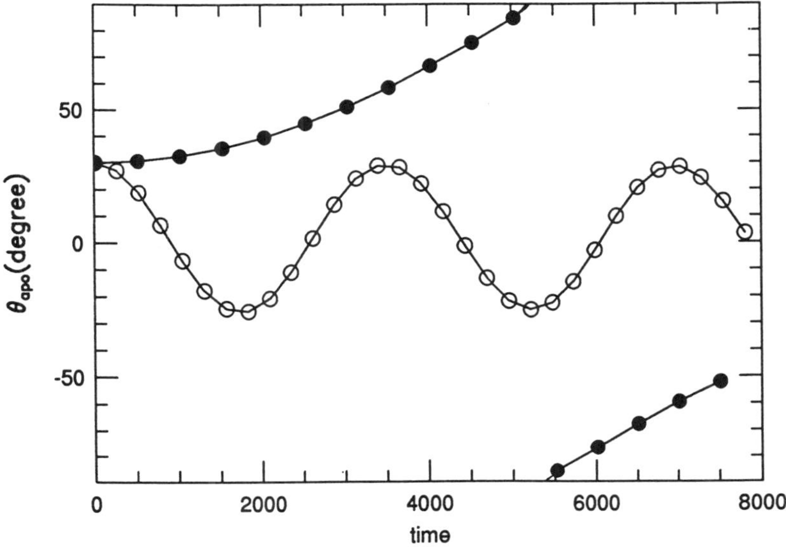


Fig. 1. The angle of apocenter relative to the bar position angle, $\theta_{\rm apo}$, for the steady perturbation. The lines with open circles and filled circles are far stars whose initial unperturbed orbits are in the abnormal region (initial $J_\theta = 0.5$) and the normal region (initial $J_\theta = 1.2$) respectively.

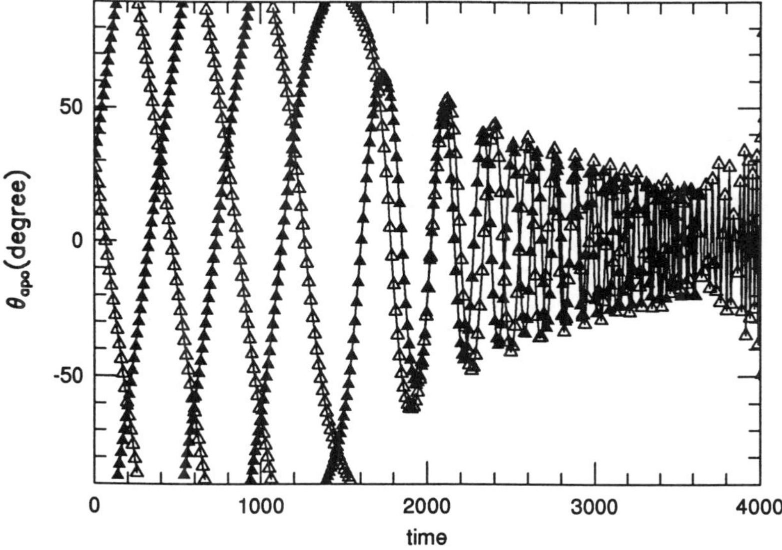


Fig. 2. The angle of apocenter relative to the bar position angle, $\theta_{\rm apo}$, for the time-dependent perturbation. The initial orbits are in the abnormal region (initial $J_\theta = 0.4$). The line with filled triangles is the orbit initially rotating faster than the bar, and with open triangles slower.

Figure 2 is similar to figure 1 but for the orbits initially in the abnormal region. The growth rate $\alpha$ of the bar potential has been taken so that $\psi_2(t = 1000)$ becomes $10\, a_0$. The initial unperturbed precession rates of orbits are 10% times faster (the line with filled triangles in the figure) or slower (open triangles) than the pattern speed of perturbation.

In the strong perturbation ovals with initial unperturbed $\Omega_i$ not close to the pattern speed would be trapped to oscillate with their elongation along that of the bar and the amplitudes of oval oscillations become smaller.

## 3. Summary

Numerical integrations of stellar orbits in a nonaxisymmetric bar-shaped potential show that the oval orbits are trapped to oscillate with their elongation along that of the bar-shaped potential if the precession rates of orbits resonate with the pattern speed of the potential and the initial orbits are in the abnormal region. As the strength of the perturbation increases, not only the oscillation of those orbits decrease, but also non-resonant orbits are trapped. Therefore we guess that a bar structure in the disk galaxy is formed as follows.

If a weak bar-like perturbation occurs in the disk, then the stars in the abnormal region will take the oval orbits with their elongation along that of the perturbation and oscillate around it. The gravity of the trapped orbits will add to the potential valley, therefore the orbits are trapped more tightly to trap more orbits. Finally, a strong bar structure may be built up.

The bar becomes more stronger, however, the position angle of apocenter tends to increase. The increase may be due to the strength of perturbation or nonself-consistency. If the strong bar-like potential breaks alignments of orbits, Lynden-Bell mechanism may not form a strong bar, which is consistent with the simulation of Athanassoula and Sellwood (1986) and the results calculated by Polyachenko (1989).

The study of orbits under a self-consistent potential will make clear the relation of Lynden-Bell mechanism to the strong bar formation.

## References

Athanassoula E., Sellwood J. A. 1986, MNRAS 221, 213
Hénon M. 1959, Ann. d'Astrophys. 22, 491
Lynden-Bell D. 1979, MNRAS 187, 101
Polyachenko V. L. 1989, SovAstronLett 15, 385

# Bi-Symmetric Instabilities of Thin Stellar Disks

Shunsuke HOZUMI
*Faculty of Education, Shiga University, 2-5-1 Hiratsu, Otsu 520, Japan*

**Abstract**

We investigate which part of thin stellar disks dominates their global stability against bi-symmetric modes by varying the halo mass distribution which is a fixed part of the disk mass distribution. The results show that the growth rates of bi-symmetric modes are well-correlated to the total mass of the halo, irrespective of its distribution. It follows, therefore, that the seed of disk instabilities exists everywhere equally in disks.

## 1. Introduction

Many numerical simulations of self-gravitating thin stellar disks have demonstrated that bar instability is one of the most serious issues in disk galaxy dynamics. To argue the stability of disks, we introduce Toomre's (1964) $Q$ parameter which is defined as

$$Q(r) = \frac{\kappa c_r}{3.36 G \Sigma}, \tag{1}$$

where $r$ is the radius, $\kappa$ the epicyclic frequency, $c_r$ the radial velocity dispersion, $G$ the gravitational constant, and $\Sigma$ the disk surface density. When $Q > 1$, disks are saved from local axisymmetric Jeans instabilities. Despite a local quantity, the $Q$ parameter is often used as a criterion of the global stability.

Recent studies have indicated that the central part of disks is more significant to the global stability than the other parts. This comes from the fact that high central $Q$ values are effective in stabilizing disks against bar modes (Athanassoula & Sellwood 1986; Hozumi et al. 1987). In fact, observations show that some galaxies have high central $Q$ values: a normal S0 galaxy NGC 1553 and an Sb spiral NGC 7184 have $Q \simeq 2.8$ (Kormendy 1984) and $Q \simeq 2.2$ (van der Kruit & Freeman 1986), respectively, near the center. As

*S. Kato et al. (eds.), Physics of Accretion Disks, 237–240.*
© 1996 OPA (Overseas Publishers Association) Amsterdam B.V.

a reference, the $Q$ value in the solar neighborhood is estimated to be 1.7 (Toomre 1974).

Although the central part of disks comes to acquire special interest regarding the global stability, there is no definite work on which part of disks dominates the global stability. Then, in this paper, we examine on the basis of $Q$ distributions which part of disks determines the global stability. For this purpose, we have constructed disk models whose $Q$ distributions can be varied in a specified part without such an extra length scale as a bulge component.

## 2. Models and Method

The mass models used are the isochrones (Hénon 1959). We follow Kalnajs's (1976) method to make equilibrium distribution functions of directly rotating stars, $F^+(\varepsilon, j)$, where $\varepsilon$ and $j$ are the energy and angular momentum of a star per unit mass, respectively. Then, the equilibrium distribution function, $F_0$, is given by

$$F_0(\varepsilon, j) = \begin{cases} (1/2)F_0^+(\varepsilon) + F_1^+(\varepsilon, j) & (j \geq 0), \\ (1/2)F_0^+(\varepsilon) & (j < 0), \end{cases} \tag{2}$$

where the functions $F_0^+(\varepsilon)$ and $F_1^+(\varepsilon, j)$ are derived from the expansion of $F^+(\varepsilon, j)$ as

$$F^+(\varepsilon, j) = F_0^+(\varepsilon) + F_1^+(\varepsilon, j). \tag{3}$$

It is to be noted from equation (1) that we can raise $Q$ values of a specified part, if we fix the corresponding part of the surface density. In so doing, we must represent the fixed part by integrals of motion. Then, it should be reminded that in general, $F_0^+(\varepsilon)$ governs stars near the central part of disks and $F_1^+(\varepsilon, j)$ those of the outer part.

First, if we need models with high central $Q$ values, we have only to modify equation (2) as

$$F_0(\varepsilon, j) = \begin{cases} (1/2)\alpha_{\rm in} F_0^+(\varepsilon) + F_1^+(\varepsilon, j) & (j \geq 0), \\ (1/2)\alpha_{\rm in} F_0^+(\varepsilon) & (j < 0). \end{cases} \tag{4}$$

Second, we can likewise raise $Q$ values of the outer part of disks if we use the following $F_0$:

$$F_0(\varepsilon, j) = \begin{cases} (1/2)F_0^+(\varepsilon) + \alpha_{\rm out} F_1^+(\varepsilon, j) & (j \geq 0), \\ (1/2)F_0^+(\varepsilon) & (j < 0). \end{cases} \tag{5}$$

Last, we can raise $Q$ values of whole disks if we use the following $F_0$:

$$F_0(\varepsilon, j) = \begin{cases} \alpha_{\rm w}[(1/2)F_0^+(\varepsilon) + F_1^+(\varepsilon, j)] & (j \geq 0), \\ (1/2)\alpha_{\rm w} F_0^+(\varepsilon) & (j < 0). \end{cases} \tag{6}$$

The parameters $\alpha_{\rm in}$, $\alpha_{\rm out}$, and $\alpha_{\rm w}$ are the indicator of self-gravity of the models, and $0 \leq \alpha_{\rm in}, \alpha_{\rm out}, \alpha_{\rm w} \leq 1$. Figure 1 shows the $Q$ distributions of the models which are determined from the values of the parameters $\alpha_{\rm in}$, $\alpha_{\rm out}$, and $\alpha_{\rm w}$, given the model parameter $n$ appearing in the equilibrium distribution functions. We adopt $n = 8$ and $n = 10$ as basic models.

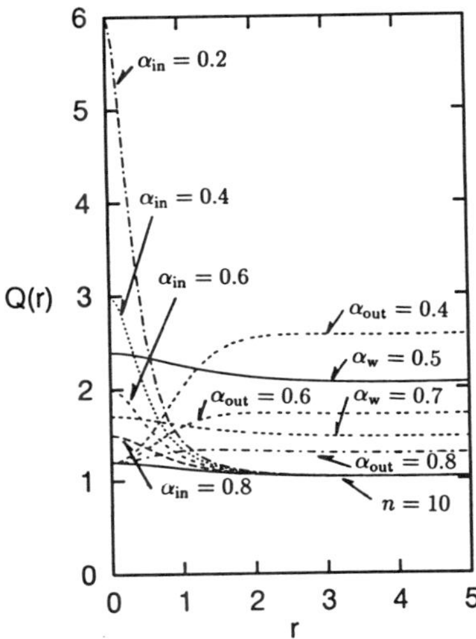


Fig. 1. Distributions of $Q$ values for the models based on the isochrone disk with $n = 10$.

The prescription mentioned above is favorable in that there is no change in shearing rate as well as no new length scale. In addition, the equilibrium distribution functions have no such peculiar feature as those used in Athanassoula & Sellwood (1986) which become double peaked near the disk center in order to make central $Q$ values raised.

Using those equilibrium distribution functions, we obtain the most unstable global modes by integrating numerically the linearized collisionless Boltzmann equation. We restrict ourselves to bi-symmetric modes which are usually the most unstable ones among the various numbers of arms of spiral modes. The details of the method are described in Hozumi et al. (1987).

## 3. Results and Conclusions

The results for the $n = 10$ based models are shown in figure 2 as an example. We summarize our results as follows:

(1) Basically, the growth rates are well-correlated to the total fixed mass, that is, the total mass of the halo, irrespective of the $Q$ distributions. In particular, the colder model ($n = 10$) has a tighter correlation between the growth rate and the fixed mass. Thus, we conclude that the seed of disk instabilities should exist everywhere equally in disks.

(2) If we are allowed to extrapolate the relation between the growth rate and the fixed mass, both $n = 8$ and $n = 10$ based models show that the disks would be stabilized with the fixed mass of about 0.4 times the total mass. This may be compared with observations that the ratio of dark to disk mass within the optical limit of spiral galaxies would be nearly unity.

(3) We have found that the pattern speeds with the high central $Q$ models show the sequence which deviates from those with the other models (figure 2b). Therefore, the origin of those modes may be different from the others.

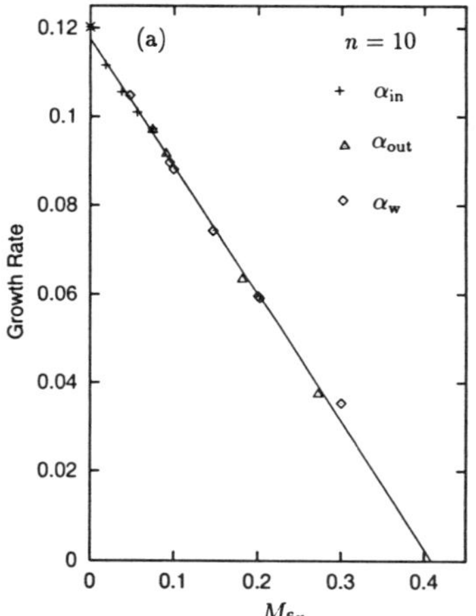


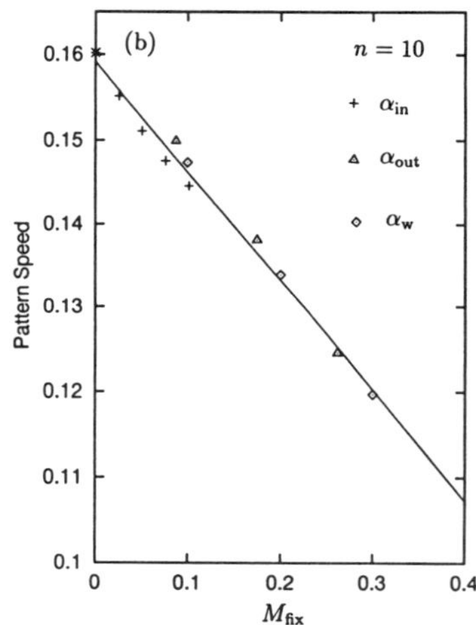


Fig. 2. Growth rates (a) and pattern speeds (b) plotted against fixed masses in units of total mass for the models based on the isochrone disk with $n = 10$. Solid lines show the least-squares fits to the data.

## Acknowledgments

The author would like to thank Dr. T. Fujiwara for many stimulating discussions.

## References

Athanassoula E., Sellwood J.A. 1986, MNRAS 221, 213
Hénon M. 1959, AnnAstrophys 22, 126
Hozumi S., Fujiwara T., Nishida M. T. 1987, PASJ 39, 447
Kalnajs A. J., 1976 ApJ 205, 751
Kormendy J. 1984, ApJ 286, 116
Toomre A. 1964, ApJ 139, 1217
Toomre A. 1974, Highlights of Astronomy 3, 457
van der Kruit P. C., Freeman K. C. 1986 ApJ 303, 556

# Dynamical Stability of Gaseous Disks

Shogo INAGAKI
*Department of Astronomy, Faculty of Science, Kyoto University, Sakyo-ku, Kyoto 606-01, Japan*

**Abstract**

Stability analysis using an energy principle is reviewed. The variatinal principle by Katz and Lynden-Bell is put into the energy principle and a sufficient condition for stability is derived.

## 1. Introduction

Stability analysis of cold disks was done by Hunter(1963, 1965) and warm disks by Takahara (1976), Iye (1976), and Aoki et al. (1979). All of these works are based on eigen-mode analysis. Global eigen-mode anlysis is difficult, so there is only a few simple disk models whose stabilities have been analysed. If the problems are formulated with a variational principle, approximate solutions are easily obtained and linear series analyses (Inagaki & Hachisu 1978) may become possible.

## 2. Stability of Barotropic Flows

### *2.1. Lagrangian Formulation*

Let $\boldsymbol{w}$ be the velocity of the vortex line. The Lagrangian variable $\alpha^k$ moving with the velocity $\boldsymbol{w}$ is defined by

$$\alpha^k(\boldsymbol{r} + \boldsymbol{w}dt, t + dt) = \alpha^k(\boldsymbol{r}, t). \tag{1}$$

We define $\boldsymbol{W}$ with the velocity of the fluid, $\boldsymbol{v}$, and $\boldsymbol{w}$ by

$$\boldsymbol{v} = \boldsymbol{w} + \boldsymbol{W}. \tag{2}$$

The vorticity is defined by

$$\boldsymbol{\omega} = \nabla \times \boldsymbol{v}. \tag{3}$$

*S. Kato et al. (eds.), Physics of Accretion Disks, 241–246.*
© 1996 OPA (Overseas Publishers Association) Amsterdam B.V.

It is noted that the mass preservation,

$$\frac{\partial \rho}{\partial t} + \nabla \cdot (\rho \boldsymbol{w}) = 0, \tag{4}$$

and the vorticity preservation,

$$\frac{\partial \boldsymbol{\omega}}{\partial t} + \nabla \times (\boldsymbol{\omega} \times \boldsymbol{w}) = 0, \tag{5}$$

are satisfied.

We define the action by

$$\begin{aligned} A &= \int_{t_0}^{t_1} L dt \\ &= \int\int \left[\frac{1}{2}w^2 - \left(\frac{1}{2}W^2 + \epsilon + \Phi\right)\right] \rho d^3x dt. \end{aligned} \tag{6}$$

(7)

Here, we shall adopt the Clebsch form of the representation of the velocity field,

$$\boldsymbol{v} = \alpha \nabla \beta + \nabla \nu. \tag{8}$$

Then we have

$$\boldsymbol{\omega} = \nabla \alpha \times \nabla \beta \tag{9}$$

and

$$\boldsymbol{\omega} \cdot \nabla \alpha = 0, \qquad \boldsymbol{\omega} \cdot \nabla \beta = 0. \tag{10}$$

We define $\mu$ by

$$\boldsymbol{\omega} \cdot \nabla \mu = \rho. \tag{11}$$

Then $\rho$ is represented in terms of a Jacobian,

$$\rho = \frac{\partial(\alpha, \beta, \mu)}{\partial(x, y, z)}. \tag{12}$$

Thus we can consider the Lagrangian variable $(\alpha, \beta, \mu)$ as $\alpha^k$.

From the condition of the stationarity of the action $A$,

$$\Delta A = 0, \tag{13}$$

we obtain Euler's equation,

$$\frac{\partial \boldsymbol{v}}{\partial t} + (\boldsymbol{v} \cdot \nabla)\boldsymbol{v} + \nabla(h + \Phi) = 0 \tag{14}$$

(Katz & Lynden-Bell 1985).

### *2.2. Lagrangian Formulation*

In particle mechanics if the action is given by

$$A = \int \left[\frac{1}{2}\dot{x}^2 - V(x)\right] dt, \tag{15}$$

the condition of the stationarity of the potential,

$$\Delta V = 0, \tag{16}$$

gives the condition for equilibrium and

$$\Delta^2 V > 0 \tag{17}$$

is the sufficient condition for stability of the equilibrium point.

In our case the action is given by

$$\begin{aligned} A &= \int\int \left[\frac{1}{2}w^2 - \left(\frac{1}{2}W^2 + \epsilon + \Phi\right)\right] \rho d^3x dt \\ &= \int \left[\int \frac{1}{2}w^2 \rho d^3x - E(t)\right] dt, \end{aligned} \tag{18}$$

(19)

where

$$E = \int \left(\frac{1}{2}W^2 + \epsilon + \Phi\right) \rho d^3x. \tag{20}$$

Thus we can expect that

$$\Delta E = 0 \tag{21}$$

gives the condition for stationary flow,

$$(\boldsymbol{W} \cdot \nabla)\boldsymbol{W} + \nabla(h + \Phi) = 0, \tag{22}$$

and

$$\Delta^2 E > 0 \tag{23}$$

gives the sufficient condition for stability. In fact this is true (Katz et al. 1993).

### *2.3. Application to Maclaurin Disk*

The unperturbed surface density of the Maclaurin disk is

$$\Sigma_0 = \Sigma_c \sqrt{1 - \frac{R^2}{a^2}}, \tag{24}$$

where $\Sigma_c$ the central surface density, $a$ the radius of the disk, and $R$ is the radius in cylindrical coordinates. The unperturbed gravitational potential is given by

$$\Phi_0 = \frac{1}{2}\Omega_G^2 R^2 = \frac{1}{2}\frac{\pi^2 G\Sigma_c}{2a}R^2. \tag{25}$$

The rotational angular speed is

$$\Omega^2 = \Omega_G^2 - \frac{3K\Sigma_c^2}{a^2}, \tag{26}$$

where $K$ is the constant appearing in the equation of state $p = K\Sigma^3$.

The unperturbed Lagrangian coordinates $\alpha_0$ and $\beta_0$ are defined by

$$\alpha_0 = \Omega^2 R^2 \tag{27}$$

and

$$\beta_0 = \phi, \tag{28}$$

respectively. We impose a perturbation

$$\alpha = \alpha_0 + \delta\alpha, \qquad \beta = \beta_0 + \delta\beta. \tag{29}$$

Then the displacement $\boldsymbol{\xi}$ is defined by

$$\alpha^k(\boldsymbol{R}+\boldsymbol{\xi}) + \delta\alpha^k(\boldsymbol{R}+\boldsymbol{\xi}) = \alpha^k(\boldsymbol{R}). \tag{30}$$

The displacement is decomposed into a scaler potential $\eta$ and vector potential $\boldsymbol{\psi}$:

$$\boldsymbol{\xi} = a^2(\nabla\eta + \nabla\times\boldsymbol{\psi}), \tag{31}$$

where

$$\boldsymbol{\psi} = \boldsymbol{e}_z\psi. \tag{32}$$

We further decompose $\eta$ and $\psi$ as

$$\eta = \sum_{l=m}^{\infty}\sum_{m=0}^{\infty}\eta_{lm}P_l^m(\xi)e^{im\phi} + cc \tag{33}$$

$$\psi = \sum_{l=m}^{\infty}\sum_{m=0}^{\infty}\psi_{lm}P_l^m(\xi)e^{im\phi} + cc \tag{34}$$

and evaluate $\Delta^2 E$. We then find $\Delta^2 E > 0$ if

$$Q < (Q_{lm})_{\min}, \tag{35}$$

where

$$Q = \Omega^2/\Omega_G^2. \tag{36}$$

The smallest value of $(Q_{lm})_{\min}$ is found when $(l, m) = (2, 2)$ and

$$(Q_{22})_{\min} = \frac{1}{2}. \tag{37}$$

This result agrees with that of the dynamical stability analysis (Binney & Tremaine 1987).

### 2.4. Linear Series

In discussing the stability, linear series are useful in many cases (Poincaré 1902). In discussing the thermodynamic stability,

$$\Delta^2 S < 0 \tag{38}$$

is the criterion for stability. Since

$$\Delta^2 S = \Delta \left(\frac{1}{T}\right) \Delta E, \tag{39}$$

we can see the stability if we plot $E$ versus $1/T$ (Katz 1978, figure 1). In figure 1, the points with vertical tangents are critical points of stability for fixed $T$ and those with horizontal tangents are critical points for fixed $E$.

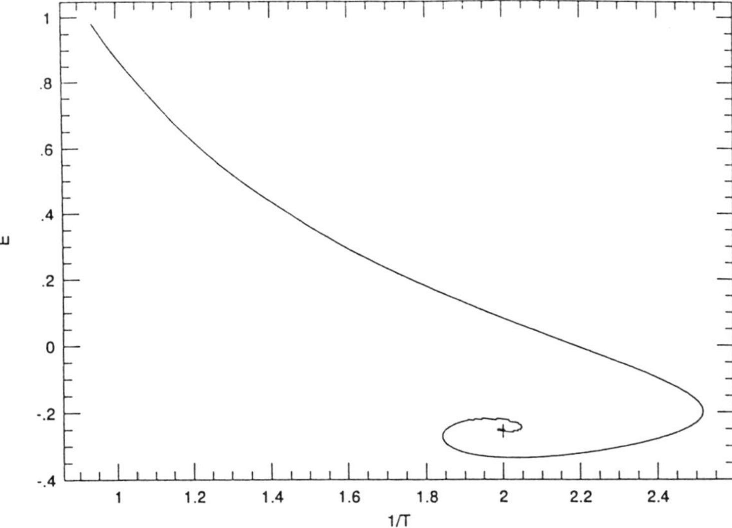


Fig. 1. Linear series for isothermal gas spheres.

Similar consideration leads us that the diagram $J$ versus $\Omega$ should be used to determine the stability of rotating figures because

$$\Delta^2 E = \Delta\Omega\Delta J. \tag{40}$$

However this is not true for the Maclaurin disks because the turning points predicts the exchange of stability with fixed angular velocity while the dynamical stability is for fixed angular momentum.

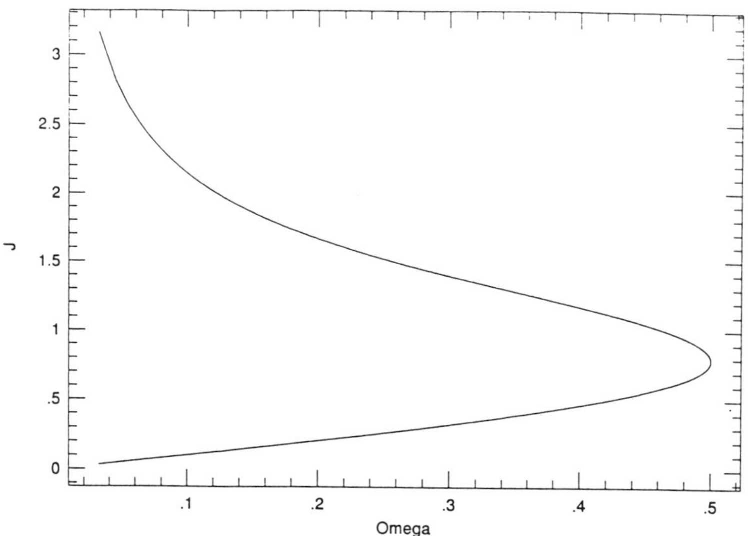


Fig. 2. Linear series for Maclaurin disks.

## 3. Conclusions

The energy principle works well but it is difficult to apply to real problems. The linear series method is questionable to apply to the dynamical stability of barotropic flows.

## References

Aoki S., Noguchi M., Iye M. 1979, PASJ 31, 737

Binney J., Tremaine S. 1987, Galactic Dynamics (Princeton University Press, Princeton) p321

Hunter C. 1963, MNRAS 126, 299

Hunter C. 1965, MNRAS 129, 321

Inagaki S., Hachisu I. PASJ 30, 39

Iye M. 1978, PASJ 30, 223

Katz J. 1978, MNRAS 183, 765

Katz J., Lynden-Bell D. 1985, Geophys. Astrophys. Fluid Dynamics 33, 1

Katz J., Inagaki S., Yahalom A. 1993, PASJ 45, 421

Poincaré H. 1902, Figures d'équilibre d'une masses fluide (Gauthier-Villars, Paris)

Takahara F. 1976, PTP 56, 1665

# Observations of Protoplanetary Disks

Masahiko HAYASHI
*SUBARU Project Office, National Astronomical Observatory*
*Mitaka, Tokyo 181, Japan*

**Abstract**

I review current understanding of disks around young stars by summarizing observational results of various disk properties with the evolutionary sequence from accreting protostars to weak-line T Tauri stars.

## 1. Evidence for Disks around Young Stars

While much indirect evidence was discovered for compact (size$\sim$100 AU) disks around young stars, only recently have we been provided with their direct evidence as images taken by the Hubble Space Telescope. Images of the Orion nebula presented by O'Dell et al. (1993) showed that many young stars inside the M42 HII region have associated elliptical nebulae or elliptical regions of high extinction seen as silhouettes against bright background light of the HII region. These elliptical nebulae or silhouettes are clear manifestation of compact disks around young stars. An image of HH 30 obtained by Burrows et al. (1995) showed a jet emanating from the vicinity of the HH 30 star, which is actually not visible as a result of being extinct by the disk viewed almost edge-on. The reflection nebula around the HH 30 star, delineating the disk itself, showed that the disk is flared.

## 2. Physical Properties

Low mass young stars are classified into three types based on their evolutionary stages, i.e., protostars, classical T Tauri stars (CTTS), and weak-line T Tauri stars (WTTS), all types of which are known to have disks. Disks around CTTS and WTTS have the radius and mass of $\sim$100 AU and 0.001–0.5 $M_{\odot}$, respectively (Beckwith et al. 1990). Since the radius is comparable to that of our solar system and the mass to the so-called "minimum mass

*S. Kato et al. (eds.), Physics of Accretion Disks, 247–252.*
© 1996 OPA (Overseas Publishers Association) Amsterdam B.V.

solar nebula," those disks may well be believed to be the site of planetary system formation. We may hence call them as protoplanetary disks.

Compact (size$\sim$100 AU) and dense ($H_2$ number density$\gtrsim 10^{10}$ $cm^{-3}$) disks around protostars are the least studied among those around the three types of young stellar objects because the disks as well as stars are deeply embedded in molecular cloud cores or, in other words, protostellar envelopes and are difficult to be observed. We know their existence from millimeter-(mm) and submillimeter (submm)-wave interferometric measurements, which showed that the protoplanetary disks around protostars might be somewhat less massive than or as massive as those around T Tauri stars[1] (Ohashi et al. 1991; Terebey et al. 1993; Moriarty-Schieven et al. 1992). Protoplanetary disks around protostars are still under growing up as being fed by surrounding, infalling envelopes that are often observed as larger disks with radius $10^3$–$10^4$ AU.

Excess emission at infrared (IR) and mm-wave for CTTS and WTTS are consistent with arising from dust particles in protoplanetary disks, while the excess from protostars originates from extended envelopes. Spectral energy distributions (SED) of CTTS and WTTS are well fitted by the superposition of blackbody radiation with temperature varying from $\sim$2000 K to $\sim$20 K. When we assume radially symmetric temperature distribution such as $T(r) \propto r^{-q}$, then the power law index $q$ can be easily obtained from SEDs at IR with the aid of the relation $\nu F_\nu = \nu^{4-2/q}$. The value of $q$ is between 0.5 and 0.75 (Beckwith et al. 1990). The $q$ value of 0.75 is the case for geometrically thin, steady accretion or reprocessing disks. Less steep temperature gradient ($q <$ 0.75) is explained if disks are flared (Kusaka et al. 1970; Kenyon & Hartmann 1987), i.e., the case when outer part of disks receive more heating radiation than the geometrically thin case. This interpretation, however, cannot be applicable to the case when disk luminosity is comparable to or larger than the extinction corrected stellar luminosity, i.e., the case for "flat spectrum" T Tauri stars, which have almost constant spectral energy density in the wavelength range 1–100 $\mu$m. Reprocessing of stellar photons by flattened accreting envelopes is proposed to consistently explain such SEDs (Kenyon et al. 1993).

At present we do not have good estimates for the density distribution of protoplanetary disks, since SED is not sensitive to the density distribution. High angular resolution ($\sim$0.1″) imaging with mm and submm interferometers is essential to probe the density distribution of protoplanetary disks.

[1]There is another group of researchers who propose "Class 0" sources, which would be in the youngest stages of evolution but could have massive disks. Observational evidence is still poor to distinguish which is the actual case.

## 3. Rotation

Line observations of disks revealed the rotation of protoplanetary disks. Weak CO ($J = 1 - 0$) emission has been unambiguously detected from disks around GG Tauri (Skrutskie et al. 1993; Kawabe et al. 1993; Dutrey et al. 1994) and DM Tauri (Handa et al. 1995; Saito et al. 1995), toward which little background emission from extended molecular gas exists. Interferometric observations are also powerful to selectively detect disk emission mixed with stronger background emission (Koerner et al. 1993). These observations have shown that protoplanetary disks are rotating such that their kinematics is consistent with the Keplerian law.

Emission and absorption lines from rotating disks show characteristic double peaked profiles (Omodaka et al. 1992; Beckwith & Sargent 1993). Line profiles of vibrational transitions of CO sometimes show such double peaked profiles. Carr et al. (1993) pointed out that the CO bandhead emission from the protostar candidate WL16 embedded in the Rho Ophiuchi molecular cloud is consistent with such a double peaked line profile, suggesting that very inner part of the disk is rotating rapidly as a result of Keplerian rotation.

## 4. Accretion

Some CTTS have the entire disk luminosity comparable to or larger than the extinction corrected stellar luminosity. This can only be explained when disks derive their energy from gravitational accretion. When the mass accretion rate onto central stars is less than $10^{-7}$ $M_\odot$ yr$^{-1}$, then the disk luminosity is dominated by reprocessing of stellar photons. Even in such a case, CTTS show evidence of accretion such as UV excess, optical veiling, near IR excess, strong emission lines, blueshifted forbidden lines, random time variation, etc., which are the major difference between CTTS and WTTS when both have disks dominated by reprocessing luminosity. It is getting more and more evident that CTTS differ from WTTS in that CTTS have viscous accreting disks while WTTS, even if they do have disks, lack the inner part of disks where the disks would otherwise be interacting with stars.

Although we cannot obtain direct evidence for viscous accretion disks around protostars because of their heavy obscuration, there are observational results suggestive of their presence. Such viscous accreting disks around protostars are surrounded by envelopes of ~0.1 pc in size, which must undergo dynamical infall due to the isothermal nature of interstellar molecular gas. Clear evidence for dynamical infall was first obtained toward HL Tauri in its flattened disk-shaped envelope of mass and radius of 0.03 $M_\odot$ and 1,400 AU, respectively (Hayashi et al. 1993). This disk-shaped envelope has an infalling velocity of ~1 km s$^{-1}$ at the radius of 1,000 AU, which is consistent with the infalling gas being accelerated by the central star of 0.55 $M_\odot$. It also shows

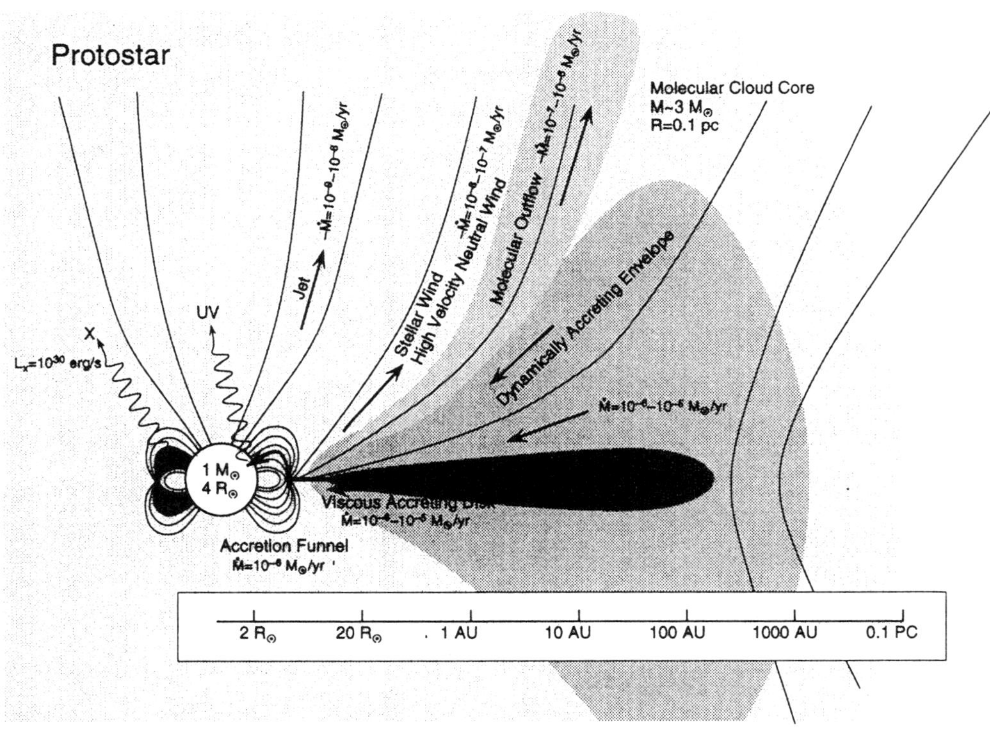


Fig. 1. Circumstellar structure around an accreting protostar

a hint of rotation at 0.2 km $s^{-1}$ at the same radius, suggesting that rotation becomes dominated inside the radius of ~40 AU and that centrifugally supported viscous accretion disk forms there (Lin et al. 1994). This radius is consistent with the recent measurements of 70 AU for the radius of the compact dust disk around HL Tauri (Lay et al. 1994).

## 5. Formation and Dissipation

Formation of protoplanetary disks takes place in the accreting protostar phase. The formation process is clearly demonstrated in the example of HL Tauri as described above such that a molecular cloud core with finite initial angular momentum dynamically collapses to form a centrifugally supported viscous accretion disk of ~100 AU in radius. If the inside-out collapse scenario works (Shu et al. 1987), then a viscous accreting disk may be initially small in radius and mass and it may grow up as the central protostar grows. This is because the infalling gas initially at larger radius has larger specific angular momentum and it eventually falls on the outermost part of the forming disk with increasing its radius and mass (see section 2).

Dissipation of protoplanetary disks was suggested from the decrease of excess IR emission following the transition from CTTS to WTTS (Strom et al. 1989; Skrutskie et al. 1990), which occurs typically 3 million years of stellar age. Mm emission from disks, on the other hand, lasts at least $10^7$ yr (Beckwith et al. 1990). These results show that a protoplanetary disk dissipates away from its inner part first leaving its outer part relatively long in the WTTS phase.

## 6. Relation between Disks and Outflows

Young stars are known to have outflows as well as circumstellar disks. It is getting clear that such outflows are closely related to accretion processes. Outflows from young stars can be classified into three groups, namely Jet/HH, SW/HVNW, and MO. Jet/HH includes radio and optical jets, Herbig-Haro objects, and ionized stellar winds. SW/HVNW means neutral stellar winds observed in [OI] or Na D lines and high velocity neutral winds detected in radio HI or CO lines. MO denotes bipolar molecular outflows. Jet/HHs are characterized by their high velocity of ~200 km $s^{-1}$, suggestive of their origin close to central stars, and small mass loss rates. While SW/HVNWs also exhibit the high velocity, they have large mass loss rates and accordingly have enough momentum to drive molecular outflows. Because it is difficult to directly drive molecular outflows from disks, MOs are gradually believed to be driven by SW/HVNWs (Shu et al. 1994). A summary for the three types of outflows is given in table 1.

All the three types of outflows are observed for embedded protostars. Only a small fraction of CTTS, possibly transient objects from the protostar to CTTS phase, have associated jets, HH objects, and molecular outflows. It is, however, evident from strong emission lines of CTTS that CTTS are all associated with ionized and neutral stellar winds. None of the three types of outflows has ever been discovered for WTTS.

Edwards et al. (1993) found excellent correlation between the presence of accretion related phenomena and SW/HVNW, suggesting that SW/HVNWs are directly related to accretion from disks onto stars and are driven at the innermost part of disks. They also pointed out that the rotation period of

Table 1. Summary of accretion and outflow phenomena. See text for the meaning of Jet/HH, SW/HVNW, and MO.

| | Protostar | CTTS | WTTS | $-\dot{M}_{outflow}$ | Observations |
|---|---|---|---|---|---|
| $\dot{M}_{accretion}$ | $10^{-6}$–$10^{-5}$ | $10^{-7}$–$10^{-6}$ | 0 | | |
| Jet/HH | Yes | 10% | No | $10^{-8}$–$10^{-9}$ | [SII], Radio |
| SW/HVNW | Yes | Yes | No | $10^{-7}$–$10^{-8}$ | [OI], HI, CO |
| MO | Yes | 10% | No | $10^{-6}$–$10^{-7}$ | CO |

CTTS takes relatively constant value of ~8 days in addition to the well known fact that T Tauri stars are slow rotators. These observational results, coupled with strong X-ray emission detected from all the three types of young stars, indicates that protostars and CTTS have stellar magnetic fields interacting with disks to drive winds, to promote accretion, and to control stellar rotation as is schematically shown in figure 1, while WTTS have magnetic fields which are free from disks because the inner part of the disks are dissipated.

**Acknowledgments**

I am grateful to Shoken M. Miyama and Nagayoshi Ohashi for fruitful discussions through which I often get inspiring ideas.

**References**

Beckwith S. V. W., Sargent A. I. 1993, ApJ 402, 280
Beckwith S. V. W., Sargent A. I., Chini R. S., Güsten R. 1990, AJ 99, 924
Burrows C. 1995, http://www.stsci.edu/pubinfo/PR/95/24.html
Carr J. S., Tokunaga A. T., Najita J., Shu F. H., Glassgold A. E. 1993, ApJL 411, L73
Dutrey A., Guilloteau S., Simon M. 1994, A&A 286, 149
Edwards S. et al. 1993, AJ 106, 372
Handa T. et al. 1995, ApJ 449, 894
Hayashi M., Ohashi N., Miyama S. M. 1993, ApJL 418, L71
Kawabe R., Ishiguro M., Omodaka T., Kitamura Y., Miyama S. M. 1993, ApJL 404, L63
Kenyon S. J., Calvet N., Hartmann L. 1993, ApJ 414, 676
Kenyon S. J., Hartmann L. 1987, ApJ 323, 714
Koerner D. W., Sargent A. I., Beckwith S. V. W. 1993, Icarus 106, 2
Kusaka T., Nakano T., Hayashi C. 1970, PTP 44, 1580
Lay O. P., Carlstrom J., Hills R. J., Phillips T. G. 1994, ApJL 434, L75
Lin D. N. C., Hayashi M., Bell K. R., Ohashi N. 1994, ApJ 435, 821
Moriarty-Schieven G. H., Wannier P. G., Tamura M., Keene J. 1992, ApJ 400, 260
O'Dell C. R., Wen Z., Hu X. 1993, ApJ 410, 696
Ohashi N., Kawabe R., Hayashi M., Ishiguro M. 1991, AL 102, 2054
Omodaka T., Kitamura Y., Kawazoe E. 1992, ApJL 396, L87
Saito M, Kawabe R., Ishiguro M., Miyama S. M., Hayashi M., Handa T., Kitamura Y., Omodaka T. 1995, ApJ 453, 384
Shu F. H., Adams F. C., Lizano S. 1987, ARA&A 25, 23
Shu F. H., Najita J., Ostriker E., Wilkin F., Ruden S., Lizano S. 1994, ApJ 429, 781
Skrutskie M. F. et al. 1993, ApJ 409, 422
Skrutskie M. F. et al. 1990, AJ 99, 1187
Strom K. M., Strom S. E., Edwards S., Cabrit S., Skrutskie M. F. 1989, AJ 97, 1451
Terebey S., Chandler C. J., André P. 1993, ApJ 414, 759

# Instabilities in Self-Gravitating Circumstellar Disks

Nobuhiro KIKUCHI[1] and Shoken M. MIYAMA[2]
*1. Department of Astronomy, University of Tokyo, Tokyo 113, Japan*
*2. National Astronomical Observatory, Mitaka, Tokyo 181, Japan*

**Abstract**

Linear and non-linear calculations of the evolution of thin self-gravitating circumstellar disks are presented in order to study the efficiency of angular momentum transfer due to non-axisymmetric instabilities. It is shown that non-axisymmetric instabilities can cause mass accretion within a time scale which is comparable to that of evolution of protostars and T Tauri stars.

## 1. Introduction

To understand processes which redistribute angular momentum within protoplanetary disks is important to understand the evolution of protoplanetary disks and initial conditions for the formation of planetary systems. Larson (1988) pointed out four mechanisms which can transport angular momentum in protoplanetary disks. They are (1) turbulent viscosity, (2) gravitational torques, (3) magnetic torques and (4) wave transport with shock dissipation. In early stage of the evolution, protoplanetary disks are likely to be relatively massive and self-gravitating because material from a collapsing molecular cloud core falls onto the disk rather than onto the central star. Therefore, gravitational torques which arise from non-axisymmetric instabilities can be an efficient mechanism for angular momentum transfer.

Since the work by Adams et al. (1989), several analyses of linear and non-linear stability of self-gravitating gaseous disks have been done (e.g. Papaloizou & Savonije 1991; Laughlin & Bodenheimer 1994; Miyama et al. 1994). In this paper we show that the growth of non-axisymmetric unstable modes into a non-linear regime can form shocks and that their dissipation can be an additional source of angular momentum transfer.

*S. Kato et al. (eds.), Physics of Accretion Disks, 253–256.*
**© 1996 OPA (Overseas Publishers Association) Amsterdam B.V.**

## 2. Basic Equations

We consider a thin, equilibrium disk orbiting around a central star. We study the stability of the disk by examining whether a small perturbation added to the equilibrium disk grows or not. The hydrodynamic equations which describe the evolution of the disk in cylindrical coordinate $(r, \phi)$ can be expressed in a weak conservation form as

$$\frac{\partial \Sigma}{\partial t} + \frac{1}{r}\frac{\partial}{\partial r}(r\Sigma u) + \frac{1}{r}\frac{\partial}{\partial \phi}(\Sigma v) = 0 \ , \tag{1}$$

$$\begin{aligned} \frac{\partial}{\partial t}(\Sigma u) &+ \frac{1}{r}\frac{\partial}{\partial r}[r(\Sigma u^2 + P)] + \frac{\cos\phi}{r}\frac{\partial}{\partial \phi}[(\Sigma v^2 + P)\sin\phi - \Sigma uv\cos\phi] \\ &+ \frac{\sin\phi}{r}\frac{\partial}{\partial \phi}[(\Sigma v^2 + P)\cos\phi + \Sigma uv\sin\phi] = -\Sigma\frac{\partial \psi}{\partial r} \ , \end{aligned} \tag{2}$$

$$\begin{aligned} \frac{\partial}{\partial t}(\Sigma v) &+ \frac{1}{r}\frac{\partial}{\partial r}(r\Sigma uv) - \frac{\sin\phi}{r}\frac{\partial}{\partial \phi}[(\Sigma v^2 + P)\sin\phi - \Sigma uv\cos\phi] \\ &+ \frac{\cos\phi}{r}\frac{\partial}{\partial \phi}[(\Sigma v^2 + P)\cos\phi + \Sigma uv\sin\phi] = -\frac{\Sigma}{r}\frac{\partial \psi}{\partial \phi} \ , \end{aligned} \tag{3}$$

$$\frac{\partial E}{\partial t} + \frac{1}{r}\frac{\partial}{\partial r}[r(E + P)u] + \frac{1}{r}\frac{\partial}{\partial \phi}[(E + P)v] = -\Sigma\left(u\frac{\partial \psi}{\partial r} + \frac{v}{r}\frac{\partial \psi}{\partial \phi}\right) \ . \tag{4}$$

In the above equations, $\Sigma$ is the surface density, $u$ and $v$ are the velocity in the radial and azimuthal directions, respectively, $P$ is the pressure integrated in the vertical direction, and $E$ is the total energy per unit area. The equation of state is

$$P = (\gamma - 1)\left[E - \frac{\Sigma}{2}(u^2 + v^2)\right] \ , \tag{5}$$

where $\gamma$ is the ratio of specific heats and is taken to be 1.4 in this paper. The gravitational potential $\psi$ is composed of two parts such as

$$\psi = \psi_{\mathrm{star}} + \psi_{\mathrm{disk}} \ . \tag{6}$$

The potential of the central star $\psi_{\mathrm{star}}$ is written as

$$\psi_{\mathrm{star}}(\boldsymbol{r}) = -\frac{GM_{\mathrm{star}}}{|\boldsymbol{r} - \boldsymbol{r}_{\mathrm{star}}|} \ , \tag{7}$$

where $\boldsymbol{r}_{\mathrm{star}}$ is the position of the central star which can be found by the conservation of the center of mass

$$M_{\mathrm{star}}\boldsymbol{r}_{\mathrm{star}} + \int_{R_{\mathrm{in}}}^{R_{\mathrm{out}}}\int_0^{2\pi} \boldsymbol{r}\Sigma r dr d\phi = \boldsymbol{0} \ . \tag{8}$$

Finally, the potential of the disk $\psi_{\rm disk}$ is given by the Poisson integral:

$$\psi_{\rm disk}(r,\phi) = -\int_{R_{\rm in}}^{R_{\rm out}} \int_0^{2\pi} \frac{\Sigma(r',\phi')r'dr'd\phi'}{\sqrt{r^2 + r'^2 - 2rr'\cos(\phi-\phi')}} \,. \tag{9}$$

## 3. Linear Stability Analysis

When the deviation from the axisymmetric initial state is small, the stability of the disk can be studied using a linear perturbation theory. The basic equations described in the previous section can be linearized with respect to a small perturbation of the form

$$\delta\Sigma(r)e^{i(m\phi-\omega t)} \,, \tag{10}$$

where $m$ is the azimuthal wave number and $\omega$ is the complex mode frequency. Given appropriate boundary conditions, linearized equations form an eigenvalue problem for $\omega$. If there is an eigenvalue for which $\mathrm{Im}(\omega) > 0$, the disk is unstable. The eigenvalue problem can be solved numerically to obtain eigenvalues and eigenfunctions. The unstable disks with some of the eigenfunctions are used as initial conditions for non-linear simulations described in the next section.

## 4. Non-Linear Simulations

The non-linear hydrodynamic equations described in section 2 are solved using a two-dimensional Eulerian hydrodynamic code. Briefly, our hydrodynamic code calculates numerical fluxes by solving a Riemann problem exactly using an algorithm described in van Leer (1979). Second order of accuracy is achieved using a technique proposed by Harten et al. (1987). The calculations presented here employ $128 \times 128$ meshes.

We consider a disk model in which the disk contains 20% of the system's total mass. The entropy distribution of the initial disk is assumed to be uniform in two-dimensional sense such that $P/\Sigma^\gamma = 0.1$. The minimum value of Toomre's $Q$ is about 2. A reflecting boundary condition is employed at both inner and outer radius of the disk. At the beginning of the calculation, the axisymmetric equilibrium disk is perturbed by the most unstable $m = 2$ mode obtained by the linear stability analysis with maximum amplitude $\delta\Sigma/\Sigma = 10^{-3}$. The subsequent evolution is followed up to $t = 30$ in non-dimensional unit corresponding to about 5 rotation periods at the outer radius of the disk. Equidensity contours at $t = 15.0$ are plotted in figure 1, which shows the development of two-armed spiral structure and the formation of shock waves. At the end of the calculation, mass which lost angular momentum accumulates near the inner boundary as shown in figure 2.

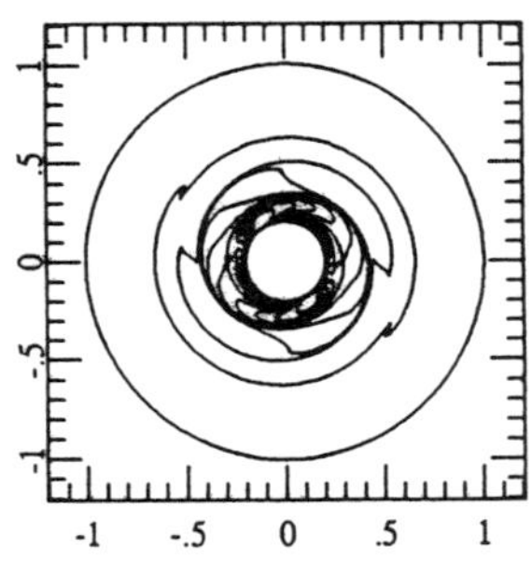


Fig. 1. Equidensity contours at $t = 15.0$.

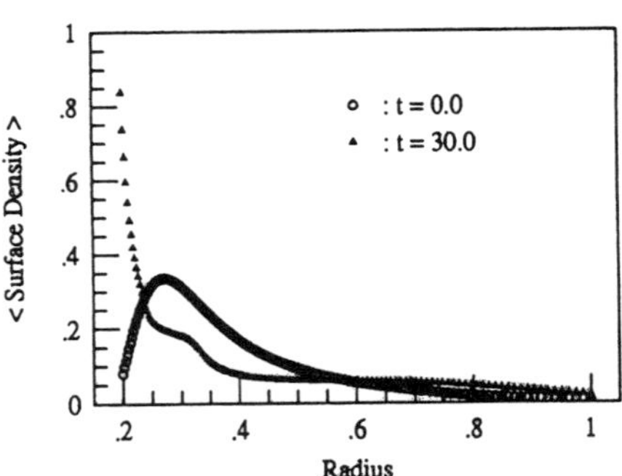


Fig. 2. Initial and final (azimuthally averaged) distribution of surface density are plotted by open circles and filled triangles, respectively.

In order to estimate mass accretion rate due to non-axisymmetric instabilities, we carried out another calculation using an inner boundary condition which allows mass accretion through the inner boundary. The mass accretion rate could be estimated to be about $10^{-3} \times \sqrt{GM^3/R_{\text{out}}}$, where $M$ is the total mass of the system and $R_{\text{out}}$ is the outer radius of the disk. This implies that if $M = 1M_\odot$ and $R_{\text{out}} = 100$ AU, the time scale of angular momentum transfer is about $10^5$ years, which is comparable to the evolutional time scale of protostars. Therefore, non-axisymmetric instabilities can be an efficient mechanism for angular momentum transfer when the disk is self-gravitating.

## Acknowledgements

N. K. was supported by Grant-in-Aid for Encouragement of Young Scientists of the Japanese Ministry of Education, Science and Culture.

## References

Adams F. C., Ruden S. P., Shu F. H. 1989, ApJ 347, 959

Harten A. et al. 1987, JComptPhys 71, 231

Larson R. B. 1988, in The Formation and Evolution of Planetary Systems, ed H. A. Weaver, L. Danly (Cambrige University Press) p31

Laughlin G., Bodenheimer P. 1994, ApJ 436, 335

Miyama S. M. et al. 1994, in Numerical Simulations in Astrophysics, ed J.Franco, S. Lizano, L. Aguilar, E. Daltabuit (Cambrige University Press) p305

Papaloizou J. C., Savonije G. J. 1991, MNRAS 248, 353

van Leer B. 1979, JComptPhys 32, 101

# Similarity Solution for Gravitationally Contracting Rotating Disks

Kazuya SAIGO and Tomoyuki HANAWA
*Department of Astrophysics, School of Science, Nagoya University, Chikusa-ku, Nagoya 464-01, Japan*

## Abstract

We obtain the structure and evolution of an isothermal rotating disk that contracts dynamically owing to its self-gravity. The disk is assumed to be geometrically thin and symmetric around its rotation axis. Our solutions show similarity in the evolution of the surface density, radial velocity, and rotation velocity: their profiles keep their functional forms unchanged while the scale factor evolves with time. The surface density is nearly constant in the central part, the radial size of which decreases linearly with time. The radial and rotation velocity are proportional to the radius in the central part and approach to asymptotic values of the order of the sound speed in the outer part. Our similarity solutions explain main features of numerical simulations of dynamically collapsing rotating disks. One of our similarity solution corresponds to a non-rotating magnetized disk during its collapse, and agrees with the numerical magnetohydrodynamical simulations with an error of several tens percents.

## 1. Introduction

Rotation and magnetic field have large effects on the star formation. When a gas cloud collapses to form a star, it will form a disk perpendicular to the rotation and magnetic field. According to numerical studies thus far (see, e.g., Narita et al. 1984; Tomisaka 1995; Nakamura et al. 1995), the gravitationally contracting disk shows 'similarity' in its structure during its isothermal collapse phase: the surface density and velocity profiles keep their functional forms unchanged. The surface density is almost flat in the central part of the disk and is inversely proportional to the radial distance from the

*S. Kato et al. (eds.), Physics of Accretion Disks, 257–260.*
**© 1996 OPA (Overseas Publishers Association) Amsterdam B.V.**

center in the envelope. As the central surface density increases, the central part of the disk shrinks in the linear dimension.

Narita et al. (1984) derived the similarity equation for the gravitationally contracting disk under the thin disk approximation. It is similar to the similarity equation for a gravitationally contracting gas sphere (Larson 1969; Penston 1969) but much more difficult to solve. The former is the boundary value problem to find a solution for an integral differential equation having a singular point. Thus, only approximate solutions have been obtained thus far. The solution of Narita et al. (1984) does not pass the singular point and Nakamura et al. (1995) modified the similarity equation into a simpler differential equation to obtain approximate solutions.

In this paper we report the exact similarity solutions for gravitationally contracting disks. They are obtained numerically but their numerical error is smaller than $10^{-4}$ in the relative amplitude. We also confirmed that the similarity solutions reproduce the numerical simulations with the error of 20 % for a magnetized disk and qualitatively for a rotating disk.

## 2. Model

We consider the evolution of a geometrically thin disk assuming the symmetry around the axis and the isothermity of the gas. In the polar coordinates, $(r,\ \varphi)$, the evolution of the surface density $\Sigma$ is described as

$$\frac{\partial \Sigma}{\partial t} + \frac{1}{r}\frac{\partial}{\partial r}(r v_r \Sigma) = 0 , \tag{1}$$

where $v_r$ denotes the radial velocity. The equation of motion is described as

$$\frac{\partial}{\partial t}(v_r \Sigma) + \frac{1}{r}\frac{\partial}{\partial r}(r {v_r}^2 \Sigma) + {c_s}^2 \frac{\partial \Sigma}{\partial r} - \frac{{v_\varphi}^2 \Sigma}{r} + g\Sigma = 0 , \tag{2}$$

where $v_\varphi$ and $c_s$ denote the rotational velocity and the isothermal sound speed, respectively. The gravitational acceleration $g$ is related to the surface density by

$$g(r) = 2\pi G \int_0^\infty \int_0^\infty k\, J_1(kr)\, J_0(kr')\, r'\, \Sigma(r')\, dr' dk , \tag{3}$$

where $J_0$ and $J_1$ denote the zeroth and first order of Bessel functions, respectively.

We obtained the self-consistent solution showing the similarity,

$$\Sigma(r,\ t) = \frac{1}{2\pi G(t_0 - t)}\, \eta(x) , \tag{4}$$

$$[v_r(r,\ t), v_\varphi(r,\ t)] = c_s\, [-\xi(x),\ \omega\, x^{-1}\, \mu(x)] , \tag{5}$$

$$g(r,\, t) \;=\; \frac{c_s}{(t_0 - t)} \int_0^\infty \int_0^\infty k J_1(kx) J_0(kx') x' \eta(x')\, dx'\, dk\;, \tag{6}$$

where

$$x \;=\; \frac{r}{c_s(t_0 - t)}\;. \tag{7}$$

The nondimensional functions, $\eta$, $\xi$, and $\phi$ satisfy

$$\left[(\xi - x)^2 - 1\right] \frac{d\eta}{dx} + \eta \left[\frac{(\xi - x)^2}{x} - \gamma + \frac{\omega^2 \mu^2}{x^3}\right] = 0\;, \tag{8}$$

$$\mu \;=\; -(\xi - x)\, x\, \eta \;=\; \int_0^x x' \mu(x')\, dx'\;, \tag{9}$$

$$\gamma \;=\; \int_0^\infty \int_0^\infty k J_1(kx) J_0(kx') x' \eta(x')\; dx'\; dk\;. \tag{10}$$

Note that equation (8) has a singular point at $\xi - x = \pm 1$. We obtained the similarity solution by an iterative method for a given $\omega$, which denotes the ratio of the specific angular momentum to the disk mass. Our solution is valid for $t < t_0$ although equation (8) has solutions valid for $t > t_0$.

## 3. Results

Figure 1 shows the surface density $\eta$ as a function of $x$ for various $\omega$s. The surface density is almost flat in the central part and inversely proportional to the radial distance for any $\omega$. We could not find a similarity solution for $\omega > 0.5$. When $\omega$ is close to 0.5, the infall velocity $\xi$ is very small as shown in figure 2. We think that there exists no similarity solution for $\omega > 0.5$ since the centrifugal force increases in proportion to $\omega^2$.

The infall velocity is $\xi \simeq (1/2)x$ in the central part and nearly constant in the envelope. For $\omega = 0$ (i.e., for a magnetized disk), the singular point $(\xi - x = -1)$ is located at $x = 1.73$.

We compared our similarity solutions with numerical simulations of gravitationally contracting disks. Our solution of $\omega = 0$ coincides with the numerical simulations of Tomisaka (1995) and Nakamura et al. (1995) who followed the gravitational collapse of a magnetized gas cloud. The surface density profile agrees with each other with an error of 20% and the velocity profile with an error of 30%. Our solution of $\omega = 0.3$ reproduces the main features of Matsumoto et al. (1995) who followed the gravitational collapse of a rotating cloud.

## Acknowledgements

We thank Drs. F. Nakamura and T. Matsumoto for valuable discussions and showing us their numerical simulations before publication.

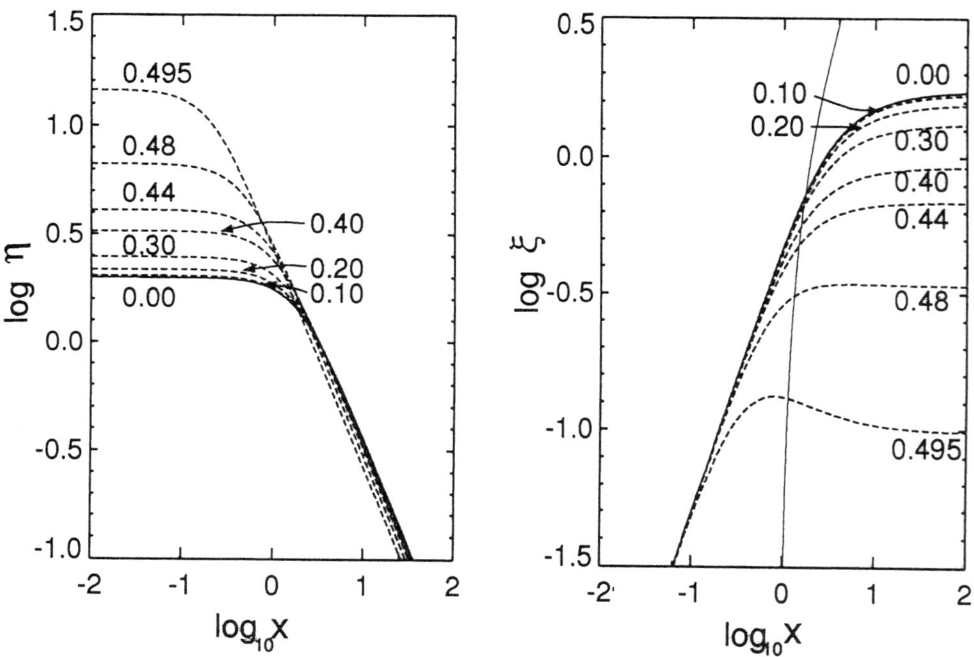


Fig. 1. The surface density profile of our similarity solutions. Each curve denotes a solution for a given $\omega$, the value of which is shown in the figure.

Fig. 2. The same as figure 1, but for the infall velocity.

## References

Larson R. B. 1969, MNRAS 145, 271
Matsumoto T., Nakamura F., Hanawa T. 1995, in preparation
Nakamura F., Hanawa T., Nakano T. 1995, ApJ 444, 770
Narita S., Hayashi C., Miyama S. M. 1984, PTP 72, 1118
Penston M. B. 1969, MNRAS 144, 425
Tomisaka K. 1995a, ApJ 438, 226
Tomisaka K. 1995b, ApJ submitted

# Magnetorotational Instability in Protoplanetary Disks

Takayoshi SANO[1] and Shoken M. MIYAMA[2]
*1. Department of Astronomy, School of Science, University of Tokyo, Bunkyo-ku, Tokyo 113, Japan*
*2. National Astronomical Observatory, Mitaka, Tokyo 181, Japan*

**Abstract**

Magnetorotational instability (Balbus-Hawley instability) in accretion disks with the dissipation of the magnetic field is investigated using local and global linear perturbation theory. When the effect of the ohmic dissipation is important, it is found that small scale perturbations are stabilized. The most unstable wavelength and the maximum growth rate become $\eta/v_A$ and $v_A^2/\eta$, respectively, where $v_A$ and $\eta$ are the Alfvén speed and the magnetic diffusivity. Sufficient conditions for instability are obtained by the global analysis which takes into account of the vertical structure of the disk. We also find the unstable modes where the amplitude of the perturbed velocity is localized only at the upper layer of the disk. The stability analysis is applied to protoplanetary disks.

## 1. Introduction

In the standard theory of accretion disks, it is thought that the turbulent viscosity plays an important role. However the origin of turbulence is still unclear. Recently, it is shown that the fully developed turbulence produced by the magnetorotational instability can transport significant amount of angular momentum (e.g., Balbus & Hawley 1991; Hawley et al. 1995).

Various infrared and radio observations reveal the existence of accretion disks around T Tauri stars, so-called protoplanetary disks, where planet formation is supposed to be taking place. These protoplanetary disks are very cold and dense so that the ohmic dissipation cannot be neglected. In this paper, we investigate the effect of the ohmic dissipation on magnetorotational instability using linear perturbation theory and discuss its application to protoplanetary disks.

*S. Kato et al. (eds.), Physics of Accretion Disks, 261–264.*
© 1996 OPA (Overseas Publishers Association) Amsterdam B.V.

## 2. Linear Analysis

Consider a magnetized axisymmetric accretion disk with the Keplerian angular velocity. The unperturbed magnetic field is assumed to be uniform and purely vertical. The continuity and momentum equations are

$$\frac{\partial \rho}{\partial t} + \nabla \cdot (\rho \boldsymbol{v}) = 0 \,, \tag{1}$$

$$\frac{d\boldsymbol{v}}{dt} + \frac{1}{\rho}\nabla\left(P + \frac{B^2}{8\pi}\right) - \frac{1}{4\pi\rho}\left(\boldsymbol{B}\cdot\nabla\right)\boldsymbol{B} + \nabla\Phi = 0 \,, \tag{2}$$

where $d/dt$ denotes the Lagrangian derivative and $\Phi$ is the gravitational potential of the central star. The induction equation is

$$\frac{\partial \boldsymbol{B}}{\partial t} - \nabla \times (\boldsymbol{v} \times \boldsymbol{B}) = \eta \nabla^2 \boldsymbol{B} \,, \tag{3}$$

where $\eta$ is the magnetic diffusivity which is assumed to be constant. The energy equation is

$$\rho T \frac{ds}{dt} = \frac{\eta}{4\pi}\left(\nabla \times \boldsymbol{B}\right)^2 \,, \tag{4}$$

where $s$ is the entropy per unit volume. The right-hand side of the equation (4) represents a loss of energy resulting from the Joule heating. We assume that the disk is isothermal, i.e., the adiabatic sound speed $c_s = \gamma P/\rho$ is constant, and in hydrostatic equilibrium along the vertical direction.

At first, we perform simple local analysis to show the effect of the ohmic dissipation briefly. We consider large-wavenumber axisymmetric Eulerian perturbations which are proportional to $\exp i\,(k_r r + k_z z - \omega t)$. It is assumed that unperturbed density distribution is spatially uniform. Equations (1)-(4) give the fourth order dispersion relation;

$$k^2\left\{\omega\left(\omega + i\eta k^2\right) - k_z^2 v_{\rm A0}^2\right\}^2 - k_z^2\kappa^2\left\{\left(\omega + i\eta k^2\right)^2 - k_z^2 v_{\rm A0}^2\right\} - 4\Omega^2 k_z^4 v_{\rm A0}^2 = 0 \,, \tag{5}$$

where $k^2 = k_r^2 + k_z^2$ and $v_{\rm A0} = B_0/\sqrt{4\pi\rho_0}$ is the Alfvén speed, and $\kappa^2$ is the square of the epicyclic frequency. The subscript zero denotes quantities at the equilibrium. In derivation of the equation (5) we adopt the Boussinesq approximation.

The dispersion relation (5) is characterized by a magnetic Reynolds number, $R_{\rm m} = v_{\rm A0}^2/\eta\Omega$. The unstable growth rate is shown as a function of $R_{\rm m}$ in figure 1. When $R_{\rm m} > 1$, characteristics of the unstable modes are quite similar to those of ideal MHD. If $R_{\rm m} < 1$, small scale perturbations are stabilized due to the effect of the ohmic dissipation and the most unstable wavelength becomes $\eta/v_{\rm A0}$. This scale is determined by the balance of unstable growth rate and resistive dumping rate. The maximum unstable growth rate is inversely proportional to the magnetic diffusivity and reduced significantly.

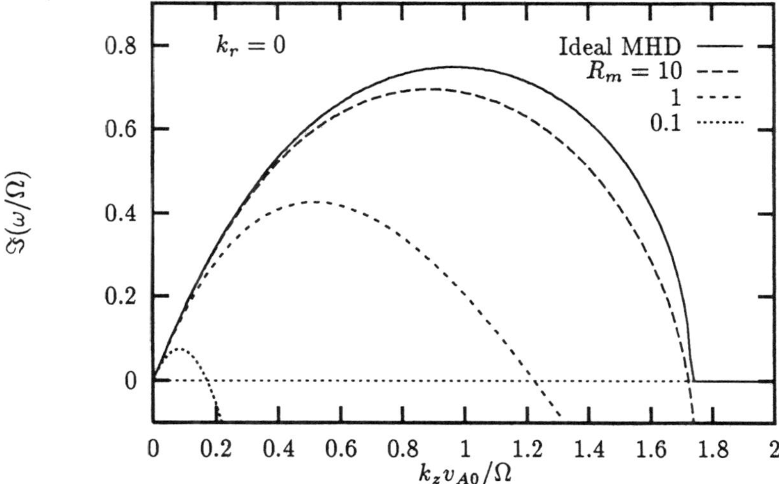


Fig. 1. $R_{\rm m}$ dependence of the unstable growth rate. For the case of $k_r = 0$ the system is most unstable.

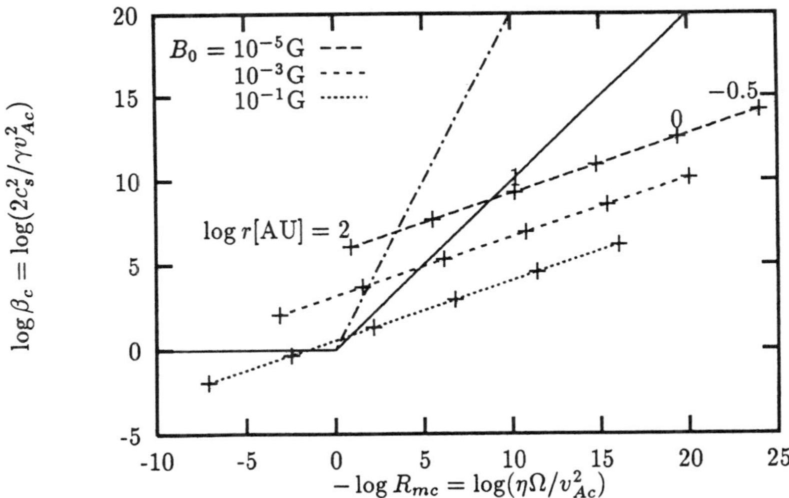


Fig. 2. The disk structure with various strength of the magnetic field in the $\beta_{\rm c}$ and $R_{\rm mc}$ diagram. Disk model is the minimum mass solar nebula and the magnetic diffusivity is calculated by Umebayashi & Nakano (1988). Solid line denotes equation (6), the criterion of the unstable mode.

Next, global analysis taking into account of the vertical structure of the disk is performed. In radial direction, we use local approximation again because we would like to consider a geometrically thin disk structure. Since the case of $k_r = 0$ would be most unstable from the result of local analysis, we consider only this case. The horizontal and vertical motions are decoupled in this case and only the horizontal displacements which concern with the unstable modes are calculated. We consider the hot halo at $|z| > z_b$, where $c_s \rightarrow \infty$, $\rho \sim P/c_s^2 \rightarrow 0$, and the perturbed magnetic field connects with the solution in the halo as the boundary condition.

This system is characterized by two parameters, $\beta_c = 2c_s^2/\gamma v_{Ac}^2$ and $R_{mc} = v_{Ac}^2/\eta\Omega$, where $v_{Ac} = B_0/\sqrt{4\pi\rho_c}$ is the Alfvén speed at the midplane of the disk. The stability criterion is approximately given by the requirement that the shortest wavelength of the unstable modes should be less than the scale height, $H = \sqrt{2}c_s/\sqrt{\gamma}\Omega$. This condition can be expressed as,

$$\beta_c \gtrsim 1 \qquad \text{and} \qquad \beta_c R_{mc} \gtrsim 1 \, . \tag{6}$$

We find the unstable modes in which the amplitude of the perturbed velocity is localized only at the upper layer of the disk, where local unstable wavelength is shorter than the scale height. The region where these modes are found lies in $\sqrt{\beta_c} \lesssim R_{mc}^{-1} \lesssim \beta_c$ and $\beta_c > 1$ in the $\beta_c$-$R_{mc}$ diagram.

## 3. Application to Protoplanetary Disks

In this section, we consider the application to protoplanetary disks. We take the minimum mass solar nebula as a disk model (Hayashi et al. 1985) and use results of Umebayashi & Nakano (1988) as the electrical conductivity $\sigma_c$, i.e., $\eta = c^2/4\pi\sigma_c = 7.2 \times 10^{23} \, (r/1\text{AU})^{-5} \, \text{cm}^2\text{s}^{-1}$. Two parameters, $\beta_c$ and $R_{mc}$, are estimated as a function of the radius and the strength of the magnetic field,

$$\beta_c = \frac{2c_s^2}{\gamma v_{Ac}^2} = 2.1 \times 10^4 \left(\frac{B_0}{0.1\text{G}}\right)^{-2} \left(\frac{r}{1\text{AU}}\right)^{-13/4} \left(\frac{\gamma}{5/3}\right)^{-1} \left(\frac{L}{L_\odot}\right)^{1/8} \left(\frac{M_*}{M_\odot}\right)^{1/2} , \tag{7}$$

$$R_{mc}^{-1} = \frac{\eta\Omega}{v_{Ac}^2} = 2.5 \times 10^{11} \left(\frac{B_0}{0.1\text{G}}\right)^{-2} \left(\frac{r}{1\text{AU}}\right)^{-37/4} \left(\frac{L}{L_\odot}\right)^{-1/8} \left(\frac{M_*}{M_\odot}\right) . \tag{8}$$

From the stability criterion (**??**), we conclude that the outer part of the disk, $r \gtrsim r_c = 14$ AU, is unstable if the magnetic field strength is less than 0.1 gauss. The critical radius, $r_c$, is independent of the field strength. For the radius from 14 AU to 53 $(B_0/10^{-5}\text{G})^{-8/61}$ AU, the amplitude of the unstable modes is localized at the upper layer of the disk. It implies that angular momentum transport might occur only in this layer in nonlinear stage.

## References

Balbus S. A., Hawley J. F. 1991, ApJ 376, 214

Hawley J. F., Gammie C. F., Balbus S. A. 1995, ApJ 440, 742

Hayashi C., Nakazawa K., Nakagawa Y. 1985, in Protostars and Planets II, ed. D. C. Black & M. S. Matthews (Univ. of Arizona Press, Tuscon), p1100

Umebayashi T., Nakano T. 1988, PTPS No. 96, 151

# Nonlinear Interaction of Global Modes in Protostellar Disks

Vladimir KORCHAGIN[1] and Gregory LAUGHLIN[1,2]
*1. National Astronomical Observatory, Mitaka, Tokyo 181, Japan*
*2. Dept. of Physics, University of Michigan, Ann Arbor, MI 48109, USA*

## Abstract

Nonlinear interactions of global spiral modes play an important role in disk dynamics. Within the context of a second-order, semi-analytic analysis, we have derived a formalism describing the nonlinear interaction of two linearly unstable modes in a massive protostellar disk. Specifically, we have considered the nonlinear generation of linearly stable single-armed spirals. We compare the results of our non-linear analysis with hydrodynamic simulations, illustrating that this "three-wave" process is indeed an important effect.

## 1. Introduction

The development of a rapidly growing $m$=1 mode within a protostellar disk could provide an attractive mechanism for realizing binary formation. Eccentric modes are also useful for eliciting mass and angular momentum transport through the disk itself. Furthermore, one-armed spirals may also be important in other astrophysical contexts, such as the accretion disks surrounding the central sources which drive quasars..

Previous simulation-based investigations of disk stability have indicated that within the linear regime, gaseous disks surrounding a central gravitating mass generally become unstable to higher order modes (i.e., $m = 2, 3, 4$, etc.) while remaining stable to $m$=1 waves. However, disks which are *linearly stable* to the growth of $m$=1 disturbances often exhibit very rapid development of $m$=1 patterns once higher-order modes have grown to significant amplitudes.

Here, we explain the nonlinear generation of such $m$=1 patterns within the context of the three-wave coupling of global spiral modes. The three-wave effect is a resonance phenomena, arising from second-order terms in

*S. Kato et al. (eds.), Physics of Accretion Disks, 265–268.*
© 1996 OPA (Overseas Publishers Association) Amsterdam B.V.

a perturbation expansion of the hydrodynamic equations. The basic idea is that a mode with azimuthal mode number $m$=1 and rotational frequency $\omega_{p_1} = \omega_{p_{n+1}} - \omega_{p_n}$ is produced from two separate, linearly unstable modes with $m = n+1$ and $m = n$. We show that such a coupling can be a very effective mechanism for generating eccentric $m = 1$ modes, and that an emergent one-armed spiral can reach the deeply nonlinear high-amplitude regime at a time coincident with the saturation of the linearly unstable modes. Our analysis illustrates the importance of nonlinear interaction of unstable modes in the overall evolution of self-gravitating disks.

## 2. Equilibrium Model and Linear Analysis

The basic equations describing the behavior of a thin self-gravitating disk are the continuity equation, the momentum equations, and Poisson's equation. In this analysis, we have selected an equilibrium configuration in which the underlying surface density distribution of the disk has a gaussian radial profile:

$$\sigma(r) = \Sigma_0 e^{-(r-R_0)^2/w}. \tag{1}$$

We have also assumed that the gas has a polytropic equation of state, with index $\gamma_p = 2$. This relatively high value was chosen to offset the increased instability of a thin disk in comparison to a full three-dimensional disk. The constant $w$ in equation (1) was set to 0.1, the radius where the surface density peaks, $R_0$, was set at $R_0 = 0.3$, the inner radius of the disk lay at $R_i = 0.2$, and the normalization constant, $\Sigma_0$, was selected so that the disk itself contained 40% of the total mass. This particular configuration has a strong overall similarity to the protostellar disks which we have obtained from axisymmetric simulations designed to follow the collapse of a hypothetical molecular cloud core.

We examined the linear stability properties of the self-gravitating disk with a matrix-eigenmode method similar to the one described by Adams et al. (1989). Specifically, the relation for the form of an unstable surface density eigenmode (with azimuthal mode number $m > 1$) is written as a second-order integro-differential equation in the perturbed surface density. This equation, together with appropriate boundary conditions, is then discretized and solved numerically to determine the global properties of the linearly unstable modes in the disk. Upon applying our mode search procedure (using 200 evenly spaced radial grid points) to the equilibrium model disk specified by equation (1), we found no significantly unstable $m$=1 modes.

The model disk is, however, unstable to higher $m$ patterns. The fastest growing $m$=3 and $m$=4 disturbances turn out to be the most unstable eigenmodes present in the disk. In units where $G$, the disk's outer radius, and the combined mass of the disk and star are all unity, the eigenvalue for the principal $m$=3 mode is $(\Omega_p = 2.66, \gamma = 0.83)$, and that for the $m$=4 mode is $(\Omega_p = 2.48, \gamma = 0.96)$. These three and four-armed spirals both grow at nearly twice the rate of the most unstable $m$=2 mode, whose eigenvalue is located at $(\Omega_p = 2.68, \gamma = 0.44)$.

## 3. Nonlinear Three-Wave Analysis

After we determined the linear stability properties of the model equilibrium disk, we went on to examine the properties of the one-armed global mode which springs from the nonlinear resonance interaction of the linearly unstable three and four armed spirals. Specifically by taking the resonance condition into account, we derived a set of governing equations which describe the development of the $m$=1 global mode in the disk:

$$\frac{\partial}{\partial t}\Big(\frac{\sigma_1}{\sigma}\Big) + i(\Omega_{p_1} - \Omega)\frac{\sigma_1}{\sigma} + \Big(\frac{1}{r} + \frac{1}{\sigma}\frac{d\sigma}{dr}\Big)u_1 + \frac{du_1}{dr} - \frac{i}{r}v_1 = e^{[(\gamma_n+\gamma_{n+1})t]}F_\sigma(r) \quad (2)$$

$$\frac{\partial u_1}{\partial t} + i(\Omega_{p_1} - \Omega)u_1 - 2\Omega v_1 = -\frac{d}{dr}\Big(\tilde{\Psi}_1 + \tilde{\Psi}_\star + h_1\Big) + e^{[(\gamma_n+\gamma_{n+1})t]}F_r(r) \quad (3)$$

$$\frac{\partial v_1}{\partial t} + i(\Omega_{p_1} - \Omega)v_1 + \frac{\kappa^2}{2\Omega}u_1 = \frac{i}{r}\Big(\tilde{\Psi}_1 + \tilde{\Psi}_\star + h_1\Big) + e^{[(\gamma_n+\gamma_{n+1})t]}F_\phi(r) \quad (4)$$

The *generating functions* $F_\sigma(r), F_r(r)$ and $F_\phi(r)$ are somewhat complicated expressions involving the known eigenfunctions of the unstable $m = n$ and $m = n + 1$ modes. The exact form of these generating functions, as well as the details of the calculations are given in Laughlin & Korchagin (1995).

## 4. Results

Equations (2)–(4) describe how the growing $m = n$ and $m = n + 1$ modes dictate the response of the disk's $m$=1 component to an arbitrary perturbation. In general, the solution converges rapidly into a non-winding, slowly rotating, one-armed eigenmode.

In order to see how the results from the semi-analytic three-wave analysis pertain to the evolution of a non-idealized system, we have computed several fully non-linear hydrodynamic simulations for the model disk. In the first calculation, the equilibrium disk model was corrupted with an $m$=1 perturbation. As was expected, given the results of the linear analysis, this perturbation was stable. In the second calculation, the disk was seeded with equal-strength $m$=3 and $m$=4 perturbations. In sharp contrast to the first simulation, rapid growth of an $m$=1 mode was observed. The resultant single-armed mode had the expected growth rate, as well as a qualitatively appropriate radial structure when compared with the semi-analytic calculations. Again, for the details, see Laughlin & Korchagin (1995).

Figure 1 plots the development of the *global* Fourier amplitudes in the disk for the $m$=1, 3, and 4 modes. The overall growth rate of the $m$=1 component agrees quite well with the quantitative prediction (shown as a straight line) up to a time $T \approx 6$. At later times in the $m$=3+4 simulation, saturation in all three of the mode components is observed, and the disk has entered the *deep non-linear regime*. The figure also shows that the $m$=1 mode does not grow at all when the disk is started with a pure $m$=1 perturbation.

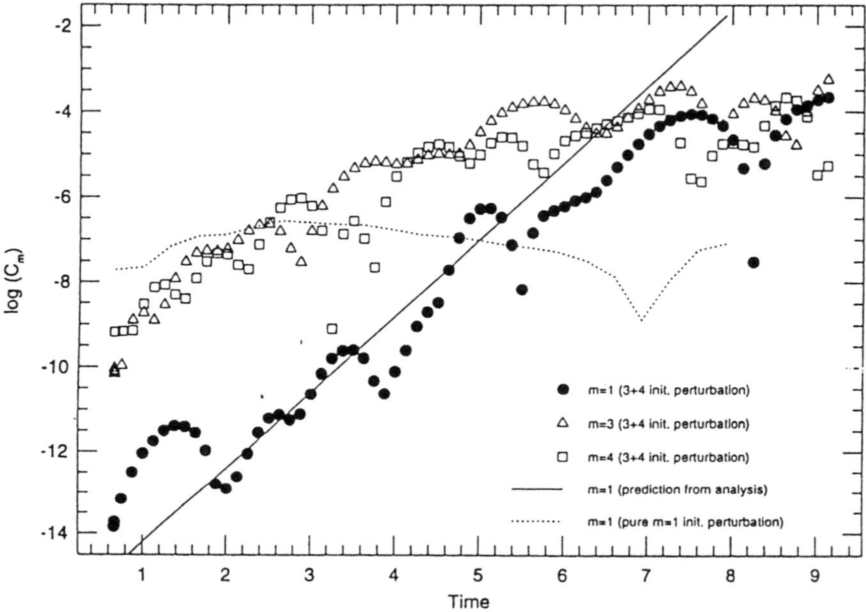


Fig. 1. Amplitude behaviour for m=1, 3, and 4 modes.

## 5. Summary

Our work has led us to the conclusion that disk evolution from an initial equilibrium configuration consists of three distinct phases. The earliest phase is the linear regime. The second regime is the three-wave era summarized here. The three-wave phase comes to an abrupt end when the amplitude of the $m$=1 mode reaches the strength of the $m = n$ or $m = n + 1$ modes. The end of the three-wave era marks the start of the epoch of mode saturation – the deep nonlinear regime. The deep non-linear regime is characterized by higher-order non-linear effects. It is also the period during which mass and angular momentum transport processes begin to significantly alter the overall properties of the disk.

It seems likely that the key to elucidating the saturation mechanism may lie in understanding how third-order effects manifest themselves in the disk.

## References

Adams F. C., Ruden S. P., Shu F. H. 1989, ApJ 347, 959
Laughlin G., Korchagin V. 1995, ApJ in press

# Equilibrium Structures of Self-Gravitating Disks

Masa-aki HASHIMOTO[1], Yoshiharu ERIGUCHI[2], Kenzo ARAI[3], and Ewald MÜLLER[4]

*1. Department of Physics, Kyushu University, Ropponmatsu, Fukuoka 810, Japan*

*2. Department of Earth Science and Astronomy, University of Tokyo, Komaba, Tokyo 153, Japan*

*3. Department of Physics, Kumamoto University, Kurokami, Kumamoto 860, Japan*

*4. Max-Planck-Institut für Astrophysik, Garching bei München, Germany*

**Abstract**

We investigate equilibrium structures of rapidly rotating toroids and disks with self-gravity using the two-dimensional code. Two rotational laws are assumed: (i) rotation having constant specific angular momentum ($j$-const law) and (ii) Keplerian rotation. Equilibrium sequences are calculated with the specified central temperature of the toroid and the mass of the central object. Configuration of hot toroids resembles to that of a thick disk. In particular, the central temperature of a toroid exceeds $10^9$ K and the luminosity exceeds the Eddington luminosity considerably. It is found that the Keplerian disks with self-gravity have thin, slim, and fat shapes which depend on the distance between the central object and the disk, and also on the mass ratio between the central object and the disk.

## 1. Structure of Self-gravitating and Rotating Hot Toroids

We have found that equilibrium sequences of self-gravitating hot toroids of the maximum temperature in the range $1.5 \times 10^7$ to $6.2 \times 10^9$ K can exist. Our computation shows that there exist equilibrium models for which the maximum temperature of the toroid can well exceed $10^9$ K if $R_{\rm in}/R_{\rm out} \sim 10^{-3}$ and if the energy generation rate depends little on the temperature, where $R_{\rm in}/R_{\rm out}$ is the ratio of distances from the inner and outer edge of the toroid

*S. Kato et al. (eds.), Physics of Accretion Disks, 269–272.*
© 1996 OPA (Overseas Publishers Association) Amsterdam B.V.

to the rotational axis. The value for $R_{\rm in}/R_{\rm out}$ is typical for thick accretion disks.

From the nucleosynthetic point of view, these temperatures imply that nucleosynthesis from hydrogen burning to silicon burning can occur, and that nuclear statistical equilibrium can be reached finally. If the toroidal star becomes very hot due to the self-gravity, nuclear burning might even lead to explosive nucleosynthesis (Hashimoto et al. 1989). Thus, in some sense, our results are complementary to those obtained by Arai & Hashimoto (1992) who were mainly concerned with nucleosynthesis triggered by viscous heating, but who have neglected the effect of self-gravity in their models.

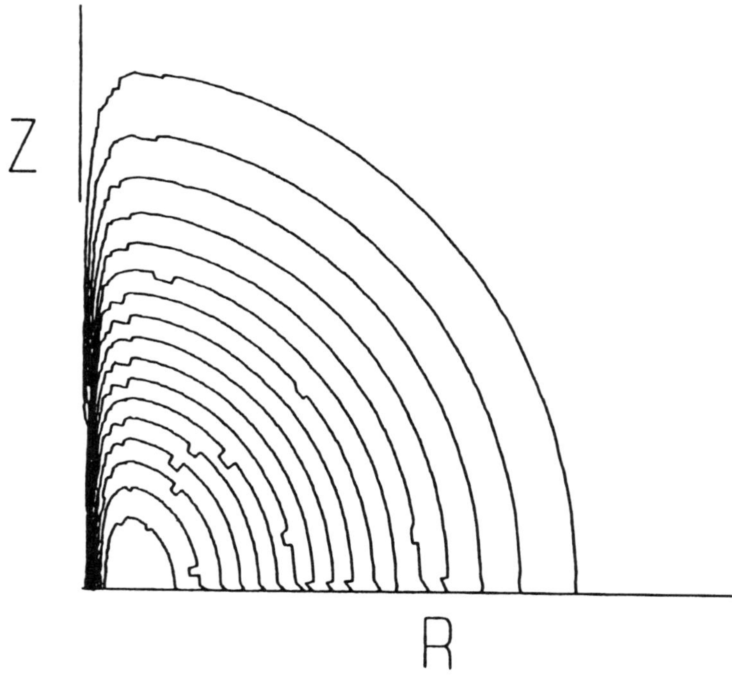


Fig. 1. Contours of constant density for the model with $j$-const law with $T_{\rm c} = 6.2 \times 10^9$ K and $R_{\rm in}/R_{\rm out} = 0.0012$. The density ratio between two adjacent contours is 1/18 of the maximum density.

Figure 1 shows that models where the temperature exceeds $10^9$ K have rather large masses, i.e., such configurations are unlikely to be formed in binary systems consisting of two stars. However, such massive accretion disks may be present around massive black holes, which are thought to exist in the center of galaxies or quasars (see, e.g., Blandford 1986). Although we have only considered *toroidal stars without a central massive object*, our conclusions concerning the existence of toroids in which nucleosynthesis is possible will not change, even if a central object is taken into account. In the latter case even higher temperatures than those in our restricted models may result, because the equilibrium state has to be maintained against both self-gravity and gravity due to the central object. Thus very high temperatures ($> 10^9$ K) can also be expected in less massive toroids compared with pure toroidal stars without central objects.

## 2. Structure of Self-Gravitating Keplerian Disks

We have computed several equilibrium sequences with the Keplerian rotation law as is shown in figure 2. Our models include self-gravity, a simplified energy generation rate, a realistic equation of state, i.e., gas pressure and radiation pressure, and a realistic description of opacity. In previous investigations, which also assumed the Keplerian rotation law, some of these ingredients have been neglected. In particular, self-gravity has not been properly included in all previous models.

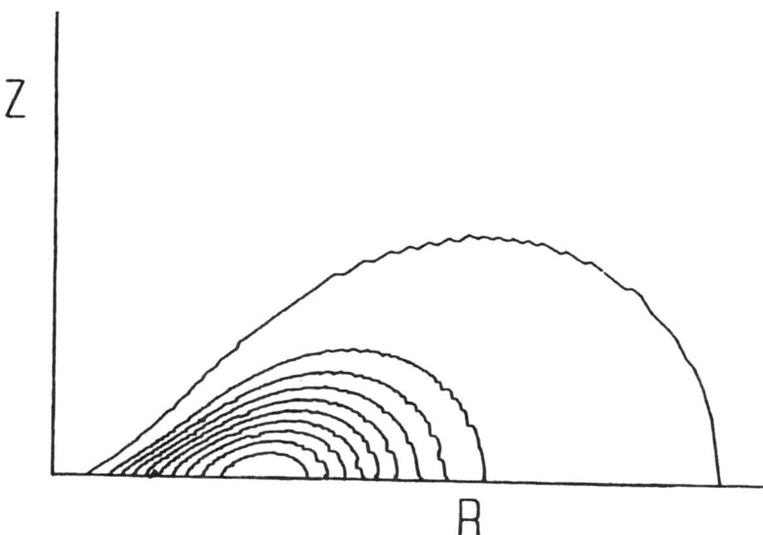


Fig. 2. Contours of constant density for configuration with $M_c = 0.6M_\odot, M_d = 0.39M_\odot, T_c = 3.0 \times 10^6$ K and $R_{in}/R_{out} = 0.05$.

The neglect of self-gravity, however, is not justified when trying to model systems with massive accretion disks consistently. Such systems have recently been suggested by observations of solar-like nebula by Kawabe et al. (1993), who reported the observation of a central star surrounded by a rather massive object, probably an accretion disk. The existence of such massive disks is further supported by hydrodynamical simulations performed by Bodenheimer et al. (1990) and Yorke et al. (1993). Our equilibrium models may help to understand these observations and simulations.

Bodenheimer et al. (1990) considered the axisymmetric collapse of a rotating $1M_\odot$ proto-stellar cloud. In their simulations they found a quasi-equilibrium structure consisting of a central part (core) of about 0.6 $M_\odot$ and an outer part (disk) of about 0.4 $M_\odot$ as seen in their figure 2. This quasi-equilibrium configuration is very similar to our model $(0.6M_\odot, 0.39M_\odot)$. The distinct accretion disk found by Bodenheimer et al. (1990) has a radial extent of about 40 AU. Our model gives $R_{\rm out} = 0.8R_\odot$, but we can scale our model by a factor $r_{\rm s} = 40\ {\rm AU}/0.8R_\odot = 1.08 \times 10^4$ to the size of their disk. On the other hand, $T_{\rm c}$ scales as $T_{\rm c}/r_{\rm s}$ and $\rho_{\rm c}$ scales as $\rho_{\rm c}/r_{\rm s}^3$ so that the central temperature and density become $T_{\rm c} \sim 300$ K and $\rho_{\rm c} \sim 3 \times 10^{-11}$ g cm$^{-3}$, respectively. For the equilibrium disk of Bodenheimer et al. (1990), the temperature ranges from 800 K to 200 K and the density ranges from $2.5 \times 10^{-9}$ to $4 \times 10^{-16}$ g cm$^{-3}$ (see their figure 2). The relative thickness of the disk in the $z$-direction also agrees rather well. We obtain a value of $z_{\rm max} \sim 0.3$, which is almost the same as that of Bodenheimer et al. (1990). Concerning the rotational curve, Bodenheimer et al. (1990) got essentially a Keplerian curve as is shown in their figure 3. Only in the outer parts, beyond 30 AU, slight deviations from the Kepler law occur, because the outer region has not yet reached an equilibrium state. Therefore, our model has essentially the same structure as that of Bodenheimer et al. (1990).

## References

Arai K., Hashimoto M., 1992, A&A 254, 191

Blandford R. D. 1986, in Quasars: IAU Symp. 119 (D. Reidel, Dordrecht) p319

Bodenheimer P., Yorke H. W., Różyczka M., Tohline J. E. 1990, ApJ 355, 651

Hashimoto M., Eriguchi K., Arai Y., Müller E. 1993, A&A 268, 131

Hashimoto M., Eriguchi Y., Müller E. 1995, A&A 297, 135

Kawabe R., Ishiguro M., Omodaka T., Kitamura Y., Miyama S. M. 1993, ApJL 404, L63.

Yorke H. W., Bodenheimer P., Laughlin G. 1993, ApJ 411, 274

# Numerical Simulations of Accretion Disks: Sources of Anomalous Viscosity

John F. HAWLEY and Steven A. BALBUS
*Department of Astronomy, University of Virginia, PO Box 3818, Charlottesville, VA 22903, USA*

## Abstract

Numerical simulations of local disk dynamics are now providing detailed information about the nature of the anomalous viscosity in accretion disks. We have found that disks are hydrodynamically stable to both linear and nonlinear amplitude perturbations. Accretion disk turbulence is produced instead by the action of a magnetohydrodynamic instability.

## 1. Disks and Anomalous Viscosity

The fundamental dynamical equation governing the behavior of hydrodynamical disks is

$$\frac{\partial \boldsymbol{v}}{\partial t} + (\boldsymbol{v} \cdot \nabla)\boldsymbol{v} = -\frac{1}{\rho}\nabla\left(P + \frac{B^2}{8\pi}\right) + \frac{\boldsymbol{B} \cdot \nabla \boldsymbol{B}}{4\pi} + \nu\nabla^2\boldsymbol{v}, \qquad (1)$$

where $\boldsymbol{v}$ is the velocity, $\rho$ the density, $P$ the pressure, $\boldsymbol{B}$ the magnetic field, and $\nu$ the physical viscosity. In the traditional approach to disk transport, magnetic effects are ignored and the tiny value of $\nu$ is bemoaned. Substantial transport is essential however, so an anomalous viscosity, huge compared with its molecular counterpart, is introduced *ad hoc*. In fact, the terms in the equations that produce the anomalous viscosity are not so mysterious: angular momentum transport is due to Reynolds and Maxwell stresses in the flow. The Reynolds stress is already present in the above equation with zero viscosity; it lies within the second term from the left. The relevant component of the Reynolds stress is $(\rho u_R u_\phi)$, where the components of $\boldsymbol{u}$ are the difference between the true velocity and Keplerian rotation. The Maxwell stress, $B_R B_\phi/4\pi$, emerges from the magnetic tension term on the right side

*S. Kato et al. (eds.), Physics of Accretion Disks, 273–278.*
**© 1996 OPA (Overseas Publishers Association) Amsterdam B.V.**

of equation (1). Thus, everything one needs to account for accretion disk evolution is, in principle, present in the basic equations of magnetohydrodynamics (MHD). Nothing further need be added.

The problem, of course, is that equation (1) is a set of three-dimensional (3D), nonlinear equation whose solutions are too difficult for the disk modeler to extract and compare with observations, at least for the present. The equations can be simplified tremendously by considering averages over vertical height and azimuth, and looking only at small deviations from a mean flow (e.g. Keplerian orbits). Then the complex nonlinearity inherent in the $(\boldsymbol{v} \cdot \nabla)\boldsymbol{v}$ and Lorentz force is reduced to an averaged net stress $w_{r\phi}$. Using dimensional and scaling arguments, Shakura & Sunyaev (1973) set this term equal to $\alpha P$, and were able to make important advances into the problem of accretion disk structure.

The $\alpha$ parameter has proven valuable as a modeling tool. What we do not have, however, is a first-principle formulation for the net anomalous stress suitable for use in $\alpha$-disk models. Instead we must infer a value of $\alpha$ from the observations. Many systems are well-modeled by $\alpha \sim 0.01$, while CVs in outburst seem to require values around 0.1. Some of the advection-dominated flows discussed in this meeting need even larger values, in excess of 1. Dense thick accretion tori, on the other hand, require a much lower value of $\alpha$.

In the absence of further information, all things may be possible. However, we are making progress in understanding the sources of the anomalous viscosity, both what seems to work, and what clearly doesn't. These results have come from the combination of analytic study and numerical simulations. In the companion paper (Balbus & Hawley in this volume) we have reviewed some of the most important analytic results. Here we will review the numerical simulations.

## 2. Hydrodynamic Disks

If the anomalous viscosity is to have its origin in hydrodynamic processes, the most obvious source is local turbulence that generates a significant positive net Reynolds stress. If hydrodynamic turbulence is to be produced and sustained, some mechanism must function to transfer energy from the mean flow into the velocity fluctuations that make up the turbulence. That is, we require an instability. As has been repeatedly noted in the disk literature, disks are locally hydrodynamical stable by the Rayleigh criterion, $dL/dR > 0$. The notion that has been with us from the inception of disk theory, is that disks might nevertheless be hydrodynamically unstable to finite-amplitude (nonlinear) perturbations. This is not without a fluid mechanical precedent: simple shear flows demonstrate just such properties.

Numerical simulations are the ideal way to investigate such a question. However, until recently such simulations have not been done. There are at least two reasons for this. First, the nonlinear instabilities in shear flows are 3D in nature, and computers capable of carrying out extended 3D simulations have only recently become generally available. The second reason appears more serious, but is, in fact, based upon a misunderstanding. This is the expectation numerical simulation codes must have too much numerical viscosity to permit the development of such instabilities.

It is well known in experimental fluid mechanics that low Reynolds number flows are often laminar, while high Reynolds number flows break down into turbulence. This has led to the common wisdom that high Reynolds number flows are always turbulent. But is such a blanket statement justified? Consider the differences between high and low Reynolds number flows. Viscosity introduces a length scale into the problem, the viscous diffusion length, and the Reynolds number is essentially the ratio of the scale of the system to this viscous length. Turbulence requires a range of wavelengths over which the fluid can self-interact without strong damping, hence the need for a "sufficiently large" Reynolds number. The Reynolds number can also directly affect dynamical instabilities in the flow. For example, if the wavelengths that would be dynamically unstable for an Euler system grow at a rate slower than the viscous diffusion time for that wavelength, the flow will be stabilized. Decreasing the viscosity will bring that instability to life. Finally, there is the possibility of a viscous instability that may arise because of the presence of a boundary layer.

In accretion disks, however, instabilities resulting from viscous boundary layers are not the issue. Any nonlinear instabilities involved with transport should result from the presence of velocity perturbations on lengthscales unaffected by viscosity. In a numerical code that solves the Euler equations, there is no explicit physical viscosity. Numerical diffusion is unavoidably present, and is significant for poorly resolved wavelengths. However, on well-resolved wavelengths the effective numerical diffusion of good numerical schemes is very small. While one cannot assign an unambiguous effective Reynolds number to such a code, experiments comparing the Euler equation PPM code to Navier-Stokes solving codes (Porter et al. 1990) suggest that Euler codes obtain more finely detailed flow features than are accessible to the highest possible Reynolds number Navier-Stokes simulations. Thus, Euler codes should be completely capable of capturing nonlinear instabilities provided only that there is a sufficient range of numerically-resolved wavelengths available.

If there are local nonlinear hydrodynamic instabilities to be found, a suitably-resolved 3D simulation should observe them. We have performed such simulations using the local shearing box approximation. In the shearing box, the local disk geometry, $(R, \phi, z)$, is reduced to a Cartesian $(x, y, z)$ frame corotating with the angular velocity $\Omega$ corresponding to the center of the box. The dynamics are those of a disk with forces from a tidal potential and a Coriolis term, although we often neglect the vertical component of gravity. The equilibrium state consists of constant density and pressure and a background angular velocity $v_y = -q\Omega x$. This represents a disk flow with $\Omega \propto R^{-q}$. For Keplerian flows, $q = 3/2$; for constant angular momentum ($L$) distributions, $q = 2$. Rayleigh unstable flows correspond to $q > 2$.

We simulated the evolution of different background angular velocity distributions, parameterized by $q$, that were perturbed at large ($> 10\%$) levels over all available wavelengths. The results are summarized in figure 1 which shows the time-evolution of the kinetic energy in the velocity perturbations. The curve labeled "Rayleigh Unstable" corresponds to a Rayleigh unstable background velocity with $q = 2.1$. Such a flow is linearly unstable and this results in an exponential growth of the velocity perturbations. In sharp contrast, the curve labeled "Keplerian" ($q = 3/2$) declines with time; the perturbations are decaying, i.e. the flow is stable.

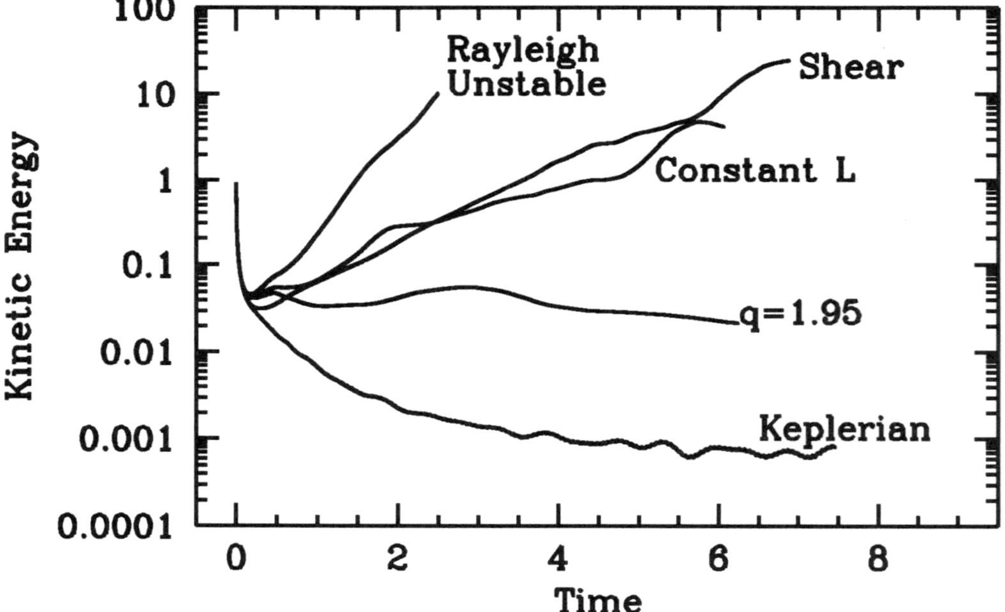


Fig. 1. Time evolution of the kinetic energy in velocity perturbations for a number of hydrodynamic simulations.

The curve labeled "Shear" is a simple shear flow without Coriolis or tidal forces. Such a flow is unstable to finite amplitude perturbations. This result is well known (e.g., Bayley et al. 1988) and the claim that disks should be nonlinearly hydrodynamically unstable is often made on the this basis. The simulation demonstrates that whatever the code's effective Reynolds number is, it is clearly large enough to find such an instability when it is present. Conversely, the complete absence of any hint of instability in the Keplerian case points to physical, not numerical, causes.

The curve labeled "Constant L" corresponds to a $q = 2$, marginally linearly stable, constant angular momentum distribution. The constant angular momentum case is particularly interesting because in such a flow the epicyclic frequency vanishes, and it is linearly stable. But an analytic study (Balbus et al. 1995) shows that a term resulting from a coupling between transport and angular momentum gradient stabilizes these disk flows by providing a sink term for the energy in the angular velocity perturbations. In the absence of this term, the flow should be no more stable than a shear layer, which also has no dynamical sink. The constant L curve shows this to be the case. This effect is further demonstrated by the curve labeled "$q = 1.95$." Although the flow has only a very small angular momentum gradient, it is completely stabilized. The velocity perturbations decline with time, albeit less rapidly than in the Keplerian case. The difference in the nonlinear stability properties of the Constant L and $q = 1.95$ runs is a clear demonstration that the stabilization is physical in origin.

To summarize, accretion disks are locally hydrodynamically stable to all local perturbations, linear and nonlinear. Hydrodynamic turbulence will not arise naturally in disks. The Coriolis force can never be ignored in disks, and it provides a strong stabilizing influence. This conclusion apparently holds even if the turbulence is driven by convection. Ryu & Goodman (1992) analyzed the linear stage of convection and found angular momentum transport to be inward. Recent simulations by Stone & Balbus (1995) show that this result holds in the nonlinear regime as well. Again, the only source for angular velocity fluctuations comes from leveling out the angular momentum gradient. In disks this means transport is inward.

Where does this leave us for hydrodynamic transport mechanisms? Global spiral waves, e.g., global shock waves (Spruit 1987), remain a viable possibility for outward angular momentum transport. But such global mechanisms will not be well-described by a local alpha prescription.

## 3. Magnetized Disks

Although disks are hydrodynamically stable, that is largely irrelevant because they are magnetohydrodynamically unstable (Balbus & Hawley 1991) when $d\Omega^2/d\ln R < 0$, a condition that always applies within accretion disks. Some of the essential aspects of the linear theory of MHD stability are presented in the companion paper (Balbus & Hawley in these proceedings). Here we will briefly mention the conclusions obtained through simulations.

There are now several years worth of numerical results concerning the local nonlinear behavior of the magnetorotational instability. The simulations range from nonstratified, differentially rotating domains using the shearing box (Hawley et al. 1995a, b), to stratified local disk cross sections (Brandenberg et al. 1995, also in these proceedings; Stone et al. 1995). While the details in these simulations vary, one aspect remains consistent: magnetized disks easily produce self-sustaining, nonisotropic turbulence that transports angular momentum outward at a rate corresponding to $\alpha \geq 0.01$.

Do the present results provide anything more to the modeler than a general justification for the $\alpha$ viscosity ansatz? Yes, but significant questions remain. The simulations show that the angular momentum transport scales with the *magnetic* pressure, i.e., total stress $w_{R\phi} \approx 0.5P_{\rm mag}$. How the magnetic pressure scales to the gas or radiation pressure $P$ and other disk parameters has yet to be determined, although stratified simulations (Brandenberg et al. 1995; Stone et al. 1995) have made a start towards addressing this issue. From fairly general considerations one expects $P_{\rm mag} < P$, and the stratified simulations support this with saturation magnetic energy amplitudes of only a few percent of $P$. Although simulations with a net vertical field can have $\alpha \sim 0.1$, it may be difficult to produce a much larger $\alpha$.

Disks must now be regarded as fundamentally MHD systems. Insofar as one may be interested in the gross spatial and evolutionary properties of disks, it is probably acceptable to ignore this fact and subsume the MHD into the general rubric of $\alpha$. This is not appropriate, however, for more specific issues of disk dynamics and questions of stability. For these we must solve the full equations, which means we will have to rely on numerical simulations.

## Acknowledgements

The authors acknowledge James Stone and Charles Gammie for their contributions to the simulations and analysis summarized here. This work is supported in part by NASA grants NAG-53058, NAGW-4431, by NSF grant AST-9423187, and a metacenter grant from the Pittsburgh Supercomputing Center and the National Center for Supercomputing Applications.

## References

Balbus S. A., Hawley J. F. 1991, ApJ 376, 214
Balbus S. A., Hawley J. F., Stone J. M. 1995, ApJ submitted
Bayley B., Orszag S., Herbert T. 1988, Ann. Rev. Fluid Mech. 20, 359
Brandenberg A., Nordlund Å, Stein R. F., Torkelsson U. 1995, ApJ 446, 741
Hawley J. F., Gammie C. F., Balbus S. A. 1995a, ApJ 440, 742
Hawley J. F., Gammie C. F., Balbus S. A. 1995b, ApJ submitted
Porter D. H. et al. 1990, Ann. NY Acad. Sci. 617, 234
Ryu D., Goodman J. 1992, ApJ 388, 438
Shakura N. I., Sunyaev R. A. 1973, A&A 24, 337
Spruit H. 1987, A&A 184, 173
Stone J. M., Balbus S. A. 1995, ApJ submitted
Stone J. M., Hawley J. F., Gammie C. F., Balbus S. A. 1995, ApJ submitted

# The Origin of Turbulent Viscosity In Accretion Disks

Steven A. BALBUS and John F. HAWLEY
*VITA, Dept. of Astronomy, Univ. of Virginia, Charlottesville, VA 22903, USA*

**Abstract**

The origin of anomalous accretion disk viscosity no longer has the air of mystery it once enjoyed. In this paper, we review the linear MHD instability which disrupts disks, and argue that outward transport must be due to its nonlinear behavior. Hydrodynamical turbulence cannot, as a matter of principle, sustain significant outward transport in Keplerian disks, even though it is quite capable of doing so in shear layers.

## 1. Introduction

Accretion disk theory has long been plagued by the absence of a compelling explanation for the greatly enhanced viscous stress tensor required to account for observed source luminosities. Indeed, in review articles on the subject, it has been a traditional ritual to cite this as the central problem, add that high Reynolds number shear is known to be sensitive to nonlinear disruption but that convection and magnetic effects may be important, and then move briskly on to $\alpha$-disk phenomenology.

From the inception of disk theory, the high Reynolds number of astrophysical plasmas was often regarded as a direct and unquestionable link to turbulence and enhanced transport. This was the approach taken in the seminal disk paper of Shakura & Sunyaev (1973), who also pointed out that a contribution to the stress tensor would arise from the passive tangling of a disk magnetic field by the turbulent flow. All of this, it was implied, could be subsumed within the $\alpha$ formulation.

The impact of Shakura & Sunyaev was immediate and profound; to this day it remains the paradigm for thin Keplerian disks. But in its details, the paper raised more questions than it answered. Are disks truly unstable? If

*S. Kato et al. (eds.), Physics of Accretion Disks, 279–284.*
© 1996 OPA (Overseas Publishers Association) Amsterdam B.V.

so, is it simply a matter of high Reynolds number in a shearing flow, or is something additional needed? Are arbitrary angular momentum distributions nonlinearly unstable? Just what can one conclude from laboratory Couette flow experiments? How is the stress of an embedded magnetic field actually transmitted to the gas? Does a field leave the disk via buoyancy before it become "dynamically important"?

For nearly two decades these were points of controversy with little consensus. Now, however, many of these questions are ripe for addressing. Issues began to clarify when Balbus & Hawley (1991) demonstrated that a simple linear MHD instability was far more powerful and general than had been previously understood. The outcome of the instability was precisely the enhanced outward transport that had been sought. In fact, it became very quickly clear that *any* differentially rotating system is sharply destabilized by a weak magnetic field. Keplerian disks are but one example of a general phenomenon.

In this review, we will try to put forth the case that the central problem really has been solved: enhanced accretion disk viscosity is almost certainly due to MHD turbulence, at least when there is any sort of magnetic couple. Hydrodynamic turbulence has difficulties of a truly fundamental nature that render it unsuitable as a source of significant stress in disks. Thus, in accretion disk physics, MHD must play a central role.

## 2. The Instability in Brief

The reason that disks are destabilized by a weak magnetic field is easily understood. Consider first a uniform medium threaded by a magnetic field with Alfvén velocity $\boldsymbol{u}_{\mathrm{A}}$. We introduce plane wave fluid displacements $\boldsymbol{\xi}_i$ of wavenumber $\boldsymbol{k}$. Incompressible displacements obey the simple wave equation

$$\frac{d^2\boldsymbol{\xi}_i}{dt^2} = -(\boldsymbol{k}\cdot\boldsymbol{u}_{\mathrm{A}})^2\boldsymbol{\xi}_i. \tag{1}$$

Next, set up the problem in the context of an accretion disk. Let $(R, \phi, z)$ be a standard cylindrical coordinate system, and let $\Omega(R)$ be the angular velocity at radius $R$. For the rotating disk system, we work in a local co-moving frame, and add a Coriolis force $-2\boldsymbol{\Omega}\times\boldsymbol{u}$ and a tidal force $-d\Omega^2/d\ln R\,\xi_R$. Separating equation (1) into its radial and azimuthal components, and assuming the WKB displacements are in the disk plane, we obtain the local equations:

$$\frac{d^2\xi_R}{dt^2} - 2\Omega\frac{d\xi_\phi}{dt} = -\left(\frac{d\Omega^2}{d\ln R} + (\boldsymbol{k}\cdot\boldsymbol{u}_{\mathrm{A}})^2\right)\xi_R, \tag{2}$$

$$\frac{d^2\xi_\phi}{dt^2} + 2\Omega\frac{d\xi_R}{dt} = -(\boldsymbol{k}\cdot\boldsymbol{u}_{\mathrm{A}})^2\xi_\phi. \tag{3}$$

Equations (2) and (3) lead to something remarkable. Although both epicycles and Alfvén waves are strictly oscillatory, their combination produces an instability. This is perhaps best appreciated by noting that the last two equations are identical to those that obtain for two orbiting point masses connected by a spring with spring constant $(\boldsymbol{k} \cdot \boldsymbol{u}_{\rm A})^2$. In this mechanical system, angular momentum is taken from the more rapidly rotating inner mass and transferred to the less rapidly rotating outer mass. The inner (outer) mass loses (gains) angular momentum, it falls (rises) to a yet lower (higher) orbit, and the separation widens. This is the heart of the instability, in either its mechanical or MHD guise.

We shall not pause here to discuss the linear dispersion relation or its technical properties; these may be found elsewhere (Balbus & Hawley 1991, 1992b; Lin & Papaloizou 1996). However, we shall write down the most general stability criteria for axisymmetric adiabatic disturbances in a differentially rotating magnetized system, as these are somewhat less well known. They were first encountered by Papaloizou & Szuszkiewicz (1992), who used a variational method for systems with vanishing angular velocity gradients along field lines, and by Balbus (1995), who used local techniques applicable to more general configurations. These criteria are based on the assumption that the field is subthermal. In a standard notation, they are:

$$-\frac{3}{5\rho}\left[\frac{\partial P}{\partial z}\frac{\partial \ln P\rho^{-5/3}}{\partial z}+\frac{\partial P}{\partial R}\frac{\partial \ln P\rho^{-5/3}}{\partial R}\right]+\frac{\partial \Omega^2}{\partial \ln R}>0, \tag{4}$$

$$-\frac{\partial P}{\partial z}\left(\frac{\partial \Omega^2}{\partial R}\frac{\partial \ln P\rho^{-5/3}}{\partial z}-\frac{\partial \Omega^2}{\partial z}\frac{\partial \ln P\rho^{-5/3}}{\partial R}\right)>0. \tag{5}$$

Equations (4) and (5) replace the classical Høiland criteria, which they closely resemble except for the substitution of angular *velocity* gradients for the angular *momentum* gradients. Indeed, because they neglect magnetic fields, the original Høiland criteria have extremely limited applicability.

For a convectively stable thin Keplerian disk, the maximum growth rate of linear instabilities corresponds to the local Oort A-value $(1/2)\ |d\Omega/d\ln R|$ (Balbus & Hawley 1992a). Instability occurs in this case when $d\Omega^2/dR < 0$, a condition routinely satisfied by astrophysical disks. For a purely toroidal field, nonaxisymmetric disturbances must be considered and these too prove unstable (Balbus & Hawley 1992b); similar growth rates are obtained (Terquem & Papaloizou 1995; Ogilvie & Pringle 1995), but at much larger poloidal wavenumbers.

## 3. Fluctuations

### 3.1 Relation to α-Formalism

Among the nonlinear outcomes of the instability are finite amplitude velocity fluctuations. The key to understanding enhanced disk transport lies with the interaction between these fluctuations and the mean flow of the disk.

Let $\boldsymbol{u} = \boldsymbol{v} - R\boldsymbol{\Omega}$, the difference between the local true and circular velocities. The radial angular momentum flux may then be written

$$R\rho(R\Omega + u_\phi)\, u_R - \frac{R}{4\pi} B_\phi\, B_R. \tag{6}$$

Denoting locally averaged quantities by angle brackets $\langle\rangle$, the net flux becomes

$$F_L = R\langle \rho R\Omega u_R\rangle + R\langle \rho u_\phi u_R - \rho u_{\mathrm{A}\,\phi} u_{\mathrm{A}\,R}\rangle, \tag{7}$$

where the $u_{\mathrm{A}\,i}$ are Alfvén velocity components. The standard α-route amounts to assuming that the density and velocity fluctuations are uncorrelated. Assuming that the net flux constant vanishes in steady state (i.e., that it vanishes at the stellar surface), and integrating over disk thickness then gives

$$\dot{M} = \frac{2\pi\Sigma}{\Omega}\, \langle u_\phi u_R - u_{\mathrm{A}\,\phi} u_{A\,R}\rangle, \tag{8}$$

where $\Sigma$ is the mass column density and $\dot{M} = -2\pi\Sigma R\langle u_R\rangle$. Thus, the accretion rate is related in a simple way to correlations in the velocity fluctuations.

The energy flux $F_E$ may be handled similarly. One finds

$$F_E = -\frac{1}{2}\Sigma R^2\Omega^2\langle u_R\rangle + \Sigma R\Omega\langle u_\phi u_R - u_{\mathrm{A}\,\phi} u_{\mathrm{A}\,R}\rangle. \tag{9}$$

Using equation (8), this simplifies to

$$F_E = -\frac{3}{2}\, \Sigma R^2\Omega^2\langle u_R\rangle \propto R^{-2}. \tag{10}$$

The divergence of $F_E$ gives the energy deposited by the energy flux per unit disk area:

$$\nabla\cdot F_E = -\frac{3}{4\pi}\dot{M}\Omega^2, \tag{11}$$

which must be radiated away at each disk surface. This is precisely the classical result of α-disk theory (Pringle 1981), recovered here with no *ad hoc* viscosity assumptions. Furthermore, by combining equations (8) and (11), we have the interesting fluctuation-dissipation relation:

$$Q = \frac{3}{4}\Sigma\Omega\langle u_\phi u_R - u_{\mathrm{A}\,\phi} u_{\mathrm{A}\,R}\rangle, \tag{12}$$

where $Q$ is $-\nabla\cdot F_E$, the disk surface luminosity. Thus, equilibrium fluctuations are related in a very simple way to the energy dissipated and radiated from the disk surface. This is something *not* obtainable via α-theory.

*3.2 Nonlinear Stability and Transport.*

Fluctuation theory can do much more: simply by writing the $u$-equations of motion in conservative form, it is possible to read off some basic properties of nonlinear stability, and to understand aspects of transport. Straightforward arguments suggest that unmagnetized Keplerian disks are stable to finite amplitude disturbances, and that convective turbulent transport flows *inward* down the angular momentum gradient, if it flows at all.

We consider, in turn, a shear layer, an unmagnetized accretion disk, and a magnetized accretion disk. The first and the last are known to be unstable, and it is highly instructive to see how their fluctuation equations differ from that of a hydrodynamical disk.

First consider Cartesian shear flow. The undisturbed flow velocity $V$ points in the $Y$-direction, and depends only upon $X$. Departures from equilibrium satisfy the exact equations:

$$\frac{1}{2}\frac{\partial}{\partial t}\left\langle \rho u_X^2 \right\rangle + \nabla \cdot \left\langle \frac{1}{2}\rho \boldsymbol{u} u_X^2 \right\rangle = -\left\langle u_X \frac{\partial P}{\partial X} \right\rangle - \text{LOSSES}, \tag{13}$$

$$\frac{1}{2}\frac{\partial}{\partial t}\left\langle \rho u_Y^2 \right\rangle + \nabla \cdot \left\langle \frac{1}{2}\rho \boldsymbol{u} u_Y^2 \right\rangle = -\frac{dV}{dX}\left\langle \rho u_X u_Y \right\rangle - \left\langle u_Y \frac{\partial P}{\partial Y} \right\rangle - \text{LOSSES}, \tag{14}$$

where LOSSES generically represent viscous dissipation. Sustaining highly dissipative turbulence requires an external source, which is represented in equation (14) by the coupling of the $XY$ component of the stress tensor ($\alpha$ by any other name) to the mean flow gradient. The dominant balance in steady state is given by pairing off this term with the losses, which leads directly to transport down the velocity gradient. Equation (13) has no direct source term, but neither does it have a transport-coupled sink. Once created by finite amplitude disturbances, turbulence is sustained.

Contrast this with a hydrodynamical disk. The local counterparts to equations (13) and (14) are

$$\frac{1}{2}\frac{\partial}{\partial t}\left\langle \rho u_R^2 \right\rangle = 2\Omega \langle \rho u_R u_\phi \rangle - \left\langle u_R \frac{\partial P}{\partial R} \right\rangle - \text{LOSSES}, \tag{15}$$

$$\frac{1}{2}\frac{\partial}{\partial t}\left\langle \rho u_\phi^2 \right\rangle = -\frac{\kappa^2}{2\Omega} \langle \rho u_R u_\phi \rangle - \left\langle \frac{u_\phi}{R} \frac{\partial P}{\partial \phi} \right\rangle - \text{LOSSES}, \tag{16}$$

where $\kappa^2$ is the standard epicyclic frequency. The only approximation used to derive these equations is that $u \ll R\Omega$. The coupling between the transport and the mean flow is now by way of the angular *momentum* gradient, and outward transport is an uphill sink! Furthermore, inward transport would produce a sink as well because of the Coriolis couple in equation (15). The result, one well-confirmed by numerical experiments (Balbus et al. 1995), is the relentless decay of any imposed finite amplitude perturbations.

What changes when a magnetic field is introduced into the disk? Everything. By combining the induction and dynamical equations, we may derive the hydromagnetic analogues of equations (15) and (16),

$$\frac{1}{2}\frac{\partial}{\partial t}\left\langle \rho u_R^2 + \rho u_{\mathrm{A}\,R}^2 \right\rangle = 2\Omega\langle \rho u_R u_\phi \rangle - \cdots, \tag{17}$$

$$\frac{1}{2}\frac{\partial}{\partial t}\left\langle \rho u_\phi^2 + \rho u_{\mathrm{A}\,\phi}^2 \right\rangle = -2\Omega\langle \rho u_R u_\phi \rangle - \frac{d\Omega}{d\ln R}\langle \rho(u_R u_\phi - u_{\mathrm{A}\,R} u_{\mathrm{A}\,\phi})\rangle - \cdots, \tag{18}$$

where $\cdots$ represent pressure gradients and dissipative losses. Here, the $R\phi$ component of the full stress tensor couples to the angular velocity, and outward transport is once more a source to sustain magnetic and velocity fluctuations. The adverse Coriolis couple in equation (18) is smaller than the angular velocity couple in both the linear and nonlinear regime (Hawley et al. 1995). But the Coriolis couple in equation (17) is a source term, and the resulting feedback causes a runaway linear instability in a magnetic disk.

Thus, we are able to understand important qualitative properties of disk turbulence by way of the transparent physics of the nonlinear fluctuation equations. A presentation of the numerical verification of these ideas may be found in the accompanying companion article (Hawley & Balbus in this volume).

## Acknowledgements

The authors would like to thank the organizers of the conference for their hospitality. This work is supported in part by NASA grants NAG-53058, NAGW-4431, and by NSF grant AST-9423187, and by a Metacenter grant from the Pittsburgh Supercomputing Center and the National Center for Supercomputing Applications.

## References

Balbus S. A. 1995, ApJ 453, 380
Balbus S. A., Hawley J. F. 1991, ApJ 376, 214
Balbus S. A., Hawley J. F. 1992a, ApJ 392, 662
Balbus S. A., Hawley J. F. 1992b, ApJ 400, 610
Balbus S. A., Hawley J. F., Stone J. M. 1995, ApJ submitted
Hawley J. F., Gammie C. F., Balbus S. A. 1995, ApJ 440, 742
Lin D. N. C., Papaloizou J. C. B. 1996, AnnRevA&A 33, 505
Ogilvie G. I., Pringle J. E. 1995, MNRAS in press
Papaloizou J. C. B., Szuszkiewicz E. 1992
Pringle J. E 1981, AnnRevA&A 19, 137
Shakura N. I., Sunyaev R. A. 1973, A&A 24, 337
Terquem C., Papaloizou J. C. B. 1995, MNRAS in press

# Dynamo-Generated Turbulence in Disks: Value and Variability of Alpha

Axel BRANDENBURG[1], Åke NORDLUND[2], Robert F. STEIN[3], and Ulf TORKELSSON[4]

*1. Nordita, Blegdamsvej 17, DK-2100 Copenhagen Ø, Denmark*
*2. Theoretical Astrophysics Center, Blegdamsvej 17, DK-2100 Copenhagen Ø, Denmark*
*3. Department of Physics and Astronomy, Michigan State University, East Lansing, MI 48824, USA*
*4. Sterrenkundig Instituut, Postbus 80000, 3508 TA Utrecht, The Netherlands*

## Abstract

Dynamo-generated turbulence seems to be a universal mechanism for angular momentum transport in accretion disks. We discuss the resulting value of the viscosity parameter alpha and emphasize that this value is in general not constant. Alpha varies with the magnetic field strength which, in turn, can vary in an approximately cyclic manner. We also show that the stress does not vary significantly with depth, even though the density drops by a factor of about 30.

## 1. Introduction

Differentially rotating disks with angular velocity decreasing outwards are susceptible to a magnetic shear (or Balbus-Hawley) instability (Velikhov 1959; Chandrasekhar 1960, 1961; Balbus & Hawley 1991). This occurs not only in accretion disks, but also in major parts of galactic disks. Balbus & Hawley (1992) suggested that the growth rate of the instability is given by the Oort $A$-value, $-\frac{1}{2}\partial\Omega/\partial\ln r \sim \Omega$. The magnetic field grows until the energy density becomes a tenth (or more) of the thermal energy density. In this way a weak magnetic field produces turbulent fluid motions (Hawley et al. 1995, Matsumoto & Tajima 1995). These motions are practically always strong

*S. Kato et al. (eds.), Physics of Accretion Disks, 285–290.*
© 1996 OPA (Overseas Publishers Association) Amsterdam B.V.

enough (i.e., the magnetic Reynolds number is large enough) that dynamo action becomes possible so that an initial seed magnetic field is amplified further (Brandenburg et al. 1995a; Hawley et al. 1996; Stone et al. 1996). It is difficult to imagine how this process can be avoided (except maybe in protoplanetary disks where there is a lack of charge carriers). In other words, non-magnetic disks are probably rare.

The standard accretion disk model is nonmagnetic, but Shakura & Sunyaev (1973) did anticipate the importance of magnetic fields for transporting angular momentum in disks. Campbell (1992) produced a thin disk model that is inherently magnetic, with field strengths comparable to the thermal equipartion value. However, the value of $\alpha$ is not constant, but depends on the magnetic field strength. Now, such information can be extracted from three-dimensional simulations of dynamo-generated turbulence.

## 2. The Model

The Balbus-Hawley instability is local and can be simulated in the framework of the shearing box approximation (Lynden-Bell & Ostriker 1967), where the radial boundary conditions are periodic in a Lagrangian system following the azimuthal shear flow. The radial dependence of gravity is expanded around $r = R$,

$$\frac{GM}{r^2} = \frac{GM}{R^2}\left[1 - 2\frac{x}{R} + 3\left(\frac{x}{R}\right)^2 - ...\right], \tag{1}$$

where $x = r - R$. Normally only the linear term is retained. In a local frame of reference rotating with angular velocity $\Omega_0 = (GM/R^3)^{1/2}$ the centrifugal force and the radial components of gravity and Coriolis force balance,

$$-\frac{GM}{R^2}\left(1 - 2\frac{x}{R}\right) + \Omega_0^2 R\left(1 + \frac{x}{R}\right) + 2u_y^{(0)}\Omega_0 = 0, \tag{2}$$

which means that $u_y^{(0)} = -\frac{3}{2}\Omega_0 x$. In practice we solve the basic hydromagnetic equations for the deviations from the linear shear flow $u_y^{(0)}$.

The shearing box approximation is now frequently being used in local disk simulations. The disadvantage is that curvature terms such as $B_\phi^2/r$ and $B_r B_\phi/r$ are ignored. These terms can be important not only for the stability of the flow (Knobloch 1992), but also for breaking the symmetry between inflow and outflow. Brandenburg et al (1996) restored such curvature terms and included also the quadratic terms in equation (1). They found that the mean accretion rate $\dot{M}$ is consistent with the standard formula (Frank et al. 1992)

$$\dot{M} = 3\pi\alpha c_s H\Sigma, \quad \text{where} \quad \Sigma = \int \rho dz. \tag{3}$$

Furthermore, it turns out that both $\dot{M}$ and $\alpha$ vary cyclically over a timescale of about 30 orbits (for further details, see Brandenburg et al. 1996). In the following we discuss the average of $\alpha$ and its variations.

## 3. The Average Value of the Viscosity Parameter $\alpha$

The horizontal components of the Reynolds and Maxwell stress, that are crucial in accretion disk theory, are parametrized in terms of a turbulent viscosity $\nu_t$,

$$\tau_{xy} \equiv \langle \rho u_x u_y - B_x B_y/\mu_0 \rangle = -\nu_t \langle \rho \rangle r \frac{\partial \Omega}{\partial r}, \tag{4}$$

where $\nu_t$ is expressed in the natural units for disks; the sound speed, $c_s$, and the vertical disk scale height, so $\nu_t = \alpha c_s H$. Since $c_s = \Omega H/\sqrt{2}$ we have $\nu_t = \alpha \Omega H^2/\sqrt{2}$.

There are certain properties of local simulations that may affect the value of $\alpha$: stratification, field strength, and distance from the central object. First, in the absence of stratification, the vertical scale height is formally infinite and $\alpha = \sqrt{2}\nu_t/(\Omega H^2)$ goes to zero, even though the stress remains finite. The reason is that the Balbus-Hawley instability operates on scales smaller than $\lambda_{\rm BH} = v_{\rm A}/\Omega$, where $v_{\rm A}$ is the Alfvén speed. In real disks, $\lambda_{\rm BH} < H$, but in a simulation with no (or very weak) stratification $\lambda_{\rm BH}$ is limited by the size of the box. In that case, a better parametrization of the stress would be to normalize by the height of the box $L_z$, i.e., $\alpha_L = \sqrt{2}\nu_t/(\Omega L_z^2)$. In figure 1 we compare two runs of Torkelsson et al. (1996) with stratification ($L_z/H \approx 2$) and with almost no stratification ($L_z/H = 0.2$). The average viscosity parameters are $\alpha \approx 0.005$ and $\alpha_L \approx 0.017$, so $\alpha_L \approx 3.2\alpha$. In other words, given the "conversion factor" $\alpha_L/\alpha$, one could (in principle) estimate $\alpha$ even if there is no stratification in the simulation.

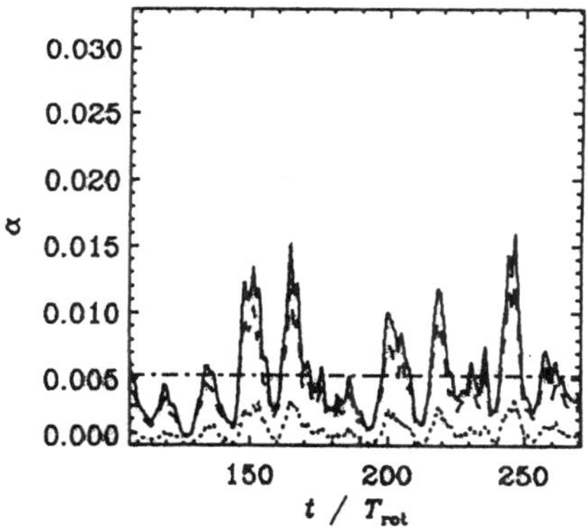


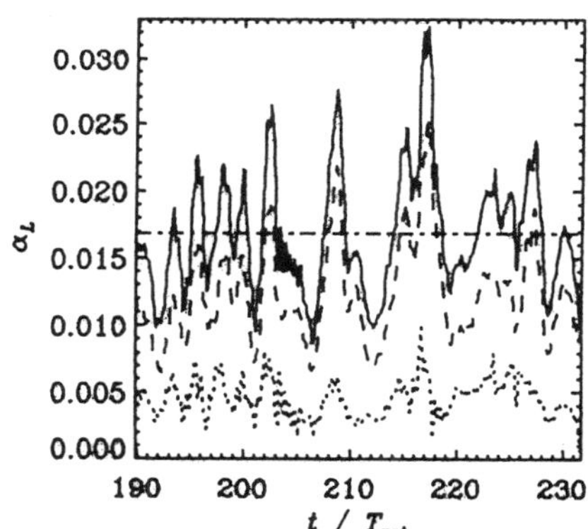


Fig. 1. Evolution of $\alpha$ and $\alpha_L$. The dotted (dashed) line refers to the contribution of the Reynolds (Maxwell) stress.

Obviously, the value of $\alpha$ depends on the definition of $H$. For isothermal stratification, the density in the equilibrium state goes like $\rho = \rho_0 \exp(-z^2/H^2)$, but it is not uncommon to define a scale height by writing $\rho = \rho_0 \exp(-z^2/2\tilde{H}^2)$, where $\tilde{H} = H/\sqrt{2}$ (e.g., Frank et al. 1992). Note that now $c_s = \Omega\tilde{H}$ and therefore $\tilde{\alpha} = \nu_t/(\Omega\tilde{H}^2)$. With this definition, the two values $\tilde{\alpha}_L$ and $\tilde{\alpha}$ are closer together ($\tilde{\alpha} = 0.008$ and $\tilde{\alpha} = 0.012$), suggesting perhaps some advantage in using the second definition, because then $\tilde{H}$ is closer to the "effective" disk height than $H$. However, in order to be consistent with previous work, we continue to use the old definition.

Note that with the new definition $\tilde{\alpha}$ would be larger than $\alpha$ by a factor of $\sqrt{2}$. However, there is yet another definition of alpha: the ratio of stress to pressure, which yields a value that is again 3/2 times larger, so altogether 2.1 times larger than our definition. Thus, it is important to keep this in mind when comparing work of different authors. We also note that the values obtained by Hawley et al. (1996) and Stone et al. (1996) agree with those of Brandenburg et al. (1995a, 1996) within a factor of two or less.

## 4. Variability of $\alpha$

It is important to realize that $\alpha$ is not constant. Variability of $\alpha$ is caused not only by turbulent motions which are inherently time-dependent, but especially by the slow changes of the large-scale magnetic field. In our model a large scale magnetic field is generated by some kind of (large scale) dynamo process. In figure 2 we compare contours of the total stress and the azimuthal field $\langle B_y \rangle$ in a $z$-$t$ diagram.

The long-term evolution of the magnetic field on a time scale of 30 orbits is perhaps somewhat reminiscent of the solar cycle. In the case of an accretion disk around a compact object of $1M_\odot$ and at a distance of $10^6$ km (typical of dwarf novae systems) 30 orbits would correspond to about 20 min. However, the cyclic variations are the result of a local simulation. Applied at different radii, we would have interference of different frequencies from different radii. In the absence of simulations in global geometry we cannot say what kind of variability (if any) can be expected in reality. What is important, however, is that $\alpha$ and $\langle \boldsymbol{B} \rangle$ are closely connected. Torkelsson et al. (1996) and Brandenburg et al. (1996) proposed the following fit formula

$$\alpha(\langle \boldsymbol{B} \rangle) = \alpha^{(0)} + \alpha^{(B)} \langle \boldsymbol{B} \rangle^2 / B_0^2, \tag{5}$$

where $B_0 = \langle \mu_0 \rho c_s^2 \rangle^{1/2}$ is the equipartition field strength based on the thermal energy, and $\alpha^{(B)}$ is around 0.5, which is much larger than the quiescent value $\alpha^{(0)}$, which is around 0.001.

It is conceivable that an improved $\alpha$-disk model would yield a self-consistent description of the disk structure and the large scale magnetic field $\langle \boldsymbol{B} \rangle$. Equation (5) could be an important ingredient of such a model.

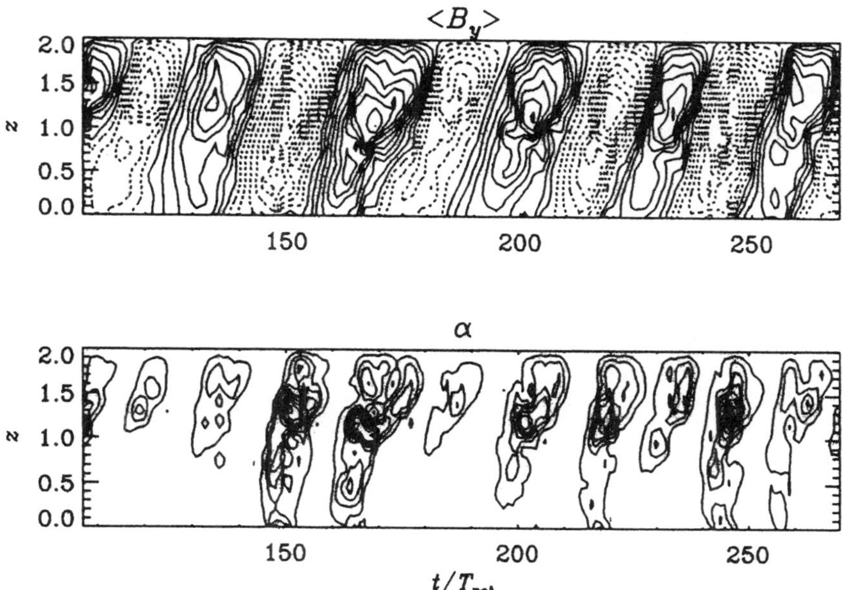


Fig. 2. Comparison of the azimuthal field $\langle B_y \rangle$ and the horizontal components of the total magnetic stress in a $z$-$t$ diagram. Dotted contours refer to negative values. Note that maxima of $\alpha$ coincide with those of $\langle B_y \rangle$.

## 5. The Vertical Dependence of $\alpha$

Originally the concept of $\alpha$-viscosity was employed within the framework of vertically integrated models. However, $\alpha$ has also been used to model the "subgrid-scale" viscosity of models that resolve the vertical dependence explicitly (e.g., Kley et al. 1993). If one assumes a constant $\alpha$, and if $c_s$ is approximately constant (isothermal disk), then equation (4) might suggest that the stress $\tau_{xy}$ is proportional to the density and would fall off towards the surface. This is however not confirmed by the simulations; see figure 3. We find that $\tau_{xy}$ is *not* proportional to $\rho(z)$ or $c_s(z)$. Indeed, a better approximation might be to use the vertically averaged stress:

$$\tau_{xy} \sim \alpha c_s \Sigma. \tag{6}$$

This new description is likely to affect the vertical disk structure of $\alpha$-disks and in that way some conclusions regarding instabilities associated with the ionization state of the disk, like dwarf nova instabilities.

The models presented here lack radiation transport and are therefore nearly isothermal. Preliminary steps toward models with surface cooling have been presented by Brandenburg et al. (1995b). Their models show narrow convection zones right beneath the cooled surface.

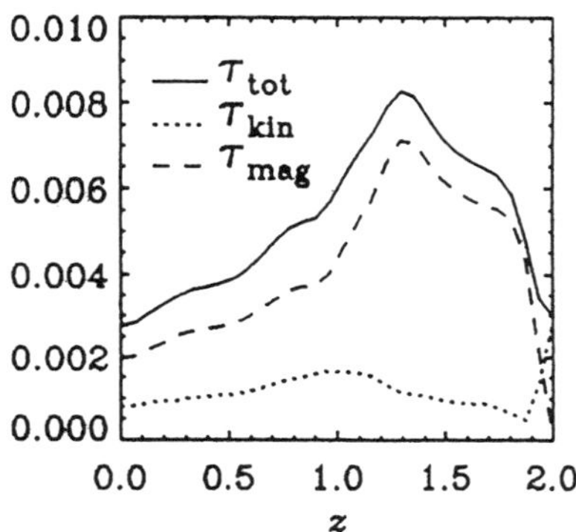


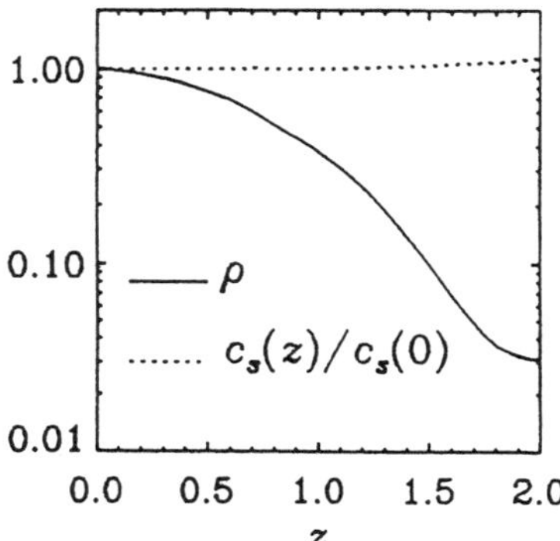


Fig. 3. Dependence of $\tau_{xy}(z)$, normalized by $\langle\rho\rangle c_s H$ (left). Note that $\tau_{xy}$ is not proportional to $\rho(z)$ or $c_s(z)$ (right).

## References


Balbus S. A., Hawley J. F. 1991, ApJ 376, 214

Balbus S. A., Hawley J. F. 1992, ApJ 400, 610

Brandenburg A., Nordlund Å., Stein R. F., Torkelsson, U. 1995a, ApJ 446, 741

Brandenburg A., Nordlund Å., Stein R. F., Torkelsson U. 1995b, Lecture Notes in Physics 462, 385

Brandenburg A., Nordlund Å., Stein R. F., Torkelsson U. 1996, APJL in press

Campbell C. G. 1992, Geophys. Astrophys. Fluid Dyn. 63, 179

Chandrasekhar S. 1960, ProcNatAcadSci 46, 253

Chandrasekhar S. 1961, Hydrodynamic and Hydromagnetic Stability (Dover Publications Inc., New York)

Frank J., King A. R., Raine D. J. 1992, Accretion power in astrophysics (Cambridge Unversity Press, Cambridge)

Hawley J. F., Gammie C. F., Balbus, S. A. 1995, ApJ 440, 742

Hawley J. F., Gammie C. F., Balbus, S. A. 1996, APJ in press

Kley W., Papaloizou J. C. B., Lin D. N. C. 1993, ApJ 416, 679

Knobloch E. 1992, MNRAS 255, 25p

Lynden-Bell D., Ostriker J. P. 1967, MNRAS 136, 293

Matsumoto R., Tajima T. 1995, ApJ 445, 767

Shakura N. I., Sunyaev R. A. 1973, A&A 24, 337

Stone J. M., Hawley J. F., Gammie C. F., Balbus, S. A. 1996, ApJ in press

Torkelsson U., Brandenburg A., Nordlund Å., Stein R. F. 1996, AstrophysLetter&Comm in press

Velikhov E. P. 1959, Sov. Phys. JETP 36, 1398

# SPH Simulations of Accretion Disks and Narrow Rings

Sarah MADDISON[1], James MURRAY[2], and Joe MONAGHAN[1]
*1. Mathematics Department, Monash University, Clayton, 3168, Australia*
*2. CITA, University of Toronto, Ontario, M5S 1A7, Canada*

## Abstract

We modelled a two dimensional viscous SPH (Smoothed Particle Hydrodynamics) disk and compared our results with that of the Lynden-Bell & Pringle (1974) theory. After several rotations a non-axisymmetric instability developed at the inner disk edge. To investigate further, we reduced the model to a single ring of SPH particles, which is actually inconsistent with SPH as it assumes a smoothed initial state. The stability analysis confirms what the simulations show.

## 1. Introduction

As a test case for a two dimensional SPH code developed to study the viscous evolution of accretion disks we modelled the spreading of a narrow viscously shearing ring of matter. We compared these simulations with semi-analytical results based upon Lynden-Bell & Pringle's (1974) analysis of a nonself-gravitating geometrically-thin accretion disk. They consider a massless 2D viscous disk in Keplerian orbit around a central mass.

To obtain an expression for the surface density evolution, $\Sigma(R,t)$, of such a disk, one can combine the equations of mass and angular momentum conservation with a prescription for the viscosity (we follow the work of Pongracic 1988). Using a prescribed surface density $\Sigma$, one then obtains a second order ordinary differential equation, which can be solved using Bessel functions and a Greens solution. For an initially Gaussian surface density the following is obtained:

$$\Sigma(R,t) = \frac{1}{R^{3/4}\sqrt{12\pi\nu t}} \int_o^\infty R'^{3/4} \exp\left\{ -\left(\frac{R_c - R'}{l}\right)^2 - \frac{(R-R')^2}{12\nu t}\right\} dR' , \tag{1}$$

*S. Kato et al. (eds.), Physics of Accretion Disks, 291–294.*
© 1996 OPA (Overseas Publishers Association) Amsterdam B.V.

where $R, R_c$, and $l$ are the radial position, radius of maximum density, and the Gaussian width, $t$ being time and $\nu$ the viscosity.

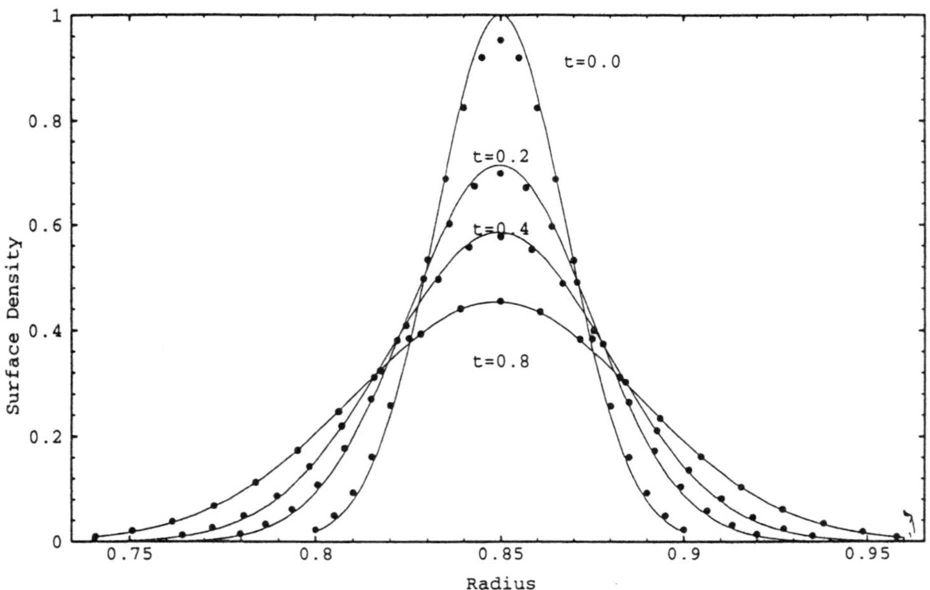


Fig. 1. Time evolution of the surface density of a Gaussian disk. The solid line denotes the theoretical solution at the times shown. The points show the corresponding SPH results.

See figure 1 for a comparison of the theoretical and simulated spread of an annulus with a Gaussian density profile. The best results were obtained when the particles were set up in concentric rings so as to give a uniform number density throughout the annulus. After several rotation periods the disk started to break up. (For full details, see Maddison et al. 1995, hereafter refered to as MMM.) The viscosity spreads the rings to the point where the innermost ring became somewhat separated from the remaining rings. This ring then started to lose shape and break up. This effect marched its way through the remaining disk. One would like to know if this is a genuine instability or an artifact of the numerical method. In the following we give a partial answer by analysing the stability of a single ring.

## 2. Stability Analysis

The radial and azimuthal equations of motion for a particle in Keplerian orbit around a central mass are given by

$$\ddot{r} - r\dot{\phi}^2 = -\frac{\mathcal{G}M}{r^2} + F_r, \tag{2}$$

$$r\ddot{\phi} + 2\dot{r}\dot{\phi} = F_\phi, \tag{3}$$

where $r, \phi, F_r$, and $F_\phi$ are the radial and azimuthal positions, radial and azimuthal force per unit masses, respectively, $\mathcal{G}$ is the gravitational constant, and $M$ is the mass of the central object.

The scaled, linearized perturbed equations of motion become

$$\ddot{q} - 3q - 2\dot{\sigma} = F_r = -\frac{g}{2}\sum_j (\dot{\sigma}_j - \dot{\sigma}_i)\sin\zeta W_{ij}, \tag{4}$$

$$\ddot{\sigma} + 2\dot{q} = F_\phi = \frac{g}{2}\sum_j (\dot{\sigma}_j - \dot{\sigma}_i)(1 + \cos\zeta)W_{ij}, \tag{5}$$

where $g = \alpha hc/\epsilon$, $\zeta = \Theta_j - \Theta_i$, and $W$ is the interpolation kernel. Applying normal mode analysis, we substitute

$$q = A\exp\{i(k\Theta + \omega\tau)\}, \tag{6}$$

$$\sigma = B\exp\{i(k\Theta + \omega\tau)\}, \tag{7}$$

where $k$ and $\omega$ are the wavenumber and frequency, into equations (4) and (5) and find

$$(-\omega^2 - 3)A - 2i\omega B = gB\omega\Psi, \tag{8}$$

$$-\omega^2 B + 2i\omega A = -gBi\omega\Phi, \tag{9}$$

where $\Phi = \Phi(\zeta, k, W)$, $\Psi = \Psi(\zeta, k, W)$, and $\zeta = \Theta_j - \Theta_i$. See MMM for a full explanation of the terms used and equation derivations. Equations (8) and (9) lead to a quartic in $\omega$, the frequency, satisfied by $\omega = 0$, leaving

$$\omega^3 - ig\Phi\omega^2 + (2ig\Psi - 1)\omega - 3ig\Phi = 0. \tag{10}$$

For the inviscid case, $\Psi = \Phi = 0$, $\omega = 0, \pm 1$. We therefore look for viscous solutions of the form:

$$\omega = \begin{cases} \gamma_1 \\ \gamma_2 + 1 \quad \text{where } |\gamma_i| \ll 1\,. \\ \gamma_3 - 1 \end{cases} \tag{11}$$

We find that $\gamma \pm 1$ is always stable, and that $\gamma$ is always unstable (again see MMM for full details). Since we are assuming $\gamma \ll 1$, we require $\alpha c \ll 1$. The short wavelength perturbations grow the most rapidly.

For the results of the single ring SPH simulation, see figure 2.

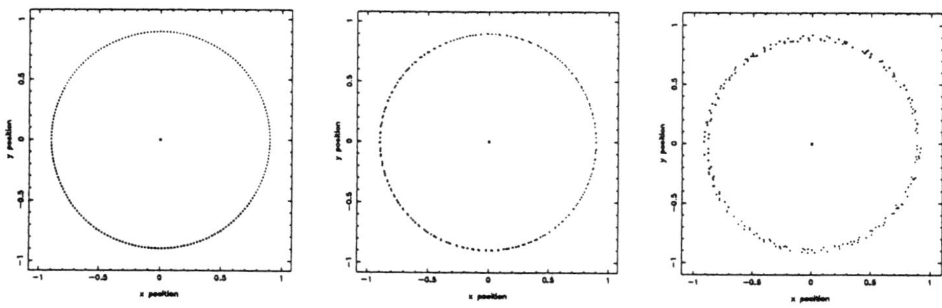

Fig. 2. The onset of viscous instability in a single ring system. The three figures are at 0.2, 4.1, and 5.8 rotational periods.

## 3. Discussion

Deviations from the standard theory of thin disks occur at the inner edge as particles move onto elliptic orbits. Lyubarskij et al. (1994) show that viscous disks can become unstable if the viscous drag on fluids in elliptical orbits is a maximum at apastron. This is the case for a standard $\alpha$ disk and this is also the case for particles in our simulations.

A single ring is the equivalent to a delta function as an initial condition, which SPH will smooth. This instability shows the robustness of SPH. To avoid this viscous instability one should increase the number of rings so that the separation remains less than $2h$. This of course increases the computational expense. A spatially varying $h$ may help overcome this instability, however viscosity is proportional to $h$, so varying $h$ introduces a varying viscosity. Tests have so far shown that constant $h$ also gives better results for the surface density evolution.

## References

Lynden-Bell D., Pringle J. E. 1974, MNRAS 168, 603
Lyubarskij Y. E., Postnov K. A., Prokhorov M. E. 1994, MNRAS 266, 583
Maddison S. T., Murray J. R., Monaghan J. J. 1995, PASA in press
Pongracic H., 1989, PhD Thesis, Monash University, Australia

# A Model of Magnetic Turbulence in Accretion Disks

Shoji KATO
*Department of Astronomy, Faculty of Science, Kyoto University, Sakyo-ku, Kyoto 606-01, Japan*

## Abstract

A model of hydromagnetic turbulence in accretion disks is presented using a second-order closure modeling. We are interested only in quantities averaged in the vertical direction. The escape of magnetic fields from the disk is taken into account phenomenologically. The transport equations of both the Reynolds stress tensor and the Maxwell tensor are closed and solved numerically. The results are shown briefly.

## 1. Introduction and Hydrodynamic Turbulence

Evaluation of turbulent viscosity in accretion disks is one of important problems remained in accretion-disk theories. Here we present our attempts to derive models of turbulence in disks using a closure modeling of turbulence. The purpose is not only to understand parameter dependences of turbulent stress tensor, but also to derive expressions for turbulent stress tensor which can be applied to time-dependent or strongly inhomogeneous disks.

A primitive way to evaluate turbulent stress tensor is to resort to a mixing length theory. The $r\varphi$-component of stress tensor, $t_{r\varphi}$, in rapidly rotating disks with angular velocity $\Omega$ is then modeled as $t_{r\varphi} = -\eta r d\Omega/dr$, where the turbulent viscosity $\eta$ is given by $\eta = \alpha \rho c_s H$, $\rho$ being the density, $c_s$ the speed of sound, $H$ the half-thickness of disks, and $\alpha$ a dimensionless constant smaller than unity. If the Keplerian disk is considered, we have $t_{r\varphi} = (3/2)\alpha p$, where $p$ is the pressure. This is the standard model of viscosity in accretion disks (Shakura & Sunyaev 1973).

The primary deficiencies of the above model based on the mixing length theory are (i) the need to produce an ad hoc specification of turbulent length and (ii) the neglect of nonlocal and history effects on the stress tensor.

*S. Kato et al. (eds.), Physics of Accretion Disks, 295–300.*
© 1996 OPA (Overseas Publishers Association) Amsterdam B.V.

To improve the deficiencies we must look to more detailed tubulent theories. There are three major methods of modeling of turbulence: (1) Large-eddy simulations, (2) one-point closure modeling and (3) two-point closure modeling (or spectral models). Among them the one-point closure modeling is proper to applying to accretion disks, since the modeling is rather simple and can be applied to systems with complicated structures. It still keep some important properties of turbulence. For example, the deficiency (ii) of diffusion-type stress tensor can be easily resolved if the one-point closure modeling is proceeded to the stage of a second-order closure (e.g., Kato 1994).

Let us now consider an incompressible turbulence. From the fluctuation component of the Navier-Stokes equation we obtain a transport equation of $\langle u_i u_j \rangle$ in the form

$$\frac{d}{dt}\langle u_i u_j \rangle = \Big( -\langle u_i u_k \rangle \frac{\partial U_j}{\partial x_k} - \langle u_j u_k \rangle \frac{\partial U_i}{\partial x_k} \Big) + \Pi_{ij} - \frac{\partial C_{ijk}}{\partial x_k} - \epsilon_{ij}, \tag{1}$$

where $u_i$ is the fluctuating component of velocity over the mean velocity $U_i$ and $\langle \ \rangle$ is the ensemble average. The terms in the parentheses on the right-hand side represent the production of stress tensor by interaction between shear and stress tensor itself, which is denoted hereafter by $C^{\mathrm{P}}_{ij}$. The term $\Pi_{ij}(=\langle p(\partial u_i/\partial x_j + \partial u_j/\partial u_j)\rangle)$ represents redistribution of $\langle u_i u_j \rangle$ by pressure, $-\partial C_{ijk}/\partial x_k$ is the diffusion of turbulence due to inhomegeneity of turbulence, and finally $\epsilon_{ij}$ is the dissipation rate of $\langle u_i u_j \rangle$ into thermal energy.

The term $\Pi_{ij}$ is a third-order quantity. Theories of turbulent closure modeling tell us that it is reasonable to approximate it as (e.g., Kato 1994)

$$\Pi_{ij} = -C_1 \kappa \Big( \langle u_i u_j \rangle - \frac{1}{3}\langle u^2 \rangle \delta_{ij} \Big) + C_2 \frac{\langle u^2 \rangle}{4} \Big( \frac{\partial U_i}{\partial x_j} + \frac{\partial U_j}{\partial x_i} \Big), \tag{2}$$

where $\kappa$ is the epicyclic frequency. The constants $C_1$ and $C_2$ are parameters specifying the turbulence. Since we concern only with the vertically-integrated or averaged quantities, the diffusion term may be omitted. Energy dissipation of turbulent energy occurs mainly in short wavelength region, where the turbulence will tend to be isotropic. Hence, $\epsilon_{ij}$ will be approximated as $\epsilon_{ij} = (1/3)\epsilon\delta_{ij}$, where $\epsilon$ is the dissipation rate of turbulence.

The time evolution of $\langle u_i u_j \rangle$ is then expressed in terms of first-order quantities and $\epsilon$. By eliminating $\epsilon$ from these approximate equations of time evolution of $\langle u_i u_j \rangle$, we obtain $\langle u_i u_j \rangle / \langle u^2 \rangle$ as functions of the mean flow. If the cylindrical coordinates are adopted, the $r\varphi$-component of $\langle u_i u_j \rangle$ in a steady state becomes

$$\frac{\langle u_r u_\varphi \rangle}{\langle u^2 \rangle} = -\frac{\kappa}{3\Omega} \frac{C_1(1-3C_2/4)}{(C_1^2+4)\kappa^2/\Omega^2 + (2/3)(d\ln\Omega/d\ln r)^2} \frac{d\ln\Omega}{d\ln r}. \tag{3}$$

This expression is obtained by Narayan et al. (1994) and Kato & Inagaki (1994) by describing turbulence in terms of a Boltzmann equation and by

Kato (1994) by a second-order closure modeling (which belongs to the one-point closure modeling) mentioned above. An important result shown by equation (3) is the existence of an upper limit of $\langle u_r u_\varphi \rangle / \langle u^2 \rangle$ for changes of $C_1$ and $C_2$. In the case of the Keplerian disk with $\gamma = 5/3$, we have $\alpha \equiv \rho \langle u_r u_\varphi \rangle / p \leq 0.0745 \langle u^2 \rangle / c_s^2$, where $c_s$ is the speed of sound. That is, in the subsonic turbulence the upper limit of $\alpha$ is 0.0745, which seems to be smaller than a typical $\alpha$-value required to dwarf novae accretion disks from observations. The presence of the upper limit of $\alpha$ in a relatively small value forces us to consider hydromagnetic turbulence.

## 2. Hydromagnetic Turbulence and Numerical Results

The second-order closure modeling discussed in the previous section is now extended to hydromagnetic turbulence. The gas is assumed to be incompressible as before. In addition, no global magnetic field is assumed, i.e., $\boldsymbol{B}_0 = 0$. The hydromagnetic turbulence in disks will generate a global magnetic field by dynamo processes, since we can expect $\langle \boldsymbol{u} \times \boldsymbol{b} \rangle \neq 0$ in disks due to the presence of helical motions. A global magnetic field generated by dynamo processes affects the structure of turbulence. In addition, an external global field may be carried by accretion motions. A consideration of the effects of the global magnetic field on the turbulence, however, is very complicated. Hence, a first step to clarify the turbulent structure in disks is to study in detail, as a limiting case, the interaction between the gas turbulence and the magnetic turbulence while neglecting the effects of any global magnetic field. The magnetic field divided by $(4\pi\rho)^{1/2}$ is denoted by $\boldsymbol{b}$ hereafter.

The basic equations are the Navier-Stokes equation with the Lorentz force and the induction equation, with subsidiary relations div $\boldsymbol{u}$ = div $\boldsymbol{b} = 0$. By taking the ensemble average of these equations after multiplying by the fluctuating components, we can derive transport equations for correlation functions, such as $\langle u_i u_j \rangle$, $\langle b_i b_j \rangle$, and $\langle u_i b_j \rangle$. The transport equation of the Reynolds stress tensor is (e.g., Kato & Yoshizawa 1993)

$$\left(\frac{\partial}{\partial t} + U_k \frac{\partial}{\partial x_k}\right) \langle u_i u_j \rangle = C_{ij}^{\mathrm{P}} + \Pi_{ij} - \frac{\partial}{\partial x_k} C_{ijk}^{\mathrm{DF}} - \epsilon_{ij}^{\mathrm{G}} - S_{ij}. \tag{4}$$

The first four terms on the right-hand side represent, in turn, those of *Production*, *Redistribution*, *Diffusion*, and *Dissipation*. The final term $S_{ij}$ represents *Transport* of $\langle u_i u_j \rangle$ to $\langle b_i b_j \rangle$. The expression for $C_{ij}^{\mathrm{P}}$ is the same as that already given in equation (1). See Kato & Yoshizawa (1993) for detailed expressions for other terms. The transport equation of the Maxwell stress tensor is obtained similarly as

$$\left(\frac{\partial}{\partial t} + U_k \frac{\partial}{\partial x_k}\right) \langle b_i b_j \rangle = \left(\langle b_j b_k \rangle \frac{\partial}{\partial x_k} U_i + \langle b_i b_k \rangle \frac{\partial}{\partial x_k} U_j\right) - \frac{\partial}{\partial x_k} F_{ijk}^{\mathrm{DF}} - \epsilon_{ij}^{\mathrm{M}} + S_{ij}. \tag{5}$$

The terms in the parentheses on the right-hand side represent *Production* of the Maxwell stress tensor by interaction with the shear. The subsequent terms represent *Diffusion*, *Dissipation*, and *Transport*. The $S_{ij}$ term appears in both equations (4) and (5), but having opposite signs, representing the most important interaction between $\langle u_i u_j \rangle$ and $\langle b_i b_j \rangle$.

To close equations (4) and (5) by second-order quantities, the third-order terms on the right-hand sides of these equations are approximated by lower-order quantities. This is made by resorting to the *Two-Scale Direct Interaction Approximation* (TSDIA) [see Yoshizawa (1985) for details]. After this modeling, we write down equations (4) and (5) explicitely in terms of the cylindrical coordinates ($r$, $\varphi$, $z$). The disk is assumed to have no motion except for rotation, i.e., $\boldsymbol{U} = (0,\ \Omega(r),\ 0)$. Then, from the components of equations (4) and (5) we have, as main equations,

$$\dot{\langle u_r^2 \rangle} = 4\Omega\langle u_r u_\varphi \rangle + \Pi_{rr} - S_{rr} - \epsilon_{\rm G}, \quad \dot{\langle b_r^2 \rangle} = S_{rr} - 2\beta\langle b_r^2 \rangle - \epsilon_{\rm M},$$

$$\dot{\langle u_r u_\varphi \rangle} = 2\Omega\langle u_\varphi^2 \rangle - \frac{\kappa^2}{2\Omega}\langle u_r^2 \rangle + \Pi_{r\varphi} - S_{r\varphi}, \quad \dot{\langle b_r b_\varphi \rangle} = r\frac{d\Omega}{dr}\langle b_r^2 \rangle + S_{r\varphi} - 2\beta\langle b_r b_\varphi \rangle,$$

$$\dot{\langle u_\varphi^2 \rangle} = -\frac{\kappa^2}{\Omega}\langle u_r u_\varphi \rangle + \Pi_{\varphi\varphi} - S_{\varphi\varphi} - \epsilon_{\rm G}, \quad \dot{\langle b_\varphi^2 \rangle} = 2r\frac{d\Omega}{dr}\langle b_r b_\varphi \rangle + S_{\varphi\varphi} - 2\beta\langle b_\varphi^2 \rangle - \epsilon_{\rm M}. \tag{6}$$

The left-side equations are for hydrodynamic turbulence, while the right-side ones are for the magnetic turbulence. The dots attached to terms on the left-hand side of each equation represent the time derivative. In the above equations, $\Pi_{ij}$, $S_{ij}$, $\epsilon_{\rm G}$, and $\epsilon_{\rm M}$ are (Kato & Yoshizawa 1995)

$$\Pi_{ij} = -C_1^\Pi\Omega\Big(\langle u_i u_j \rangle - \frac{1}{3}\langle u^2 \rangle\delta_{ij}\Big) - C_2^\Pi\Omega\Big(\langle b_i b_j \rangle - \frac{1}{3}\langle b^2 \rangle\delta_{ij}\Big) + C_0^\Pi r\frac{d\Omega}{dr}\langle u^2 \rangle,$$

$$S_{ij} = C_1^S\Omega\langle u_i u_j \rangle - C_2^S\Omega\langle b_i b_j \rangle, \quad \epsilon_{\rm G} = \frac{1}{3}\epsilon_{ii}^{\rm G}, \quad \epsilon_{\rm M} = \frac{1}{3}\epsilon_{ii}^{\rm M}. \tag{7}$$

The last term of $\Pi_{ij}$ exists only for $\Pi_{r\varphi}$ and $\Pi_{\varphi r}$. In deriving the above equations we have assumed that the magnetic field escapes from disks by magnetic buoyancy and the Parker instability. The rate of escape of $\langle b_i b_j \rangle$ is written phenomenologically as $\beta\langle b_i b_j \rangle$, where $\beta$ is the escape rate. The speed by which the magnetic field escapes will be on the order of the Alfvén speed in gas-dominated disks. Hence, we write $\beta$ as $\beta \equiv v_{\rm esc}/H = X(\langle b^2 \rangle^{1/2}/c_{\rm s})\Omega$. Hereafter the dimensionless parameter $X$ is used instead of $\beta$. The dimensionless parameters involved in the above modeling are eight; $C_1^S$, $C_2^S$, $C_0^\Pi$, $C_1^\Pi$, $C_2^\Pi$, $\nu_{\rm G}$, $\epsilon_{\rm M}$, and $X$.

The meanings of equations (6) are briefly discussed here. Let us first consider how the gas turbulence is maintained. A summation of equations of hydrodynamic turbulence gives

$$\frac{d}{dt}\langle u^2 \rangle = \frac{d}{dt}\langle u_{ii}^2 \rangle = -2r\frac{d\Omega}{dr}\langle u_r u_\varphi \rangle - S_{ii} - 2\epsilon_{\rm G}. \tag{7}$$

This shows that turbulent kinetic energy is generated i) by a combined action of the shear and the Reynolds stress ($\langle u_r u_\varphi \rangle > 0$) and ii) by the energy input from magnetic turbulence ($-S_{ii} > 0$). The positive value of $-S_{ii}$ is expected, since $S_{ii} = \Omega(C_1^S \langle u^2 \rangle - C_2^S \langle b^2 \rangle)$, $C_1^S \sim C_2^S$, and $\langle u^2 \rangle < \langle b^2 \rangle$ (see the final results). The origin of the $r\varphi$-component of the Reynolds stress is related to the shear of the differential rotation. Even when the turbulence is isotropic, $\langle u_r^2 \rangle = \langle u_\varphi^2 \rangle = (1/3)\langle u^2 \rangle$, we have $2\Omega \langle u_\varphi^2 \rangle - (\kappa^2/2\Omega)\langle u_r^2 \rangle = -(rd\Omega/3dr)\langle u^2 \rangle > 0$. Hence, equation for time evolution of $\langle u_r u_\varphi \rangle$ shows that the differential rotation acts so as to generate $\langle u_r u_\varphi \rangle$. Finally, the pressure-strain tensor $\Pi_{ij}$ has the effect of making the generated turbulence isotropic.

We next consider the magnetic turbulence. Equations of time evolutions of $\langle b_r b_\varphi \rangle$ and $\langle b_\varphi^2 \rangle$ show that they are amplified by $b_r$ being stretched by differential rotation. (Note that $\langle b_r b_\varphi \rangle$ is generally a negative quantity.) The components $\langle b_r^2 \rangle$ and $\langle b_z^2 \rangle$ are, however, not generated by differetial rotation. They can be maintained against dissipation and escape only by energy input from the gas turbulence ($S_{rr} > 0$ and $S_{zz} > 0$ are necessary). This is possible since, unlike magnetic turbulence, in gaseous turbulence there is the pressure-strain tensor, which acts so as to make the gaseous turbulence isotropic. Hence, we can expect the situation where $\langle u_r^2 \rangle$ and $\langle u_z^2 \rangle$ are larger than $\langle b_r^2 \rangle$ and $\langle b_z^2 \rangle$, respectively. This leads to $S_{rr} > 0$ and $S_{zz} > 0$, inspite of $\langle S_{ii} \rangle < 0$.

As a typical example of steady turbulence obtained by solving the set of equations (6) we show in figure 1 the results in the case of $C_1^{\mathrm{S}} = C_2^{\mathrm{S}} = C_0^{\Pi} = C_1^{\Pi} = C_2^{\Pi} = 0.1$ and $\nu_{\mathrm{G}} = \nu_{\mathrm{M}} = 0.03$. The hydrodynamic quantities, $\langle u_r^2 \rangle / c_{\mathrm{s}}^2$, $\langle u_\varphi^2 \rangle / c_{\mathrm{s}}^2$, $\langle u_z^2 \rangle / c_{\mathrm{s}}^2$, and $\langle u_r u_\varphi \rangle / c_{\mathrm{s}}^2$, are shown in the left panel as functions of $X$. The right panel shows the hydromagnetic quantities, $\langle b_r^2 \rangle / c_{\mathrm{s}}^2$, $\langle b_\varphi^2 \rangle / c_{\mathrm{s}}^2$, $\langle b_z^2 \rangle / c_{\mathrm{s}}^2$, and $\langle b_r b_\varphi \rangle / c_{\mathrm{s}}^2$. The magnitude of $-\langle b_r b_\varphi \rangle$ is larger than $\langle u_r u_\varphi \rangle$, implying that angular momentum is transported mainly by the turbulent magnetic field as is expected.

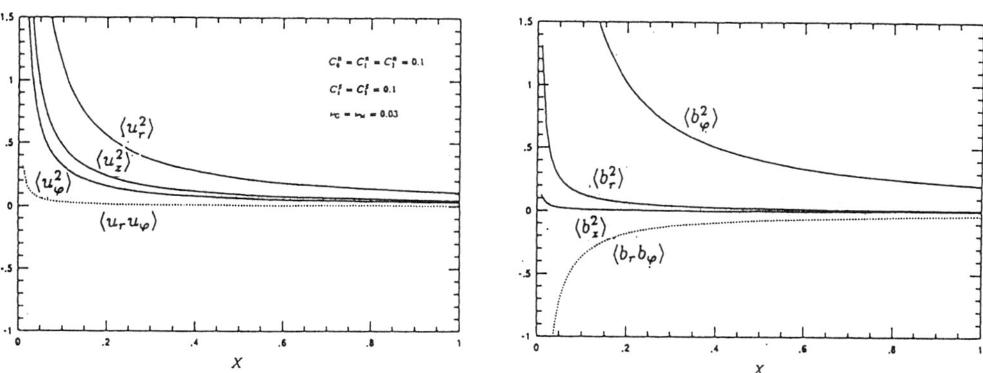


Fig.1. The left panel shows components of the Reynolds stress tensor, and the right one components of the Maxwell stress tensor. They are the values normalized by $c_{\mathrm{s}}^2$.

## 3. Discussion

We have presented, using second-order closure modeling, a model of hydromagnetic turbulence in accretion disks with no mean magnetic field ($\boldsymbol{B}_0 = 0$). The results show that steady turbulent states exist when the constants characterizing turbulence are taken reasonablly. The results are also reasonable and qualitatively in agreement with those obtained by simulations (Hawley et al. 1995; Matsumoto & Tajima 1995; Brandenburg et al. 1995).

A direct check of validity of our closure modeling can be made numerically. From exact expressions for $\Pi_{ij}$ and $S_{ij}$, we calculate their numerical values at each time in simulations. On the other hand, these values are compared with those obtained from their approximate expressions. These comparisons determine the value of the $C$'s at each time. Our modeling is relevant if these values are really universal.

A generalization of this modeling to the case of $\boldsymbol{B}_0 \neq 0$ is of importance. To do so it seems to be necessary to increase the number of the basic set of equations, which makes calculations complicated. Considerations of effects of density stratification and compressibility of the gas on structure of the turbulence are also important (Nakao & Kato 1994, 1995; Nakao 1996).

Finally we comment on *hydrodynamic* turbulence discussed in section 1. The pressure-strain tensor acts so as to make the turbulence isotropic. Hence, $\langle u_\varphi^2 \rangle$ can be maintained against the dissipation and the sink due to the Keplerian shear by energy input from $\langle u_r^2 \rangle$ through $\Pi_{\varphi\varphi}$. This is different from the arguments by Balbus & Hawley (1996, in this volume). The coefficient $C_1$ [see equation (2)] is a measure of efficiency of isotropization. Even when the coefficient is small, our model has a steady hydrodynamic turbulence, because in such case the input from $\langle u_r^2 \rangle$ to $\langle u_\varphi^2 \rangle$ decreases but both the dissipation and the sink also decrease. Non-linear effects of anisotropy of turbulence on the pressure-strain tensor, however, have been neglected in our modeling.

## References

Brandenburg A, Nordlund A., Stein R. F. 1995, ApJ 446, 741
Hawley J. F., Gammie C. F., Balbus S. A. 1995, ApJ 440, 743
Kato S. 1994, PASJ 46, 589
Kato S., Yoshizawa A. 1993, PASJ 45, 103
Kato S., Inagaki S. 1994, PASJ 46, 289
Kato S., Yoshizawa A. 1995, PASJ 47, 629
Matsumoto R., Tajima T. 1995, ApJ 445, 767
Nakao Y., Kato S. 1994, PASJ 46, 273
Nakao Y., Kato S. 1995, PASJ 47, 451
Nakao Y. 1996, PASJ submitted (see also this volume)
Narayan R., Loeb A., Kumar P. 1994, ApJ 431, 359
Shakura N. I., Sunyaev R. A. 1973, A&A 24, 337
Yoshizawa A. 1985, Phys. Fluids 28, 3313

# Vertical Structure of MHD Turbulence in Accretion Disks

Yasushi NAKAO
*Department of Astronomy, Faculty of Science, Kyoto University, Sakyo-ku, Kyoto 606-01, Japan*

## Abstract

Vertical structure of magnetohydrodynamic turbulence in accretion disks with density stratification is considered by applying one-point closure modeling of turbulence. Transport equations of turbulent kinetic energy ($K$), turbulent magnetic energy ($K_{\mathrm{W}}$) and density variance ($K_{\mathrm{D}}$) are solved simultaneously under the assumptions of isothermal disks and no-global magnetic field. Moreover, the turbulence in the disk is assumed to be homogeneous in the horizontal-plane. Imposed boundary conditions are (i) both the turbulent energy $K$ and the density variance $K_{\mathrm{D}}$ do not escape from the disk through the surface; (ii) on the other hand, escape of the turbulent magnetic energy $K_{\mathrm{W}}$ is considered phenomenologically. The results show that both $K$ and $K_{\mathrm{W}}/\rho_{\mathrm{M}}$ (magnetic energy per unit mass) increase with height by turbulent diffusion. It follows that the so-called $\alpha$-value generally increases from the equator toward the surface, though it decreases, due to the escape of magnetic fields, in a region rather close to the surface. The results further show that the magnetic energy is larger than the turbulent kinetic energy in a wide range of the disk-height, implying that the main contributor to angular momentum transport and viscous heating in disks is the magnetic fields.

## 1. Introduction

Two important problems in accretion disk theory have been remained unanswered. The one is the mechanism that makes the disk accreting, the other is the source that radiates hard X-ray photons. As for the former, many people consider that the turbulence, probably MHD turbulence, controls the

*S. Kato et al. (eds.), Physics of Accretion Disks, 301–304.*
© 1996 OPA (Overseas Publishers Association) Amsterdam B.V.

angular momentum transport. Therefore, it is quite of importance to clarify the structure of MHD turbulence in disks, especially in the disks with density stratification. Furthermore, such studies will give us some information about "hot corona models," which have been proposed to explore the source of hard X-ray radiation.

## 2. Model of Turbulence

Let us divide a physical quantity $f$ into the mean component $\langle f\rangle$ and the fluctuating one $f'$. For stationary turbulence in disks, the transport equations of $K \equiv \langle u'^2\rangle/2$ ($K$-equation) and $K_{\rm W} \equiv \langle b'^2\rangle/2$ ($K_{\rm W}$-equation) are

$$\begin{aligned}\frac{\partial K}{\partial t} &= \nu_{\rm T}\left(r\frac{d\Omega}{dr}\right)^2 - \varepsilon_{\rm K} + \frac{\langle p'\nabla\cdot\boldsymbol{u}'\rangle}{\rho_{\rm M}} + \frac{c_{\rm s}^2}{\rho_{\rm M}}\frac{d}{dz}\left(\nu_{\rm T}\frac{d\rho_{\rm M}}{dz}\right) + \frac{d}{dz}\left(\nu_{\rm T}\frac{dK}{dz}\right)\\ &+\frac{1}{\rho_{\rm M}}\frac{d}{dz}\left(\eta_{\rm T}\frac{dK}{dz}\right) - \mathcal{E} = 0, \end{aligned} \tag{1}$$

$$\frac{\partial K_{\rm W}}{\partial t} = \eta_{\rm T}\left(r\frac{d\Omega}{dr}\right)^2 - \varepsilon_{\rm W} + \frac{d}{dz}\left[\left(\nu_{\rm T} + \frac{\eta_{\rm T}}{\rho_{\rm M}}\right)\frac{dK_{\rm W}}{dz}\right] + \mathcal{E} = 0, \tag{2}$$

where $\nu_{\rm T}$ (the turbulent viscosity) and $\eta_{\rm T}$ (the turbulent magnetic diffusivity) are modeled as

$$\nu_{\rm T} = C_{\rm K}\frac{K}{\Omega} \qquad \text{and} \qquad \eta_{\rm T} = C_{\rm K}\frac{K_{\rm W}}{\Omega}, \tag{3}$$

respectively ($C_{\rm K}$ is a dimensionless constant whose value is $\sim 0.1$). Other symbols have their usual meanings: $c_{\rm s}$ is the sonic speed, $\Omega$ the angular velocity of the rotation (we assume $\Omega \propto r^{-3/2}$), $\rho_{\rm M}(z)$ mean-density stratification in the vertical ($z$) direction ($\propto \exp[-(z/H)^2/2]$), and $H$ the scale-height.

The first term on the right-hand side of equation (1) represents *the production* of $K$, the second term, $\varepsilon_{\rm K}$, *the dissipation*, the third term means an additional dissipation by compressibility effects and succeeding three terms are *the turbulent diffusions* due to the density stratification. Similarly, the terms on the right-hand side of equation (2) denote *the production* of $K_{\rm W}$, *the dissipation* and *the diffusion* in order of their appearances, respectively. The term $\mathcal{E}$ which appears in both $K$- and $K_{\rm W}$-equations with the opposite signs denotes *the energy exchange* between $K$ and $K_{\rm W}$. Energy flows from $K$ into $K_{\rm W}$ when this term is positive, and vice versa.

After constructing the transport equation of $K_{\rm D} \equiv \langle \rho'^2\rangle/2$, we must model the correlations involved in $K$-, $K_{\rm W}$-, and $K_{\rm D}$-equations to close the set of equations (see Nakao 1996 for the detail).

## 3. Results

We employ $z_s = 2.0H$ as a standard surface height. Obtained results are follows: (i) $K$ generally increases toward the disk surface; (ii) $K_W/\rho_M$ shows a gradual increase until $z$ reaches $\sim 75\%$ of the surface, but rapidly decreases toward the surface beyond that height; (iii) as $z$ increases, the conventional $\alpha$-value increases in accordance with the increases in $K$ and $K_W/\rho_M$. However, $\alpha$ can decrease near the surface, corresponding to the decrease in $K_W$ due to the escape of the energy; (iv) the energy exchange between $K$ and $K_W$ is typically in the direction from $K_W$ to $K$, especially in the region near the equator. That is, the magnetic-energy density is much larger than the energy density of gas turbulence, implying that the main contributor to both viscous heating and angler momentum transfer is the magnetic field. These are shown in figure 1.

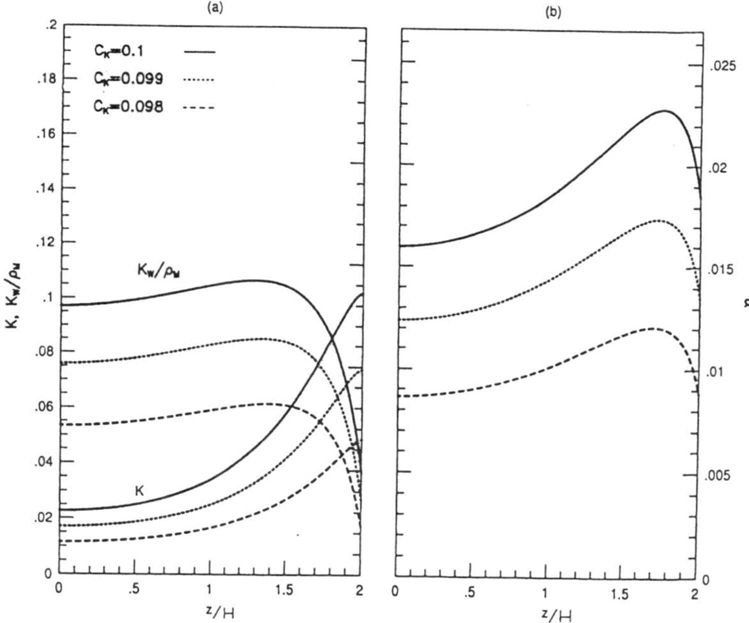


Fig. 1. (a) $z$-dependences of $K$ and $K_W/\rho_M$ (both are divided by $c_s^2$; i.e., dimensionless variables) for some parameter values of $C_K$. The same types of lines are used for $K$ and $K_W/\rho_M$ when they are a set of solutions for the same parameter. The right vertical axis is a scale for $\alpha_K$ and $\alpha_W$, which are proportional to $K$ and $K_W/\rho_M$, respectively. (b) $z$-dependences of $\alpha$ (the sum of $\alpha_K$ and $\alpha_W$). Types of lines have the same meaning as in panel (a).

The result (i) is similar to our previous works (Nakao & Kato 1994, 1995). It shows that the energy transport of $K$ (from the equator toward the surface) by turbulent diffusion occurs not only in hydrodynamic turbulence but also in MHD turbulence. The result (ii) is also connected with the turbulent transport of $K_W/\rho_M$. Escape of magnetic energy, however, brings about rapid decrease of $K_W/\rho_M$ in the surface region. The result (iii), which is a direct consequence of results (i) and (ii), gives a favorable situation for the formation of disk coronae. The result (iv) shows a good agreement with numerical simulations (e.g., Hawley et al. 1995; Matsumoto & Tajima 1995; Brandenburg et al. 1995), for the results of these simulations tell us that magnetic energy is much greater than gas turbulent energy.

A schematic diagram of energy flow is shown in figure 2.

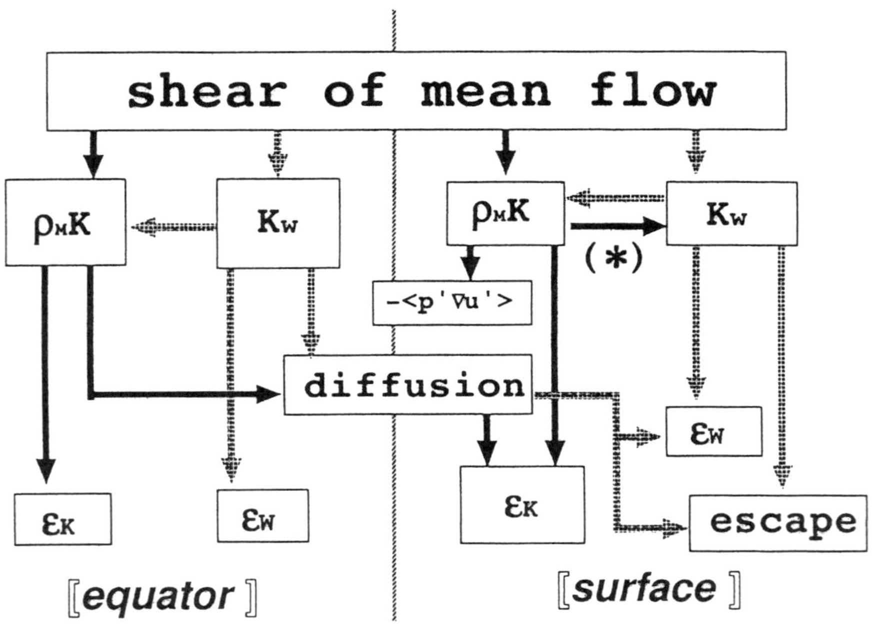


Fig. 2. A schematic diagram of the energy flow. The energy-flow of $K$ and that of $K_W$ are written by the different types of arrows. The left half represents a low $z$ region (*equator*) and the right does a high $z$ region (*surface*). The asterisk ($*$) denotes that the energy-flow is not always in such a direction in our calculations.

## Acknowledgments

The author thanks Professor S. Kato for invaluable discussions and continual encouragement during this study.

## References

Brandenburg A., Nordlund Å., Stein R., Torkelsson U. 1995, ApJ, 446, 741
Hawley J. F., Gammie C. F., Balbus S. A. 1995, ApJ 440, 742
Matsumoto R., Tajima T. 1995, ApJ 445, 767
Nakao Y., Kato S. 1994, PASJ 46, 273
Nakao Y., Kato S. 1995, PASJ 47, 451
Nakao Y. 1996, PASJ submitted

# Three Dimensional Structure of Magnetized Disks

Ryoji MATSUMOTO[1,2], Takaaki MATSUZAKI[1], Mitsuru HAYASHI[3], Toshiki TAJIMA[4], Shin MINESHIGE[5], and Kazunari SHIBATA[6]

*1. Department of Physics, Faculty of Science, Chiba University, 1-33 Yayoi-Cho, Inage-Ku, Chiba 263, Japan*

*2. Advanced Science Research Center, JAERI, Naka, Japan*

*3. Department of Physics, Graduate School of Science and Technology, Chiba University, Inage-ku, Chiba 263, Japan*

*4. Institute for Fusion Studies, University of Texas at Austin, Austin, TX 78712, USA*

*5. Department of Astronomy, Faculty of Science, Kyoto University, Sakyo-ku, Kyoto 606-01, Japan*

*6. National Astronomical Observatory, Mitaka, Tokyo 181, Japan*

## Abstract

We present the results of two-dimensional and three-dimensional simulations of magnetohydrodynamic processes in accretion disks such as (1) the escape of magnetic flux by the Parker instability, (2) generation of fluctuating magnetic fields due to the Balbus-Hawley instability, (3) jet formation in accretion disks threaded by large-scale magnetic fields, and (4) plasma heating and plasmoid ejection by magnetic reconnection. Large amplitude $1/f$ fluctuations in X-rays from black hole candidates may be explained by sporadic magnetic reconnections in advection-dominated, low-$\beta$ disks in which the magnetic pressure exceeds the gas pressure.

## 1. Introduction

Magnetic fields play important roles in accretion disks. They provide a mechanism of angular momentum transport which enables the mass accretion and gravitational energy release. In the standard theory of accretion disks (Shakura & Sunyaev 1973), the phenomenological $\alpha$ parameter was introduced in order to incorporate the angular momentum transport through the Reynolds stress $\langle \rho v_r v_\varphi \rangle$ and Maxwell stress $\langle B_r B_\varphi/(4\pi) \rangle$ ,where the bracket

*S. Kato et al. (eds.), Physics of Accretion Disks, 305–310.*
© 1996 OPA (Overseas Publishers Association) Amsterdam B.V.

denotes the spatial average. Balbus & Hawley (1991) pointed out the importance of the magnetic shearing instability (Velikhov 1959; Chandrasekhar 1961) on the generation of magnetic fluctuations in accretion disks. Subsequently, several groups have carried out local three-dimensional MHD simulations of the Balbus & Hawley instability (Hawley et al. 1995; Matsumoto & Tajima 1995; Brandenburg et al. 1995). They showed that the value of $\alpha_B = -\langle B_r B_\varphi/(4\pi P)\rangle$ in the saturation stage of the instability is between 0.001 and 1 depending on the initial magnetic field configuration.

It has been thought that the Parker instability sets a lower limit for the plasma $\beta(=P_{\rm gas}/P_{\rm mag})$ in accretion disks because magnetic flux escapes from the disk by buoyancy. In differentially rotating disks, however, the rotation and velocity shear have stabilizing effects on the Parker instability (e.g., Foglizzo & Tagger 1994). Furthermore, the growth rate of the Parker instability decreases in low-$\beta$ ($\beta < 1$) disks because pressure scale height increases. Based on the results of 2.5D MHD simulations of the Parker instability with velocity shear, Shibata et al. (1990) pointed out that once low-$\beta$ disk is formed, it can stay in low-$\beta$ state for much longer time than the dynamical time scale. If the magnetic energy in low-$\beta$ disk is suddenly released by some instabilities, the disk may show large amplitude flare-like activities.

In the following, after presenting the results of our MHD simulations of the Parker instability, jet formation, and magnetic reconnection, we discuss the possibility that advection-dominated disks are low-$\beta$ disks.

## 2. Parker Instability in Accretion Disks

Matsumoto et al. (1988) performed two-dimensional local MHD simulations of the Parker instability in nonuniform gravitational field $g(z) = -GMz/(r_0^2+z^2)^{3/2}$ which mimics the vertical gravity in accretion disks. The growth rate of the Parker instability is $0.2-0.5v_A/H$ where $H$ is the pressure scale height and $v_A$ is the Alfvén speed (e.g., Matsumoto et al. 1988). Since the growth rate of the Balbus & Hawley instability is $0.2-0.7c_s/H$ in Keplerian disks where $c_s$ is the sound speed (e.g., Matsumoto & Tajima 1995), the Parker instability can become important when $\beta \sim 1$.

We carried out three-dimensional local MHD simulations of the Parker instability in shear rotating plasma. In a frame rotating with the disk with angular speed $\Omega$, we take local Cartesian coordinates with $x$ in the radial direction, $y$ in the azimuthal direction, and $z$ in the vertical direction. The initial state consists of the isothermal gas layer in magnetohydrostatic equilibrium with horizontal magnetic field. The plasma $\beta$ at $t=0$, $\beta_0$, is assumed to be uniform. The velocity shear is given by $v_y = -2Ax$. We use the normalization $\rho(z=0) = c_s = \Omega_{\rm k} = 1$, where $\Omega_{\rm k}$ is the Keplerian angular speed. A characteristic scale height is defined as $H = c_s/\Omega_{\rm k} = 1$. We computed only the upper half plane and imposed symmetric boundary conditions at the midplane ($z = 0$). The azimuthal boundaries are periodic. The radial boundaries are treated by using sliding periodic condition (Hawley et al. 1995). The upper boundary is a free boundary where waves can transmit.

Figure 1 shows the results of three-dimensional MHD simulations of the Parker instability in horizontal magnetic fields with $\beta_0 = 1$ and $H/r_0 = 1/4$. Sinusoidal velocity perturbations with wavelength $\lambda_y = 20H$ are imposed at $t = 0$. Figure 1a shows the non-rotating case ($\Omega = A = 0$). Magnetic field lines are overlayed on density isosurfaces. As the instability grows, dense regions are created in the pocket of undulating magnetic field lines. Figure 1b shows the case when the uniform rotation ($\Omega = \Omega_k$) is included. Although the growth rate of the instability is reduced owing to the rotation, the Parker instability still takes place. Figure 1c shows the case when both Kepler rotation ($\Omega = \Omega_k$) and Keplerian velocity shear ($A = 0.75\Omega_k$) exist. The region around the midplane subject to the Balbus & Hawley instability. Near the surface, long wavelength magnetic loops are created by the Parker instability. The magnetic fluctuation $\delta B^2$ grows almost exponentially and saturates when $\delta B^2/(8\pi) \sim P$. The magnetic viscosity in the saturation stage is $\alpha_B \sim 0.05$ (figure 1d).

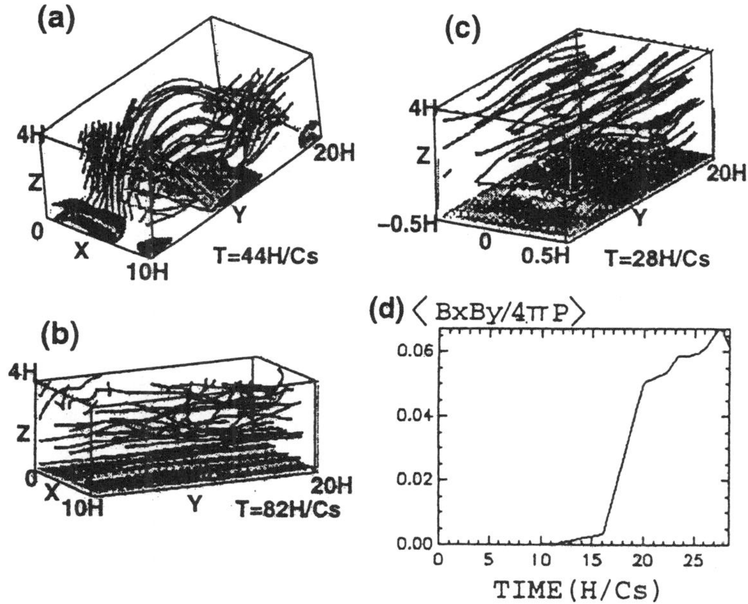


Fig. 1. Results of three-dimensional local MHD simulations of the Parker instability. Solid curves show magnetic field lines. The model parameters are $\beta_0 = 1$, $H/r_0 = 1/4$ and (a) $\Omega = A = 0$, (b) $\Omega = \Omega_k$, $A = 0$, (c) $\Omega = \Omega_k$, $A = 0.75\Omega_k$, (d) Time variation of $\alpha_B = -\langle B_x B_y/(4\pi P)\rangle$.

## 3. Global MHD Simulations of Jet Formation

When the disk is threaded by large-scale magnetic field, the nonlinear torsional Alfvén waves generated by the disk rotation entrain the matter in the low density part of the disk and create spinning bipolar jets (Uchida & Shibata 1985; Shibata & Uchida 1986).

Figure 2 shows the results of global 3D MHD simulation of the jet formation from a rotating torus threaded by uniform large-scale magnetic field. This simulation is an extension of our previous 2.5D axisymmetric simulation of jet formation (Matsumoto et al. 1996). The equilibrium torus is $n = 3$ polytrope with constant specific angular momentum $L_0$. The model parameters are $A_1 = c_{s0}^2/(\gamma v_{\varphi 0}^2) = 0.05$ and $A_2 = v_{A0}^2/v_{\varphi 0}^2 = 10^{-3}$ where the subscript 0 denotes quantities at the pressure maximum of the torus ($r_0 = L_0^2/GM$), and $v_{\varphi 0}$ is the rotation speed at $r = r_0$, respectively. We assume isothermal spherical static halo with $\alpha = (GM/RT_{\rm halo})/r_0 = 1$. The initial plasma $\beta$ at $(r, z) = (0, r_0)$ is $\beta_h = 2.0$.

Small amplitude $m = 2$ non-axisymmetric perturbations are imposed for $v_\varphi$ at $t = 0$. Numerical results show that bipolar jets with maximum speed $v_{\rm jet} \sim 2.0 v_{\varphi 0}$ are created. The surface layer of the disk accretes faster than the equatorial part because magnetic braking most effectively affects that layer. Accretion proceeds along two spiral channels which correspond to the surface avalanche flow appeared in 2.5D simulations (Matsumoto et al. 1996). In figure 2b, we can see non-axisymmetric loop-like structure which indicates the growth of the Parker instability. Figure 2c shows that low-$\beta$ regions are created in the innermost part of accretion disks where toroidal magnetic fields are accumulated.

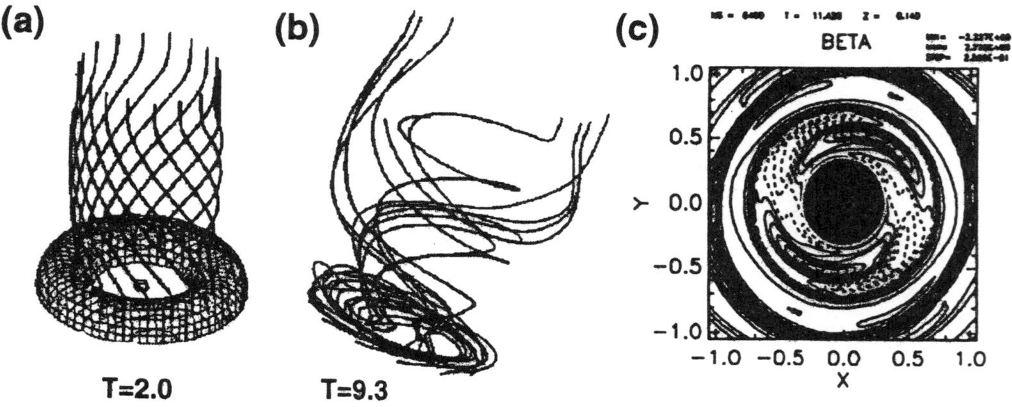


Fig. 2. Results of three-dimensional global MHD simulations of jet formation. (a) density isosurface and magnetic field lines at $t = 2.0 r_0/v_{\varphi 0}$, (b) magnetic field lines at $t = 9.3 r_0/v_{\varphi 0}$, (c) isocontours of plasma $\beta$ in the disk at $z = 0.14 r_0$. The dashed curves show the region where $\beta < 1$.

## 4. Magnetic Reconnection in Accretion Disks

Figure 3 shows the magnetic field lines in the local resistive MHD simulation of Keplerian disks with $\beta_0 = 100$ (Matsumoto & Tajima 1995). We adopted anomalous resistivity in which the resistivity has non-zero value only in the region where the current density exceeds a threshold $J_c$. Typical magnetic Reynolds number in the diffusion region is $R_m \sim 100$. As the Balbus & Hawley instability grows, magnetic field lines are tangled. The resulting magnetic reconnection helps saturating the growth of magnetic fluctuations.

Magnetic reconnection can be the origin of flares in the Sun, stars, and in accretion disks. Zylstra (1988) and Ghosh & Lamb (1991) showed that when the magnetic field of the central star is twisted by the rotation of the disk, equilibrium solution disappears when the critical twist is exceeded. By 2.5D axisymmetric MHD simulation, we showed that magnetic reconnection takes place and ejects plasmoids (figure 4). Hayashi et al. (1996) applied this mechanism to hard X-ray flares observed in protostars (Koyama et al. 1994).

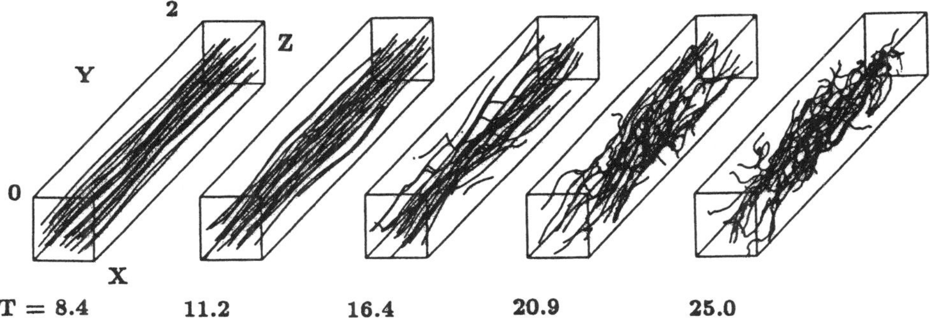


Fig. 3. Time evolution of magnetic field lines in a Keplerian disk with initially azimuthal magnetic field.

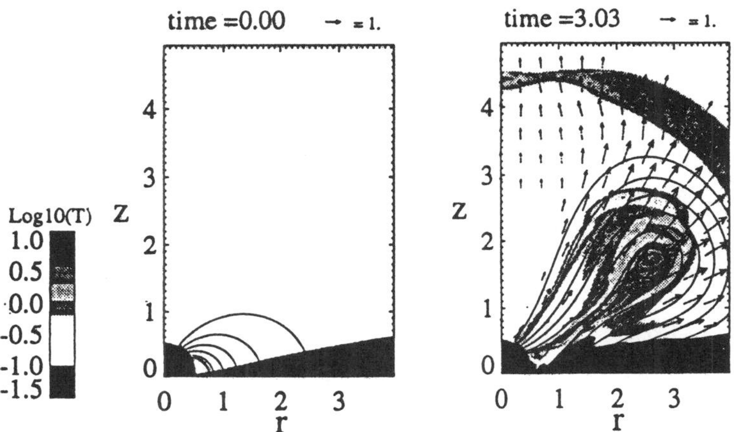


Fig. 4. The result of 2.5D MHD simulation of the interaction between the dipole magnetic field of the central star and the Keplerian disk.

## 5. Magnetic Model of X-ray Time Variabilities in Accretion Disks

Mineshige et al. (1994) proposed a self organized criticality (SOC) model of $1/f$ fluctuations of X-rays in black hole candidates. Here we note that the amplification of magnetic fields by shear rotation, escape of magnetic flux by buoyancy, and dissipation of magnetic energy by magnetic reconnection lead the disk to the marginally stable state where the magnetic fluctuation level is determined by the balance between these processes. Small deviation from the marginal state can produce big flares as expected from the SOC model.

In the low-state, black hole candidates show sporadic X-ray variations whose amplitude is the order of the total X-ray luminosity. If the time variation is caused by the release of magnetic energy, magnetic pressure should be dominant in low-state disks. Mineshige et al. (1995) proposed a model that low-state disks are low-$\beta$ disks. Since advection continuously generates radial component of magnetic fields, we expect that the saturation level of magnetic field strength will be higher in advection-dominated, low-state disks. We would like to study the magnetic structure of such disks in near future.

## References

Balbus S. A., Hawley J. F. 1991, ApJ 376, 214

Brandenburg A., Nordlund A., Stein R.F., Torkelsson U. 1995, ApJ 446, 741

Chandrasekhar S. 1961, Hydrodynamic and Magnetohydrodynamic Stability, (Oxford University Press, Oxford)

Foglizzo T., Tagger M., 1994, A&A 287, 297

Ghosh P., Lamb F. K., 1991, in Neutron Stars: Theory and Observation, ed. Ventura & Pines, (Kluwer Academic Publishers), p363

Hayashi M.R., Shibata K., Matsumoto R. 1996, in preparation

Hawley J. F., Gammie C. F., Balbus S. A. 1995, ApJ 440, 742

Koyama K. et al. 1994, PASJ 46, L125

Matsumoto R., Horiuchi T., Shibata K., Hanawa T. 1988, PASJ 40, 171

Matsumoto R., Tajima T., 1995, ApJ 445, 767

Matsumoto R., Uchida Y., Hirose S., Shibata K., Hayashi M.R., Ferrari A., Bodo G., Norman C. 1996, ApJ in press

Mineshige S., Ouchi N. B, Nishimori H. 1994, PASJ 46, 97

Mineshige S., Kusunose M., Matsumoto R., 1995, ApJL 445, L43

Shakura N. I., Sunyaev R. A. 1973, A&A 24, 337

Shibata K., Tajima T., Matsumoto R. 1990, ApJ 350, 295

Shibata K., Uchida Y. 1986, PASJ 38, 631

Uchida Y., Shibata K. 1985, PASJ 37, 515

Velikhov E. P. 1959, Soviet Phys. JETP Lett. 35, 1398

Zylstra G. 1988, Ph.D. thesis, University of Illinois at Urbana-Champaign

# Mass Flux and Terminal Velocities of Magnetically Driven Jets from Accretion Disks: Steady and Nonsteady Solutions

Takahiro KUDOH and Kazunari SHIBATA
*National Astronomical Observatory of Japan, 2-21-1, Osawa, Mitaka, Tokyo, 181, Japan*

**Abstract**

We solve 1.5-dimensional steady and axisymmetric magnetohydrodynamical (MHD) equations to study basic properties of astrophysical jets from accretion disks. We find that the mass flux depends on the poloidal magnetic field of the disk when the toroidal component of the magnetic field is dominant at the disk, although it is independent of the magnetic field when the the poloidal component is dominant there. We also perform time-dependent 1.5D MHD numerical simulations of the astrophysical jets to understand the relation between the nonsteady and steady jets. It is found that the nonsteady jets have the same properties of the steady magnetically driven jets, even if the closed boundary conditions, which are not suitable for obtaining the steady states, are assumed at the equatorial plane of the disk (i.e., the foot point of the jet) in the simulations.

## 1. Introduction

Magnetically driven jets from accretion disks have been worked as the most promising models for astrophysical jets which are observed in young stellar objects and active galactic nuclei (Blandford 1993). The magnetohydorodynamical (MHD) models of jet formation have been studied by many authours (e.g., Blandford & Payne 1982; Sakurai 1987; Contopoulos & Lovelace 1994; Li 1995), based on the steady and axisymmetric MHD wind which was first solved by Weber & Davis (1967) for the solar wind. By using the 1.5-dimensional approximation, we investigated the mass flux and terminal velocity of the magnetically driven jet, including a thermal energy of the

*S. Kato et al. (eds.), Physics of Accretion Disks, 311–314.*
© 1996 OPA (Overseas Publishers Association) Amsterdam B.V.

jet in order to obtain the mass flux (Kudoh & Shibata 1995). We also perform time-dependent 1.5D MHD numerical simulations of the jets, which will enable us to make the comparison between the steady and nonsteady jets.

## 2. Steady solutions

A method of solving the 1.5D steady MHD equations is summarized in Sakurai (1985). The geometry of the poloidal magnetic field line is given by the same equation that was used in Cao & Spruit (1994). The asymptotic geometry of this magnetic field line becomes paraboloidal; $z \propto r^2$. The absolute value of the poloidal field ($B_p$) is assumed to be, $B_p = B_{p0}(r_0/r)^2$, for simplicity, where $B_{p0}$ is the poloidal magnetic field at the foot point of the jet ($r = r_0$). The parameters we have used in this paper are

$$E_{\rm th} = \left(\frac{a_0}{V_{\rm K0}}\right)^2/\gamma \simeq 6.5 \times 10^{-4}\left(\frac{T_0}{10^3{\rm K}}\right)\left(\frac{M}{M_\odot}\right)^{-1}\left(\frac{r_0}{15R_\odot}\right) \quad , \tag{1}$$

$$E_{\rm mg} = \left(\frac{V_{{\rm A}p0}}{V_{\rm K0}}\right)^2 \simeq 3.8 \times 10^{-3}\left(\frac{B_{p0}}{10{\rm G}}\right)^2\left(\frac{M}{M_\odot}\right)^{-1}\left(\frac{n_0}{10^{13}{\rm cm}^{-3}}\right)^{-1}\left(\frac{r_0}{15R_\odot}\right) \quad , \tag{2}$$

where $a_0 = (\gamma P_0/\rho_0)^{1/2}$, $V_{{\rm A}p0} = (B_{p0}^2/4\pi\rho_0)^{1/2}$ , $V_{\rm K0} = (GM/r_0)^{1/2}$, $P_0$, $\rho_0$, and $n_0$, are sound, poloidal Alfvén, Keplerian velocities, pressure, mass density, and number density at the equatorial plane ($r = r_0$, $z = 0$), and $M_\odot$, and $R_\odot$ are the solar mass and radius, respectively.

We found that the mass flux depends on the poloidal magnetic field at the disk ($\dot{M} \propto B_{p0}$) when the toroidal component of the magnetic field become dominant at the disk, although it is independent of the magnetic field when the the poloidal component is dominant there (figure 2b and d). Since the Michel's scaling law is almost satisfied in magnetically driven jet (figure 2a), the terminal velocities depend on the poloidal magnetic field as $v_\infty \propto B_{p0}^{2/3}$ for $|B_\phi/B_p|_0 \ll 1$ and $B_{p0}^{1/3}$ for $|B_\phi/B_p|_0 \gg 1$ (figure 2c).

When $E_{\rm mg}$ is small, the strong acceleration takes place after the flow passes through the Alfvén point. The magnetic pressure plays an important role for acceleration of the flow in the weak magnetic field regime.

## 3. Time-dependent simulations

Using the same poloidal filed line configuration, we studied time-dependent 1.5D simulations assuming the initial and boundary conditions (the symmetrical boundary condition at the equatorial plane) similar to those of Shibata & Uchida (1986) to understand the relation between the nonsteady and steady jets. The numerical methods we used in this paper are CIP (Yabe & Aoki 1991) and MOC-CT schemes (Evans & Hawley 1988; Stone & Norman 1992).

Figure 1a shows the time evolution of the poloidal velocity. This figure shows that there are two velocity peaks. The leading one is the front of the nonlinear torsional Alfvén wave (*coronal-jet*). The following peak corresponds to the jet which is directly ejected from the disk (*disk-jet*). The first peak becomes the fast shock and the second one becomes the slow shock. The density evolution (figure 1b) shows that the mass which is initially in the disk is ejected to the corona.

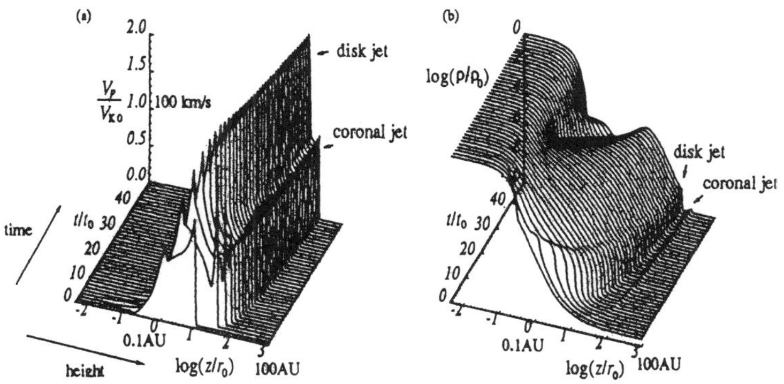


Fig. 1. The example of the time variation of (a) the poloidal velocity and (b) the density. The velocities are normalized by the Keplerian velocity at the foot point of the jet. The density is normalized by the initial value at the foot point. The spatial axis, $z/r_0$ is taken as a logarithmic scale.

The dependences of the normalized maximum mass flux and the toroidal component on $E_{\rm mg}$ are consistent with the results of the steady solution (figure 2b and d). The Michel's scaling law is also satisfied in figure 2a, the dependence of terminal velocity on $E_{mg}$ is also consistent with that of the steady solution (figure 2c). Our results of the nonsteady jets correspond to the case of the weak magnetic field regime. The differences of the absolute values are caused by the different boundary conditions at the equatorial planes between nonsteady and steady solutions.

## 4. Discussions

On the basis of above results, we can estimate the following physical quantities at the foot point of the jet in the disk from the mass loss rate observed in T Tauri winds and/or fast neutral winds ($\dot{M} \sim 10^{-8} - 10^{-7} M_{\odot}{\rm yr}^{-1}$); $n_0 \sim 10^{13}{\rm cm}^{-3}$, $T_0 \sim 1500$ K, and $B_{p0} \sim 10$G at $r_0 \sim 15 R_{\odot}$ for the jet/wind whose terminal velocity is $\sim$ 100 km s$^{-1}$. Our results also explain the velocity of the jet and the associated mass flux found in nonsteady 2.5D MHD simulations (Shibata & Uchida 1986; Matsumoto et al. 1996).

We thank for Dr. T. Sakurai for fruitful discussions, and Dr T. Yabe for helpful comments about computational methods.

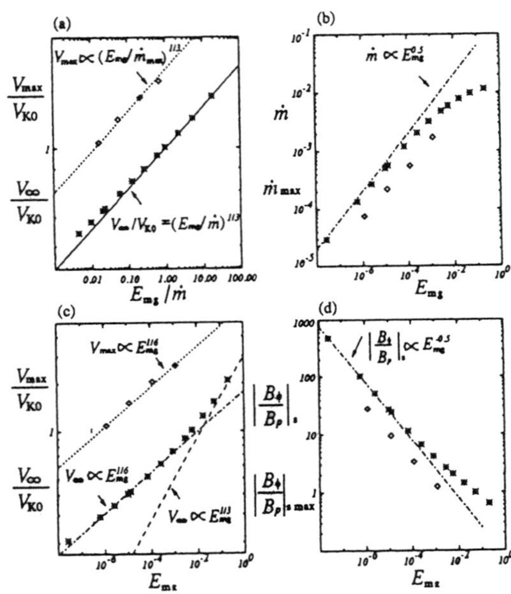

Fig. 2. (a) The maximum poloidal velocities of the disk jets are plotted as a function of $E_{mg}/\dot{m}_{max}$. The dotted line shows $v_{max} \propto (E_{mg}/\dot{m}_{max})^{1/3}$. The straight line shows the Michel's solution. (b) The maximum mass flux are plotted as a function of $E_{mg}$. (c) $v_{max}$ versus $E_{mg}$. The dotted line shows $v_{max} \propto E_{mg}^{1/6}$ (d) The maximum value of $|B_\phi/B_p|$ at the slow point versus $E_{mg}$. The results of the nonsteady solutions are plotted as "diamonds". The results of the steady solutions of the same parameters are also plotted as "asterisks".

## References


Blandford R. D. 1993, in Astrophysical jets, eds D. Burgarella, M. Livio, C. P. O'Dea (Cambridge: Cambridge Univ. Press), p15

Blandford R. D., Payne D. G., 1982, MNRAS 199, 883

Cao X., & Spruit H. C. 1994, A&A 287, 80

Contopoulos J., Lovelace R. V. E. 1994, ApJ 429, 139

Evans C. R., Hawley J. F. 1988, ApJ 332, 659

Kudoh T., Shibata K. 1995, ApJL 452, L41

Li Z-Y. 1995, ApJ 444, 848

Matsumoto R. et al. 1996, ApJ in press

Sakurai T. 1985, A&A 152, 121

Sakurai T. 1987, PASJ 39, 821

Shibata K., Uchida Y. 1986, PASJ 38, 631

Stone J. M., Norman M. L. 1992, ApJS 80, 791

Weber E. J, Davis L., Jr. 1967, ApJ 148, 217

Yabe T., Aoki T. 1991, CompPhysComm 66, 219

# Time-Dependent Accretion Flow onto a Black Hole: Formation of Shock Waves and Jets

Koji NOBUTA and Tomoyuki HANAWA
*Department of Astrophysics, School of Science, Nagoya University, Chikusa-ku, Nagoya 464-01, Japan*

## Abstract

We examine time-dependent accretion flows onto a black hole as a model of Active Galactic Nuclei (AGN) using two-dimensional numerical hydrodynamical simulations. In our simulations we follow the evolution of subsequentially-accreting gas rings, each of which mimics a blob spiraling onto a central black hole. The simulations show accretion flows with shock waves and jets. The shock waves are expected to cause thermal X-ray and non-thermal $\gamma$-ray flares. In other words our models simulate simultaneous X-ray and $\gamma$-ray flares as observed in Mkn 421.

## 1. Introduction

A great number of observations have revealed that Active Galactic Nuclei have the 'core-jets structure' in the radio map and strong time variability in the X-ray luminosity. Moreover, recent observations showed that a BL Lac object Mkn 421 has simultaneous X-ray and $\gamma$-ray flares (Kerrick et al. 1995). One of the possible theoretical interpretations for these observations is as follows. A time dependent gas flow accretes onto a massive black hole and forms shock waves. Through shock waves the gas is heated and emits X-rays or $\gamma$-rays. A part of the heated gas escapes away along the rotation axis to evolve into jets. We present here a time-dependent accretion flow model that forms shock waves and jets.

## 2. The Time-Dependent Accretion Flow Model

*S. Kato et al. (eds.), Physics of Accretion Disks, 315–318.*
© 1996 OPA (Overseas Publishers Association) Amsterdam B.V.

For simplicity we assumed that the system is symmetric with respect to the rotation axis. We employed the pseudo-Newtonian potential (Paczyński & Wiita 1980) and Newtonian mechanics. The gas consists of both ideal gas and black body, and accretes inviscidly and adiabatically. The mass of the central black hole is $10^7 M_\odot$. As a model of time-dependent accretion flows, we injected two gas rings with a mass of $2.6 \times 10^{28}$ g with a time interval of 0.9 hours in a steady accretion flow. The average accretion rate is set to the Eddington accretion rate, $\dot{M}_{\rm E} = 0.22\, M_\odot\, {\rm yr}^{-1}$. The half of the gas is supplied constantly and the other half intermittently in the form of subsequently accreting two gas rings.

The computation is done in the spherical polar coordinates $(r, \theta, \phi)$ with an orthogonal grid. The computation grid consists of 600 cells in the $r$-directions and 50 cells in the $\theta$-directions. The radial spatial extent is from $r = 2$ to $27.2\, r_{\rm g}$, where $r_{\rm g}$ denotes Schwarzshild radius. The spatial resolution in the $r$-direction is kept better than $\Delta r / r \leq 0.0088$. We used a total variation diminishing method in solving the equations of hydrodynamics.

## 3. Accretion Flow with Shock Waves and Jets

### *3.1. Typical Example*

Figure 1 shows a typical example of our simulations, in which shock waves and jets are formed. The contours denote the density and the arrows the velocity field. As the initial model, we superimposed a dense gas ring at $r = 14\, r_{\rm g}$ on a steady flow. At $t = 0.9$ hours a second gas ring is set in the same way (figure 1a). The first gas ring has large specific angular momentum ($\ell = 2.1\, r_{\rm g}\, c$) and forms a torus revolving the hole stationarily with the radius $r = 6\, r_{\rm g}$. The second ring has lower specific angular momentum ($1.7\, r_{\rm g}\, c$) than the first. At $t = 2$ hours the rings collide with each other and form a shock wave at the interface [figure 1(b)]. The temperature of the post-shock gas increases from $2 \times 10^5$ to $5 \times 10^5$ K. Thermal X-rays will be emitted from the shock heated gas and non-thermal $\gamma$-rays will be generated at the shock front. In other words our models simulate simultaneous X-ray and $\gamma$-ray flares. The shock wave first appears near the equatorial plane and extends toward the pole to cover the central hole. On the pole the shock wave converges and generates gas outflows that evolve into jets. Figure 1c represents the jet propagation. At $t = 4$ hours the jet reaches 17 $r_{\rm g}$. Another 3 hours later, the jet reaches 24 $r_{\rm g}$ and the shock waves disappear.

### *3.2. Jet*

Figure 1b shows the early stages of the jet formation. The post-shock flow is almost parallel to the shock front. When the shock front converges on the rotating axis the post-shock gas collides head-on and become dense and hot.

The resultant high-pressure gas can escape only along the axis. The outflow grows into a jet. Figure 2a shows the radial velocity profile of the jet along the rotation axis.

We define the jet as a gas component having positive velocity and positive total enthalpy so that the gas can flow out to a large distance. Figure 2b shows the evolution of mass of the jet. At $t \simeq 4$ hours, it has its maximum value, $6 \times 10^{26}$ g. It is about 1% of that of the gas ring. Afterwards the mass decreases. This is because the jet collides with the steady flow with negative velocity and loses its momentum.

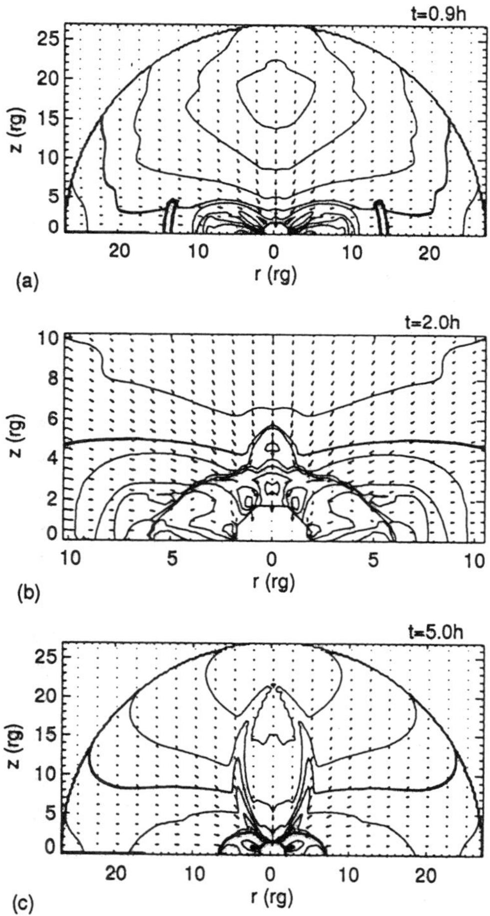


Fig. 1. Density contours and velocity vectors in the meridional plane $(r, z)$ at three stages. Contour levels are $2 \times 10^{-14}$, $5 \times 10^{-14}$, $1 \times 10^{-13}$, ... , $5 \times 10^{-11}$, $1 \times 10^{-10}$ g cm$^{-3}$. The thick contour denotes $1 \times 10^{-13}$ g cm$^{-3}$. (a) An early stage at $t = 0.9$ hours. The second gas ring was set on the steady flow. (b) The onset of the jet formation at $t = 2.0$ hours. The gas rings collide with each other and form shock waves and jet. Note that only the central region near the pole is shown. (c) A late stage at $t = 4.0$ hours. The jet reaches $r = 17\, r_g$ .

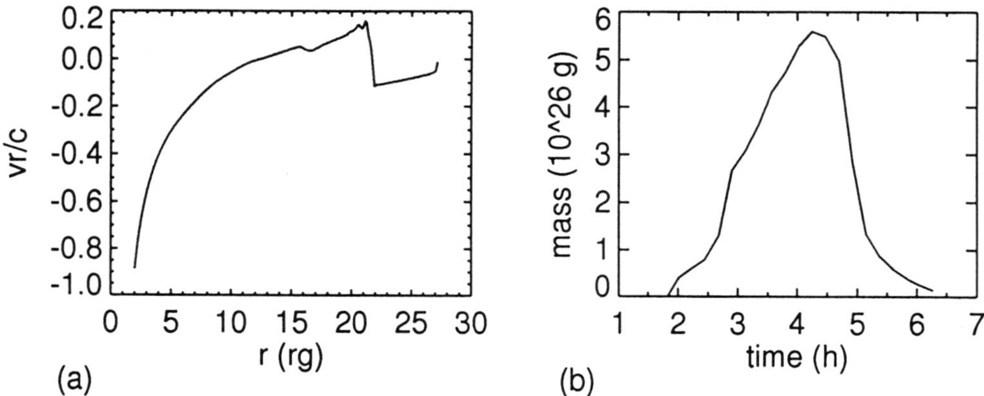


Fig. 2. (a) Radial velocity profile of the jet along the rotation axis in units of $c$ at $t = 4.0$ hours. (b) The mass of the jet as a function of time.

### *3.3. Other Models*

We constructed 4 models in which the specific angular momenta of the gas rings are different each other. They have either large ($2.2\, r_{\rm g}\, c$) or small ($1.7\, r_{\rm g}\, c$) specific angular momentum. Note that the critical specific angular momentum is $2\, r_{\rm g}\, c$ for our model. When it is smaller than the critical, the gas can be absorbed into the black hole without losing its angular momentum. It is essential for the jet formation that the first gas ring has large angular momentum and collides with the second gas ring. Otherwise it accretes onto the hole before colliding with the second gas ring.

## 4. Conclusions

We have examined time-dependent accretion flows onto a black hole using numerical simulations. As preliminary results we got an accretion flow with the torus-shape shock waves and the jets. The mass ejection rate of the jets is about 1% of the total accretion rate. Our simulations show that the large time variability is thought to be the source of X-ray and $\gamma$-ray flares.

## Acknowledgements

This reseach is supported in part by the Grant-in-Aid for Scientific Research on Priority Areas (07247212) of the Ministry of Education, Science and Culture in Japan.

## References

Kerrick A. D. et al. 1995, ApJL 438, L59
Paczyński B., Wiita P. J. 1980, A&A 88, 23

# Magnetized Accretion Disk as a Solution of Diffusion-Type Equation

Osamu KABURAKI[1] and Takahiro KUDOH[2]
*1. Astronomical Institute, Graduate School of Science, Tohoku University, Aramaki, Sendai 980-77, Japan*
*2. Department of Astronomical Science, The Graduate University for Advanced Studies, NAOJ, Japan*

**Abstract**

The behavior of *resistive*-MHD accretion disks around black holes is discussed. A diffusion-type equation is obtained from general relativistic Maxwell's equations and Ohm's law under a thin disk approximation. A plausible solution to this equation specifies the radial dependence of the accretion flow as identical with that in the standard model of viscous $\alpha$-disks. The energy extraction rate from a disk is shown to have a similar angular velocity dependence to that in the force-free or ideal-MHD calculations, suggesting that the presence of a finite resistivity does not affect this dependence.

## 1. Derivation of Diffusion-Type Equation

The MHD jet model from an accretion disk is one of the most promising way of explaining astrophysical jets. If we want to obtain a globally-consistent, stationary magnetic configuration, we have to use the *resistive*-MHD in order to avoid the piling up of the field lines anywhere (Kaburaki 1986; Lovelace et al. 1987; Königl 1989). In order to apply such a model to disks and jets around black holes in stellar-mass binary systems or in AGN, we have to treat them general relativistically. Within a thin disk approximation, we obtain below a diffusion-type equation which has the same form as derived by Lovelace et al. (1987) in the non-relativistic regime.

We abandon from the beginning to solve self-consistently the whole set of MHD equations, and assume that the velocity field is given. Then, the

*S. Kato et al. (eds.), Physics of Accretion Disks, 319–322.*
© 1996 OPA (Overseas Publishers Association) Amsterdam B.V.

problem is to find a resultant structure of the electromagnetic field. Our simplifying assumptions adopted here are as follows. The background spacetime is determined only by a central Kerr (i.e., rotating) black hole: the Kerr spacetime is described by Boyer-Lindquist coordinates $(t, r, \theta, \phi)$. The electromagnetic field and plasma flow also satisfy the stationarity and axisymmetry. The electric conductivity $\sigma$ is constant in time and space. The angular thickness of a disk is constant and small (i.e., $\Theta =$ const. $\ll 1$) so that a new variable of order unity, $\xi \equiv (\theta - \pi/2)/\Theta$, is useful to describe the latitudinal disk structure.

Reserving only the leading terms in each component of Maxwell's equation $\nabla_\nu F^{\mu\nu} = 4\pi j^\mu$ supplemented by Ohm's law $(g^{\mu\nu} + u^\mu u^\nu) j_\nu = \sigma F^{\mu\rho} u_\rho$, we obtain the same diffusion-type equations:

$$D(x)\frac{\partial^2 A_\alpha}{\partial \xi^2} - \frac{\partial A_\alpha}{\partial x} = 0 \;, \quad (\alpha = t,\ \phi)$$

for the electromagnetic potentials $A_t$ and $A_\phi$, where $x = -r$ and $D(x) = -(4\pi u^r x^2 \sigma \Theta^2)^{-1}$ ($> 0$ for the accretion flow). Once a set of solutions to these equations is found, the toroidal component of the electromagnetic field is obtained from $F_{r\theta} = (u^t \partial A_t/\partial\xi + u^\phi \partial A_\phi/\partial\xi)/u^r\Theta$.

## 2. Plausible Solution

In our treatment, the electromagnetic potential is divided into two parts: that of a global outer field and of the disk field, $A_\alpha = A_\alpha^{\mathrm{G}} + A_\alpha^{\mathrm{D}}$. The latter is produced by the current flowing in the disk and its components obeys the diffusion equations obtained above. The boundary conditions for $A_t^{\mathrm{D}}$ and $A_\phi^{\mathrm{D}}$ are that they become 0 when $\xi$ tends to $\infty$ and have an even symmetry with respect to $\xi$ (i.e., $A_\alpha(\xi) = A_\alpha(-\xi)$).

The fundamental solutions of the diffusion equation satisfy the above boundary conditions and hence they are adopted as the solutions to our equations:

$$A_t^{\mathrm{D}} = -\frac{\Phi_0}{2\sqrt{\pi}\Gamma}\exp(-\frac{\xi^2}{4\Gamma^2}) \;, \quad A_\phi^{\mathrm{D}} = \frac{\Psi_0}{2\sqrt{\pi}\Gamma}\exp(-\frac{\xi^2}{4\Gamma^2}) \;,$$

where $\Gamma^2 = \int_{x_b}^{x} D(x')dx'$, $x_b = -r_b$, and $r_b$ is the radius of disk's outer edge. Since we have assumed that $\Theta =$ const., $\Gamma$ should be nearly constant, for consistency. The weakest possible $x$-dependence of $\Gamma$ is obtained when the radial velocity varies with the radius like $u^r = u_b^r (r/r_b)^{-1} = -u_b^x (x/x_b)^{-1}$ resulting that $\Gamma^2 = -R_m^{-1}\ln(x/x_b) = -R_m^{-1}\ln(r/r_b)$, where $R_m = -4\pi\sigma\Theta^2 x_b u_b^x$.

Thus the $r$-dependence of $u^r$ is uniquely determined by the consistency requirement with our assumption, although we do not solve the Euler equation. It is interesting to see that the dependence is identical with those in the standard $\alpha$-viscosity model (e.g., Pringle 1981) and in the resistive magneto-Keplerian disk model (Kaburaki 1986).

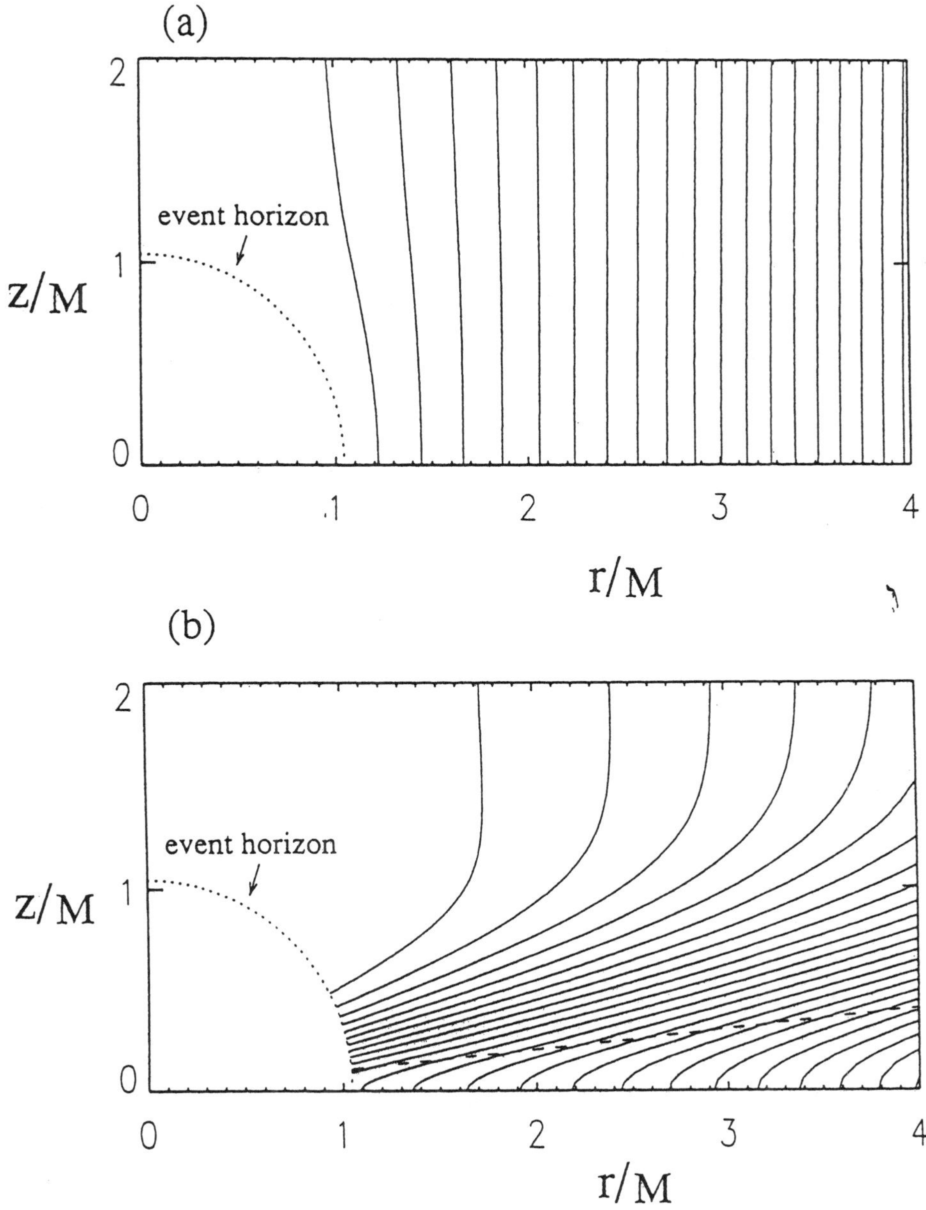


Fig. 1. Poloidal magnetic field lines (the full curves) for, (a) Wald's vacuum field with $a = 0.999$ (where $a$ is the specific angular momentum of a central black hole), and (b) the Wald field ($a = 0.999$) modified by the action of disk current. The adopted parameters are $R_{\rm m} = 1$, $\Theta = 0.1$ and $r_0 = 10M$. Although the hole has a tendency to exclude field lines in a vacuum (a), the accretion flow pushes them into the hole (b). The inclined dashed line shows the disk surface.

## 3. Energy Output from a Disk

We assume that the electromagnetic energy extracted from the disk surfaces is converted to the kinetic energy of a bipolar jet as usually expected. Our results show that the electromagnetic energy is extracted from the disk only if the condition

$$\Omega_F(\Omega - \Omega_F) > 0 \ ,$$

is satisfied, i.e., if the angular velocity of the disk, $\Omega = u^\phi/u^t$, is larger than the "angular velocity" of the electromagnetic field, $\Omega_F \equiv -A_{t,r}/A_{\phi,r} = -A_{t,\theta}/A_{\phi,\theta} = \Phi_0/\Psi_0$. The appearance of the factor $\Omega_F(\Omega-\Omega_F)$ in the energy flux is quite similar to the cases of the force-free (e.g., Blandford & Znajek 1977) and ideal-MHD (Takahashi et al. 1990) magnetospheres, suggesting that the presence of a finite resistivity does not affect this result. Nevertheless, the resistivity is essential to guarantee the conversion of gravitational energy into heat and the realization of a stationary magnetic structure, in the disk.

We assume that the global magnetic field outside the disk is something like a uniform one. The fitting of the $\theta$-component of the disk magnetic field to the outer one at the disk's outer edge fixes the value of $\Psi_0$. Similarly, the value of $\Phi_0$ is determined by the connection of the $r$-components of the disk and outer electric fields. We site the two extreme cases as an illustration.

(1) The connection to Wald's vacuum field.

In this case, the surrounding space is approximated by a vacuum so that the disk is driven by the external electric component caused by the rotating black hole. Then we obtain $\Omega_F = \Phi_0/\Psi_0 \simeq aM/r_b^3$, where $M$ and $a$ are, respectively, mass and specific angular momentum of the black hole. A powerful jet is expected to emanate mainly from a rapidly rotating (i.e., $a \sim 1$) hole. The deformation of the poloidal magnetic field lines by the accretion flow is shown in figure 1.

(2) The connection to the ideal-MHD surroundings.

In this case, the surrounding space is assumed to be plasma rich so that the electric field at the boundary is determined by the ideal MHD-condition (i.e., there is no external driving force). Then we obtain $\Omega_F = \Phi_0/\Psi_0 \simeq \Omega_b$, where $\Omega_b$ is $u^\phi/u^t$ at the boundary. Thus, the hole's rotation affects the results only through the background metric.

## References

Kaburaki O. 1986, MNRAS 220, 321

Königl A. 1989, ApJ 342, 208

Lovelace R. V. E., Wang J. C., Sulkanen M. E. 1987 ApJ, 315, 504

Pringle J. E. 1981, ARA&A 19, 137

Takahashi M., Nitta S., Tatematsu Y., Tomimatsu A. 1990, ApJ 363, 206

# Plasma Heating due to Wave Absorption in Magnetized Accretion Disks

KOUICHI HIROTANI
*National Astronomical Observatory, Mitaka, Tokyo 181, Japan*

**Abstract**

The propagation properties of magneto-gravity waves in an accretion disk penetrated by toroidal magnetic fields are studied. The disk is assumed to be composed of an equatorial cool region and slowly rotating coronae. The wavelength in the radial direction is assumed to be short compared with the typical length scale in the vertical direction. Further, the whole disk is assumed to be weakly isentropic in the vertical direction. In the equatorial cool region, the epicyclic frequency, $\kappa$, nearly equals the Keplerian angular frequency $\Omega_K$. Magneto-gravity waves are transmissive in this region, because $\omega < \kappa$ holds there. In the slowly rotating coronae, on the other hand, $\kappa$ nearly vanishes. It follows that there is the transition layer, in which $\kappa$ decreases from $\Omega_K$ to 0 with a height. In this layer there will be a height at which $\omega$ matches $\kappa$. At this height, we find that magneto-gravity waves are resonantly absorbed to heat the plasma. We discuss briefly some implications of this result with respect to the time variabilities observed from X-ray stars.

## 1. Introduction

There has been increasing interests in the subject of disk oscillations. Fukue & Okada (1990) demonstrated that low-frequency, magneto-gravity waves are localized near the equator in a disk threaded by toroidal magnetic fields. More recently, Huang & Vishniac (1993) discussed that vertically propagating gravity waves will be thermalized, as they propagate away from the equatorial plane to the region with increasing specific entropy.

In our previous work (Hirotani & Kato 1995), we demonstrated that vertically propagating gravity waves are absorbed at the height where $\omega$ matches

*S. Kato et al. (eds.), Physics of Accretion Disks, 323–326.*
© 1996 OPA (Overseas Publishers Association) Amsterdam B.V.

the epicyclic frequency $\kappa$ in a non-magnetized disk. In the present paper, we describe the vertical propagation of magneto-gravity waves in a disk penetrated by toroidal magnetic fields, and show that they are resonantly absorbed at the height where $\omega$ matches $\kappa$.

## 2. Vertical Oscillation of Disks

As an unperturbed state, we consider a steady axisymmetric disk comprising a perfect fluid rotating around a central object. We adopt a cylinderical system of coordinates $(r,\phi,z)$ which is centered on the central object; the $z$-axis is chosen as the rotation axis of the disk. The unperturbed disk can be an arbitrarily rotating configuration; however, it is assumed to have no poloidal motion. The magnetic field of the unperturbed state is supposed to have the toroidal component $B_\phi$ alone.

Adopting the short-wavelength limit in the radial direction, we finally obtain a single wave equation describing vertical oscillation of disks (Fukue & Okada 1990)

$$\frac{d^2\Phi}{dz^2} + R(z)\Phi = 0, \tag{1}$$

where the variable, $\Phi \equiv (p_1 + B_\phi b_\phi/4\pi)/\sqrt{\rho_0}$, essentially represents the energy density of oscillations; $p_1$ and $b_\phi$ refer to the perturbed pressure and magnetic field, respectively. The coefficient $R$ is defined by

$$R \equiv \frac{(\omega^2 - \omega_+^2)(\omega^2 - \omega_-^2)}{(c_s^2 + v_A^2)(\omega^2 - \kappa^2)}, \tag{2}$$

where $c_s \equiv \sqrt{\gamma P_0/\rho_0}$ and $v_A \equiv B_\phi/\sqrt{4\pi\rho_0}$ are the sound and Alfvén speeds, respectively; $\omega_\pm$ are defined by

$$\omega_\pm^2 \equiv \frac{1}{2}\left\{F \pm \sqrt{F^2 - 4[(c_s^2 + v_A^2)k_r^2 N^2 + \kappa^2 N_{ac}^2]}\right\}, \tag{3}$$

$$F \equiv \kappa^2 + (c_s^2 + v_A^2)k_r^2 + N_{ac}^2. \tag{4}$$

The acoustic cut-off frequency $N_{ac}$ is of the order of the Keplerian angular frequency, $\Omega_K$ $(\equiv \sqrt{GM/r^3})$. The explicit expression for $N_{ac}$ is not important here. Since we are concerned with nearly isentropic disk in the vertical direction, the Brunt-Väisälä frequency $N$ is supposed to be zero.

## 3. Resonant Absorption of Waves

Let us describe the oscillatory properties of the solutions $\Phi$ by means of a propagation diagram. For some fixed radius $r$ and radial wavenumber $k_r$, the

$\omega$-$z$ plane is divided into propagation region where $R(\omega, z) > 0$ and evanescent one where $R < 0$. These regions are separated by the loci of $\kappa(z)$ and $\omega_\pm(z)$.

Let us first discuss the variation of $\kappa$ with $z$. We suppose that the equatorial region of the disk is so cool that the epicyclic frequency $\kappa$ nearly equals $\Omega_K$. Above the cool disk, let there be a hot corona which rotates so slowly that $\Omega$ nearly vanishes. In such a corona, the epicyclic frequency $\kappa$ is very small. In the transition layer between the cool disk and the hot corona, a sharp temperature increase will occur. In this layer the value of $\kappa(z)$ will change sharply, as is demonstrated by the solid curve in figure 1. In this figure, the abscissa is the height $z$, while the ordinate is the frequency.

Next, let us describe the functions $\omega_\pm(z)$. In the short-wavelength limit, we have $\omega_- \approx N \approx 0$. In this case, the propagation diagram would look something like figure 1. The upper dotted curve and the lower dashed one express the functional forms of $\omega_+(z)$ and $\omega_-(z)$, respectively. Perturbations are evanescent in the hatched regions, whereas they are oscillatory in the other regions. Magneto-acoustic waves transmit in the propagation region of higher frequencies, whereas magneto-gravity waves transmit in that of lower frequencies.

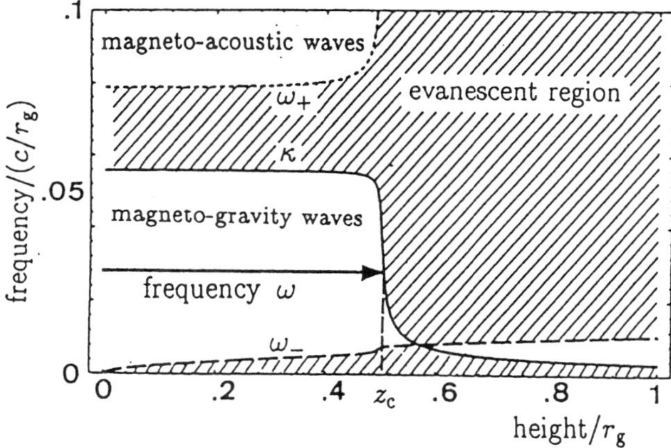


Fig. 1. A typical example of propagation diagrams in the case when the vertical entropy gradient is small. Magneto-gravity waves with frequency $\omega$ are confined in the region where $-z_c < z < z_c$ is satisfied.

This diagram indicates that the waves with a frequency $\omega$ are transmissive in the region $|z| < z_c$, where $z_c$ is the height at which $\omega$ equals $\kappa(z)$. The surface of $z = \pm z_c$ is probably located in the transition layer where $\kappa$ decreases rapidly. The wave amplitudes decay exponentially in the evanescent region ($|z| > z_c$).

It will be useful to derive the local dispersion relation. Replacing $z$-derivatives into $ik_z$ in equation (1), we obtain $k_z{}^2 = R(z)$. As a wave approaches to the singular surfaces ($z = \pm z_c$) where $R$ diverges, its wavelength in the vertical direction ($k_z{}^{-1}$) vanishes. Furthermore, the group velocity in the vertical direction also vanishes at $z = \pm z_c$. This can be clearly seen, if we differentiate the dispersion relation $k_z{}^2 = R(z)$ with respect to $k_z$. That is, we have

$$\frac{\partial\omega}{\partial k_z} = \frac{(c_s^2 + v_A^2)(\omega^2 - \kappa^2)}{2\omega^2 - (c_s^2 + v_A^2)k_z{}^2 - \omega_+{}^2 - \omega_-{}^2}\left(\frac{\omega}{k_z}\right)^{-1}. \tag{5}$$

It follows that $\partial\omega/\partial k_z$ vanishes at $z = \pm z_c$.

So far, we have seen that both the wavelength and the group velocity in the vertical direction vanish at the singular surfaces. This drives us to the conjecture that waves are absorbed at $z = \pm z_c$ (like those on a beach), when they propagate to the surfaces from the transmission side at $|z| < z_c$. We can explicitly show that magneto-gravity waves are indeed absorbed there by solving one-dimensional wave equation (1), however, its procedure is similar to what was presented in Hirotani & Kato (1995). Therefore, we omit the mathematical proof in this paper.

## 4. Discussion

Let us discuss an implication of our results with respect to X-ray observations. The vertical group velocity of a magneto-gravity wave packet, $\partial\omega/\partial k_z$, decreases as it propagates to the singular surfaces. Thus its energy accumulates in the transition layer and may develop a vorticity discontinuity there. The dissipaption of wave energy due to the formation of a "shock" (or a discontinuity) possibly occurs quasi-periodically, because the dissipation rate may vary with the phase of a wave owing to a nonlinear effect. This time-dependent dissipation of waves likely takes place, as we have seen, in the high-temperature region, such as the transition layer or the lower coronae. Higher energy X-ray photons are, therefore, expected to show larger fractional variabilities. To explore the effect of the shock quantitatively, we need a calculation in the nonlinear regime.

## Acknowledgements

The author thanks Professor S. Kato for discussions. This research has been partly supported by Research Fellowship of the Japan Society for the Promotion of Science for Young Scientists.

## References

Fukue J., Okada R. 1990, PASJ 42, 533

Huang M., Vishniac E. T. 1994, Proc. Cataclysmic Variables and Related Physics, 2nd Technion Haifa Conference, ed. O. Regev, G. Shaviv, p291

Hirotani K., Kato S. 1995, PASJ 47, 653

# Binary Black-Hole Engines in Active Galactic Nuclei

Osamu KABURAKI and Yoshiaki TANIGUCHI
*Astronomical Institute, Graduate School of Science, Tohoku University, Aramaki, Sendai 980-77, Japan*

**Abstract**

We propose an alternative model for the prime movers of active galactic nuclei (AGN), which seems to provide a new theoretical basis in considering a unified scheme of AGN. In our model, the ultimate source of AGN activities are attributed to the gravitational energy of a close-binary black-hole system which was formed by a merger process of two galaxies. The orbital motion in a central magnetic field induces an electric dipole moment on the surface of each hole. Under the electric field of the ultra low-frequency, long wave-length dipole radiation, electrons in a nearly vacuum space are accelerated to cause a cascade creation of electron-positron pair plasma. The rotational motion of the pair plasma concentrated around the orbital plane within the Roche lobe of the binary system drives a powerful bipolar jet.

## 1. Introduction

Although a single, accreting, supermassive black hole has been considered to be a prime mover of AGN (cf. Rees 1984; Antonucci 1993; Wilson & Colbert 1995), many radio galaxies and quasars tend to be found in galaxy mergers including giant elliptical galaxies (cf. Hutchings et al. 1984). If each of the two galaxies in a merger system has a black hole in its own nucleus, it is natural to postulate that the merger of the two black holes orbiting around each other would occur at a final stage of the process. Indeed, there are some such observational indications (e.g. Roos et al. 1993). Among them, the most impressive example may be the 11 year periodicity in the optical and radio luminosity variations of the archetypical, brightest BL-Lac type quasar, OJ 287 (Sillanpää et al. 1988). Therefore, it seems indispensable to reconsider the basic mechanism for the prime mover of AGN based on the binary nature of central black holes, without being constrained by the current paradigm.

*S. Kato et al. (eds.), Physics of Accretion Disks, 327–330.*
© 1996 OPA (Overseas Publishers Association) Amsterdam B.V.

## 2. Central Engines

We consider the cases in which two supermassive black holes with generally different masses, $M_1$ and $M_2$ ($M_1 \geq M_2$), are orbiting around their center of mass. Although the actual orbit of a black-hole binary may be highly eccentric (Fukushige et al. 1992), we apply the formula for circular orbits in order to estimate the emission power at an arbitrary moment. The size of the holes are represented by their Schwarzschild radii, $a_i = (2GM_i)/c^2$ $(i = 1, 2)$. According to the membrane paradigm (Thorne et al. 1986), black holes may be regarded as if they were made of ordinary matter with finite resistivity. Thus, the orbital motion across the magnetic field causes dipole type charge separations on the surfaces of the holes and resultant electric dipole radiation into surrounding nearly vacuum space.

The total power of the radiation is estimated as $P_{\rm ED} \sim 10^{46}$ $[F(q)/0.47]$ $(a/l)^9$ $[(M_1 + M_2)/10^9 M_\odot]^2$ $(B/10^4{\rm G})^2$ erg s$^{-1}$, where $a \equiv a_1 + a_2$, $l \equiv l_1 + l_2$ with $l_i$ being the distances of the holes from their center of mass, $q \equiv M_2/M_1$ and $F(q) = (1+q^8)/(1+q)^8$. We have adopted here a set of canonical values, $M_1 = 10^9 M_\odot$, $M_2 = 10^8 M_\odot$, $F(0.1) = 0.47$, and $B = 10^4$ Gauss.

A typical period of the oscillation of this dipole radiation is that of the orbital motion and its wave length is $\lambda \sim 2\pi c/\Omega$, where $\Omega$ is the orbital angular velocity. Such an ultra low-frequency and large-scale electric field is very suitable for accelerating charged particles (especially electrons) to relativistic energies. The inverse Compton scattering of the electric dipole photons by these relativistic electrons produce $\gamma$-ray photons which are energetic enough to induce a cascade production of electron-positron pair plasma. This plasma is considered to fill the Roche lobe of the binary system as far as the separation $l$ is not too large. Thus, the central engine of AGN in our model is a huge cloud of pair plasma rotating rigidly according to the rotation of the Roche lobe. As explained in the next section, the output from this central engine comes out in the form of kinetic energy of a relativistic plasma jet.

The value of the electric field at an edge of the Roche lobe, $R \sim l$, is crucial to judge whether the plasma can fill the lobe or not. It is roughly given by $E_{\rm rad} \sim \sqrt{P_{\rm ED}/cR^2} \sim \sqrt{F(q)/8}\ (a/l)^{11/2} B$, and the maximum energy of an electron accelerated by this field is $\epsilon \sim |e| E_{\rm rad}\lambda \sim 2\pi\sqrt{F(q)/8}\ |e| aB\ (a/l)^4$ where $e$ is the charge of the electron. At the immediate grazing ($l \sim a$) we obtain $\epsilon \sim 10^{24}$ eV, which is by many orders larger than the threshold for pair creation ($\sim 1$ MeV). The critical separation $l_{\rm cr}$ above which the Roche lobe cannot be filled completely is obtained from the condition $\epsilon \sim 1$ MeV. This gives roughly that $l_{\rm cr} \sim 10^3 a$ for $M_1 + M_2 = 10^9 M_\odot$.

## 3. Radio Jets

The rotation of plasma cloud in the magnetic field causes an electromotive force in the radial direction with respect to the center of mass. As a result,

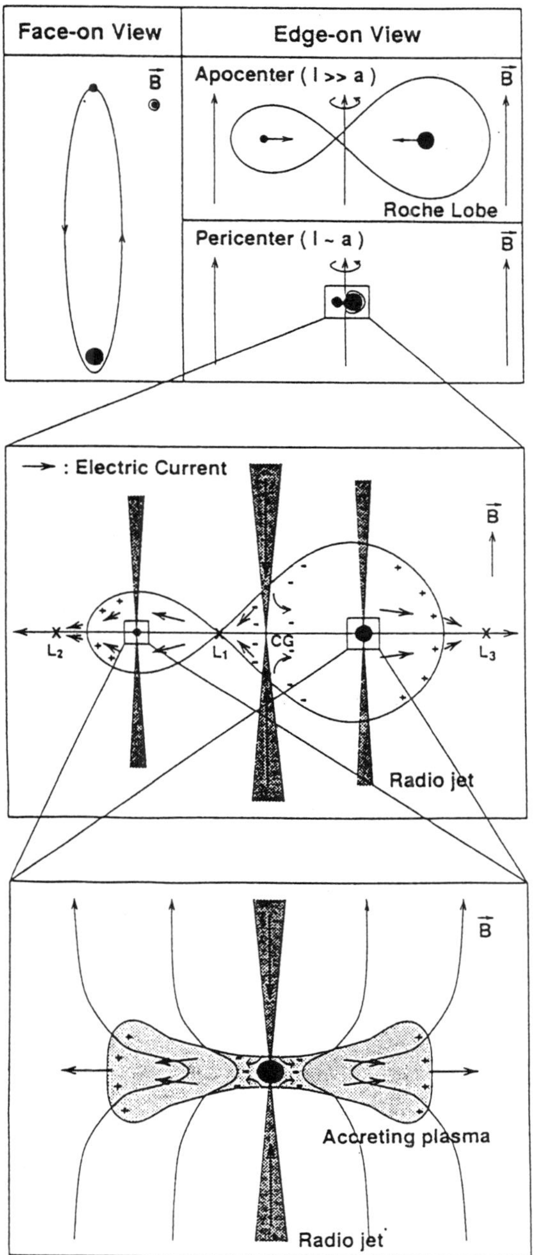


Fig. 1. An illustration of a grazing super-massive black-hole binary system. [1] (*upper panel*) 1) A face-on view of the black-hole binary with an eccentric orbit (*left*). 2) Two edge-on views of the binary at the apocenter (*upper right*) and at the pericenter (*lower right*). [2] (*middle panel*) The primary jets emanating from the center of gravity (CG) of the binary system. [3] (*lower panel*) The two secondary jets emanating from the two black holes.

inward or outward current is driven in the plasma cloud depending on the relative directions of rotation axis and magnetic field (for an idealized case, see figure 1). Owing to the requirement of current closure, a bipolar current should be developed in the polar regions. Once such a polar current is formed, the $\boldsymbol{j} \times \boldsymbol{B}$ force accelerates and pinches the conical flow of plasma as clarified in Kaburaki & Itoh (1987). This is the primary jet driven by the orbital motion. In addition to this, there may be two more secondary jets emanating from each of the binary black holes driven by the local accretion disks around them. These jets may be observed as radio jets due to their synchrotron emissions.

By assuming the radial dependence of the magnetic field like $B = B_0\,(a/r)^\alpha$, we estimate the jet power as $P_{\rm jet} = \int \boldsymbol{j} \cdot \boldsymbol{E}\, dV \sim 10^{47}\,(a/l)^{\alpha-1}(n/n^0_{\rm GJ})\,[(M_1 + M_2)/10^9 M_\odot]^2\,(B_0/10^4\ {\rm G})^2$ erg s$^{-1}$ (for $\alpha < 4$), where $n^0_{\rm GJ} = \Omega(l)B_0/(2\pi ce)$ is the Goldreich-Julian density for the magnetic field $B_0$. In the most realistic case of $\alpha = 2$, the dependence on $(a/l)$ becomes rather weak. Since the actual plasma density in the main part of the cloud seems to exceed the Goldreich-Julian value, our model can easily explain the observed powers of most intense jets.

At the final stage of a merger of binary black holes, the time scale for evolution is determined by the gravitational radiation loss, i. e., $T_{\rm GR} \sim 5 \times 10^7\ q^{-1}\,[(M_1 + M_2)/10^9 M_\odot]^{-3}[l/(3 \times 10^{17}{\rm cm})]^4$ yr, while their orbital period is $T_{\rm orb} \sim 10^2\,[(M_1 + M_2)/10^9 M_\odot]^{-1/2}\,[l/(3 \times 10^{17}{\rm cm})]^{3/2}$ yr. For a binary of canonical masses, we have $a \sim 3 \times 10^{14}$ cm, and hence $l_{\rm cr} \sim 3 \times 10^{17}$ cm and $T_{\rm GR} \sim 10^8$ yr. When the semi-major axis of a binary system becomes less than $l_{\rm cr}$, the jet activity is expected to last continuously with a periodic variation in its power output as far as the system has an eccentric orbit (e.g., OJ 287). Even when the axis is larger than $l_{\rm cr}$, one may expect recurrent phases of jet formation as far as binary's pericenter is within the critical distance. In such cases, a chain of bright knots may be observed in a radio jet.

## References

Antonucci R. 1993, ARA&A 31, 473

Fukushige T., Ebisuzaki T., Makino J. 1992, PASJ 44, 281

Hutchings J. B., Crampton D., Campbell B. 1984, ApJ 280, 41

Kaburaki O., Itoh M. 1987, A&A 172, 191

Rees M. J. 1984, ARA&A 22, 471

Roos N., Kaastra J. S., Hummel C. A. 1993, ApJ 409, 130

Sillanpää A., Haaraka S., Valtonen M. J., Sundelius B., Byrd G. G. 1988, ApJ 325, 628

Thorne K. S., Price R. H., Macdonald D. A. 1986, Black Holes: The Membrane Paradigm (Yale University Press, New Heaven)

Wilson A. S., Colbert E. J. M. 1995, ApJ 438, 62

# INDEX

**Individual objects**